Marine Ecosystems

Human Impacts on Biodiversity, Functioning and Services

Ecosystem services are emerging as a key driver of conservation policy and environmental management. Delivery of ecosystem services depends on the efficient functioning of ecosystems, which in turn depends on biodiversity and environmental conditions. Many marine ecosystems are extremely productive and highly valued, but they are increasingly threatened by human activities.

In this volume, leading researchers integrate current understanding of the effects on biodiversity, ecosystem functioning and ecosystem services caused by the range of human activities and pressures at play in coastal marine ecosystems, including fisheries, aquaculture, physical structures, nutrients, chemical contaminants, marine debris and invasive species. These reviews, combined with critical syntheses of the latest evidence and conceptual developments, make this a unique resource both for environmental managers and policy makers, and for researchers and students in marine ecology and environmental management.

TASMAN P. CROWE is Associate Dean of Science and a member of the Earth Institute and the School of Biology and Environmental Science at University College Dublin. He has undertaken research in Australia, Indonesia, Vanuatu, Ireland, the UK and continental Europe, studying individual and combined impacts of a range of stressors on marine biodiversity and ecosystem functioning, and making recommendations for management.

CHRISTOPHER L. J. FRID is Professor of Marine Biology and Head of the Griffith School of Environment at Griffith University in Queensland, Australia. His research has sought to understand how marine ecosystems function and how human impacts alter the dynamics of these systems. He has worked in the UK and throughout Europe, as well as in Ghana, Indonesia, Thailand and Australia and with many government agencies in support of marine ecosystem-based management.

ECOLOGY, BIODIVERSITY AND CONSERVATION

Series Editors
Michael Usher   *University of Stirling, and formerly Scottish Natural Heritage*
Denis Saunders   *Formerly CSIRO Division of Sustainable Ecosystems, Canberra*
Robert Peet   *University of North Carolina, Chapel Hill*
Andrew Dobson   *Princeton University*

Editorial Board
Paul Adam   *University of New South Wales, Australia*
H. J. B. Birks   *University of Bergen, Norway*
Lena Gustafsson   *Swedish University of Agricultural Science*
Jeff McNeely   *International Union for the Conservation of Nature*
R. T. Paine   *University of Washington*
David Richardson   *University of Stellenbosch*
Jeremy Wilson   *Royal Society for the Protection of Birds*

The world's biological diversity faces unprecedented threats. The urgent challenge facing the concerned biologist is to understand ecological processes well enough to maintain their functioning in the face of the pressures resulting from human population growth. Those concerned with the conservation of biodiversity and with restoration also need to be acquainted with the political, social, historical, economic and legal frameworks within which ecological and conservation practice must be developed. The new Ecology, Biodiversity, and Conservation series will present balanced, comprehensive, up-to-date and critical reviews of selected topics within the sciences of ecology and conservation biology, both botanical and zoological, and both 'pure' and 'applied'. It is aimed at advanced final-year undergraduates, graduate students, researchers and university teachers, as well as ecologists and conservationists in industry, government and the voluntary sectors. The series encompasses a wide range of approaches and scales (spatial, temporal and taxonomic), including quantitative, theoretical, population, community, ecosystem, landscape, historical, experimental, behavioural and evolutionary studies. The emphasis is on science related to the real world of plants and animals rather than on purely theoretical abstractions and mathematical models. Books in this series will, wherever possible, consider issues from a broad perspective. Some books will challenge existing paradigms and present new ecological concepts, empirical or theoretical models, and testable hypotheses. Other books will explore new approaches and present syntheses on topics of ecological importance.

Ecology and Control of Introduced Plants
Judith H. Myers and Dawn Bazely

Invertebrate Conservation and Agricultural Ecosystems
T. R. New

Risks and Decisions for Conservation and Environmental Management
Mark Burgman

Ecology of Populations
Esa Ranta, Per Lundberg and Veijo Kaitala

Nonequilibrium Ecology
Klaus Rohde

The Ecology of Phytoplankton
C. S. Reynolds

Systematic Conservation Planning
Chris Margules and Sahotra Sarkar

Large-Scale Landscape Experiments: Lessons from Tumut
David B. Lindenmayer

Assessing the Conservation Value of Freshwaters: An International Perspective
Philip J. Boon and Catherine M. Pringle

Insect Species Conservation
T. R. New

Bird Conservation and Agriculture
Jeremy D. Wilson, Andrew D. Evans and Philip V. Grice

Cave Biology: Life in Darkness
Aldemaro Romero

Biodiversity in Environmental Assessment: Enhancing Ecosystem Services for Human Well-being
Roel Slootweg, Asha Rajvanshi, Vinod B. Mathur and Arend Kolhoff

Mapping Species Distributions: Spatial Inference and Prediction
Janet Franklin

Decline and Recovery of the Island Fox: A Case Study for Population Recovery
Timothy J. Coonan, Catherin A. Schwemm and David K. Garcelon

Ecosystem Functioning
Kurt Jax

Spatio-Temporal Heterogeneity: Concepts and Analyses
Pierre R. L. Dutilleul

Parasites in Ecological Communities: From Interactions to Ecosystems
Melanie J. Hatcher and Alison M. Dunn

Zoo Conservation Biology
John E. Fa, Stephan M. Funk, and Donnamarie O'Connell

Marine Protected Areas: A Multidisciplinary Approach
Joachim Claudet

Biodiversity in Dead Wood
Jogeir N. Stokland, Juha Siitonen and Bengt Gunnar Jonsson

Landslide Ecology
Lawrence R. Walker and Aaron B. Shiels

Nature's Wealth: The Economics of Ecosystem Services and Poverty
Pieter J. H. van Beukering, Elissaios Papyrakis, Jetske Bouma and Roy Brouwer

Birds and Climate Change: Impacts and Conservation Responses
James W. Pearce-Higgins and Rhys E. Green

Marine Ecosystems

Human Impacts on Biodiversity, Functioning and Services

Edited by

TASMAN P. CROWE
University College Dublin, Ireland

CHRISTOPHER L. J. FRID
Griffith University, Queensland, Australia

CAMBRIDGE
UNIVERSITY PRESS

University Printing House, Cambridge CB2 8BS, United Kingdom

Cambridge University Press is part of the University of Cambridge.

It furthers the University's mission by disseminating knowledge in the pursuit of education, learning and research at the highest international levels of excellence.

www.cambridge.org
Information on this title: www.cambridge.org/9781107037670

© Cambridge University Press 2015

This publication is in copyright. Subject to statutory exception and to the provisions of relevant collective licensing agreements, no reproduction of any part may take place without the written permission of Cambridge University Press.

First published 2015

Printing in the United Kingdom by TJ International Ltd. Padstow Cornwall

A catalogue record for this publication is available from the British Library

ISBN 978-1-107-03767-0 Hardback
ISBN 978-1-107-67508-7 Paperback

Cambridge University Press has no responsibility for the persistence or accuracy of URLs for external or third-party internet websites referred to in this publication, and does not guarantee that any content on such websites is, or will remain, accurate or appropriate.

Contents

List of contributors		*page* ix
Part I	**Key concepts**	
1	Introduction Tasman Crowe, Melanie Austen and Christopher Frid	3
2	Ecosystem services and benefits from marine ecosystems Melanie Austen, Caroline Hattam and Tobias Börger	21
3	Assessing human impacts on marine ecosystems Christopher Frid and Tasman Crowe	42
4	Modifiers of impacts on marine ecosystems: disturbance regimes, multiple stressors and receiving environments Devin Lyons, Lisandro Benedetti-Cecchi, Christopher Frid and Rolf Vinebrooke	73
5	Impacts of changing biodiversity on marine ecosystem functioning Tasman Crowe	111
Part II	**Impacts of human activities and pressures**	
6	Marine fisheries and aquaculture Odette Paramor and Christopher Frid	137
7	Artificial physical structures Fabio Bulleri and M. Gee Chapman	167
8	Eutrophication and hypoxia: impacts of nutrient and organic enrichment Samuli Korpinen and Erik Bonsdorff	202

viii · Contents

9 **Pollution: effects of chemical contaminants and debris** 244
Emma L. Johnston and Mariana Mayer-Pinto

10 **Invasions by non-indigenous species** 274
Mads Solgaard Thomsen, Thomas Wernberg and David Schiel

Part III Synthesis and conclusions

11 **Human activities and ecosystem service use: impacts and trade-offs** 335
Melanie Austen, Caroline Hattam and Samantha Garrard

12 **Conclusions** 377
Tasman Crowe, Dave Raffaelli and Christopher Frid

Index 395

Colour plate section between pages 182 and 183.

Contributors

MELANIE AUSTEN
Plymouth Marine Laboratory, Plymouth, UK

LISANDRO BENEDETTI-CECCHI
Dipartimento di Biologi, University of Pisa, Pisa, Italy

ERIK BONSDORFF
Department of Biosciences, Åbo Akademi University, Finland

TOBIAS BÖRGER
Plymouth Marine Laboratory, Plymouth, UK

FABIO BULLERI
Dipartimento di Biologi, University of Pisa, Pisa, Italy

M. GEE CHAPMAN
School of Biological Sciences, University of Sydney, Australia

TASMAN CROWE
School of Biology and Environmental Science, University College Dublin, Belfield, Dublin, Ireland

CHRISTOPHER FRID
Griffith School of the Environment, Griffith University, Queensland, Australia

SAMANTHA GARRARD
Plymouth Marine Laboratory, Plymouth, UK

CAROLINE HATTAM
Plymouth Marine Laboratory, Plymouth, UK

EMMA L. JOHNSTON
Evolution and Ecology Research Centre, School of Biological, Earth and Environmental Sciences, University of New South Wales, Sydney, Australia

SAMULI KORPINEN
Finnish Environment Institute, Marine Research Centre, Helsinki, Finland

DEVIN LYONS
School of Biology and Environmental Science, University College Dublin, Belfield, Dublin, Ireland

MARIANA MAYER-PINTO
Evolution and Ecology Research Centre, School of Biological, Earth and Environmental Sciences, University of New South Wales, Sydney, Australia

ODETTE PARAMOR
The University of Nottingham Ningbo China, Department of Geographical Sciences, Ningbo, Zhejiang, P.R. China

DAVE RAFFAELLI
Environment Department, University of York, York, UK

DAVID SCHIEL
Marine Ecology Research Group, School of Biological Sciences, University of Canterbury, Christchurch, New Zealand

MADS SOLGAARD THOMSEN
Marine Ecology Research Group, School of Biological Sciences, University of Canterbury, Christchurch, New Zealand

ROLF VINEBROOKE
Department of Biological Sciences, University of Alberta, Edmonton, Alberta, Canada

THOMAS WERNBERG
UWA Oceans Institute and School of Plant Biology, The University of Western Australia, Crawley, Australia

Part I
Key concepts

1 · Introduction

TASMAN CROWE, MELANIE AUSTEN
AND CHRISTOPHER FRID

1.1 Marine ecosystems

The Earth is a blue planet. Seas and oceans cover over 70% of the Earth's surface and, with an average depth of over 3.2 km, the total volume of marine ecosystems is vastly greater than that of terrestrial and freshwater environments combined, comprising 98% of the total inhabitable space on the planet (Speight and Henderson, 2010). Marine ecosystems contain 31 of the 33 phyla of animals, each of which constitutes a unique and distinctive body plan, with 15 of those phyla occurring only in the sea (Angel, 1993). Approximately 250 000 marine species have been described, with an estimated 750 000 still to be discovered (Census of Marine Life, 2010). Marine creatures include the largest ever to live (blue whales), yet the energy to fuel these giants is mainly captured by microscopic plankton, rather than more substantial plants, one of a number of fundamental differences between marine and terrestrial ecosystems (Steele, 1985, 1991; Webb, 2012). Of the global annual net primary productivity, approximately 104.9 petagrams (10^{15} grams) of carbon per year, around half is produced by marine ecosystems (Field et al., 1998). Although coral reefs and beds of seaweed are, per unit area, among the most diverse and productive ecosystems on Earth, open oceans have levels of productivity akin to terrestrial deserts because life is so thinly spread, but nevertheless make the single greatest contribution to global productivity because of their size (Whittaker and Likens, 1975).

Archaeological records show that even the earliest humans exploited the marine environment for food, as a medium for transportation and as a repository for waste (Jackson, 2001; Jackson et al., 2001). Today, the marine environment provides approximately 80 million tonnes per

Marine Ecosystems: Human Impacts on Biodiversity, Functioning and Services, eds T. P. Crowe and C. L. J. Frid. Published by Cambridge University Press. © Cambridge University Press 2015.

year of fish and shellfish from capture fisheries, representing about 3% of our global animal protein supply. Marine aquaculture contributes an additional 20 million tonnes (FAO, 2012). Therefore, marine ecosystems can be seen as having an important role in global food security (Frid and Paramor, 2012). It is increasingly recognised that they contribute much more than that to human well-being, in economic, social and cultural terms. This recognition was initially crystallised when Costanza *et al.* (1997) estimated that coastal seas, open oceans, estuaries and saline marshes provide an estimated 68% of the total economic value of all ecosystem goods and services derived from the natural environment. Whilst the detail of Costanza *et al.*'s valuations has been subsequently disputed by environmental economists, their analysis did alert a much wider audience to the overwhelming importance of marine environments to human well-being. More recently, Sumaila and Cisneros-Montemayor (2010) estimated that globally 121 million people a year participate in ecosystem-based marine recreational activities, generating 47 billion USD (2003) in expenditure and supporting one million jobs.

For many years, marine ecosystems were considered to be inexhaustible sources of bounty and a convenient and resilient dumping ground. In fact, even early human societies were capable of exploitation at levels that were, in modern parlance, unsustainable. For example, Stone Age rubbish tips around the Mediterranean show that the size of the shellfish being exploited decreased over time and then the resource collapsed (Desse and Desse-Berset, 1993). It is now recognised that human activities are degrading marine ecosystems in many places through overfishing and activities that cause habitat destruction, pollution and the spread of invasive species (Millennium Ecosystem Assessment, 2005; Halpern *et al.*, 2008). Much of that degradation is out of sight and largely out of mind for much of society. Nevertheless, there is a groundswell of support for conservation and more sustainable exploitation, driven in part by high profile issues and events, such as overfishing, algal and jellyfish blooms and major oil spills, as well as improving public awareness of the beauty and fragility of the marine realm.

Given their great richness and long-standing cultural importance, a strong argument can be made that marine ecosystems deserve conservation on purely aesthetic and moral grounds. However, new arguments are being made alongside these that require a new way of thinking and a particular kind of scientific underpinning.

1.2 Ecosystem services, ecosystem functioning and biodiversity

Ecosystem services are an emerging driver of conservation policy and a component of the ecosystem approach to environmental management (Millennium Ecosystem Assessment, 2005; UNEP, 2010; UK National Ecosystem Assessment, 2011). They can be defined in terms of the contributions of ecosystems to human well-being, encompassing both tangible goods such as food and raw materials, but also more intangible services such as climate regulation, flood protection and cultural and aesthetic enrichment (de Groot *et al.*, 2010; TEEB, 2010). There is a clear distinction between ecosystem services, which are made available by ecosystems and the benefits that society chooses to derive from them. Thus, ecosystem services should be considered from an ecological perspective and benefits must be considered and measured from social, economic and health perspectives. It is the benefits which can be measured in value terms such as monetary units, as well as through other non-monetary value systems such as happiness, employment, and health improvement (Chapter 2), and thereby be more easily taken into account in environmental policy and management decisions, that can be justified in terms understood by the public and politicians. The prominence of ecosystem services in the international policy arena is underscored by the recent establishment of the Intergovernmental Panel on Biodiversity and Ecosystem Services to promote and inform the application of the concept in policy and management.

Delivery of ecosystem services depends on the efficient functioning of ecosystems (Worm *et al.*, 2006). Ecosystem functioning can be defined in a number of ways (Paterson *et al.*, 2012), but can broadly be described as the processing of energy and materials by ecosystems. Ecosystem functioning is quantified by measuring the rates of ecosystem processes. Ecosystem processes include primary and secondary production, respiration, decomposition, nutrient cycling and flows of energy through food webs (trophic dynamics). These major processes encompass functional pathways such as photosynthetic activity, nutrient fluxes and uptake, sediment mixing and stabilisation and clearance of particles from the water column. Community ecological processes, such as predation, herbivory, parasitism, mutualism and competition also influence ecosystem processes, particularly trophic dynamics, and are included in some definitions of functioning (e.g. Lawton and Brown, 1993; Martinez, 1996; Duffy 2009).

It is worth making clear distinctions between the structure of ecosystems, their properties and their functioning. The structure of an ecosystem constitutes its physical and chemical properties and the identity and relative abundance of its biological constituents (or biodiversity – see below). An ecosystem property can be defined as any aggregate structural variable describing the state of the system, such as biomass or sediment nitrogen content, and an ecosystem function as any aggregate process, such as production or grazing rate (Duffy, 2009). The stability of the structure and functioning of an ecosystem can be thought of as one of its key properties, and again comprises a number of distinct facets (see Chapter 3). Resistance to invasion is also an ecosystem property, essentially a particular aspect of stability, that has received considerable attention in this context (Stachowicz et al., 2007). Pacala and Kinzig (2002) also discuss these distinctions but consider stocks, fluxes and stability as different aspects of ecosystem functioning. These terms correlate more closely with the interdisciplinary (natural and social sciences) terminology used to quantify ecosystem services and the benefits that arise from them.

The functioning of ecosystems and the services they provide depend on biodiversity and environmental conditions (Balvanera et al., 2006; Stachowicz et al., 2007; Naeem et al., 2009). Biodiversity is another term that is used by many people to mean many things. Since being coined in 1988 (Wilson and Peters, 1988), it has found its way into the public lexicon becoming a byword for wildlife and used as a rallying call for conservation. In the broadest terms, it can be thought of as the variety of life, encompassing genes, organisms and habitats or ecosystems (United Nations, 1992) and is a key aspect of ecosystem structure. A given area that contains populations with high levels of genetic variability, many different species, many different habitat types or all three can be thought of as having high biodiversity. Although for most people, the number of species is the most obvious aspect of biodiversity, it is also worth taking account of which species are present, their relative abundances, the functional roles they play and the ways in which they are related to each other. In recent years, there has been extensive research establishing a link between the functioning of ecosystems and the biodiversity they contain. It is now widely accepted that, on average, loss of biodiversity causes reductions in the rates of many ecosystem processes, although this generalisation masks a much more complex story (see Chapter 5). There are complex feedbacks between the biodiversity in an ecosystem and the physical and chemical conditions and many of these remain unresolved.

As emphasised by a number of authors (e.g. Srivastava and Vellend, 2005; Duffy, 2009), it is important to make a clear distinction between changes in ecosystems and their consequences for society. Changes in ecosystem structure and functioning, whether positive or negative, large or small, are not necessarily either 'good' or 'bad' for society. In contrast, an ecosystem service is considered to have some positive value to human well-being, but it is not necessarily clear whether, on balance, a particular change to an ecosystem process should be considered beneficial or harmful. For example, an increase in primary production induced by nutrient inputs can be considered either positive for society (e.g. by increasing the productivity of a fishery) or negative (e.g. by degrading water quality via eutrophication). Traditionally, ecosystem scientists have primarily been concerned with characterising patterns and processes of ecosystem change in purely objective terms. As emphasised in this book, they are increasingly engaging with social scientists and economists to explore the equally complex challenge of assessing their consequences for society (e.g. Barbier *et al.*, 2011; Isbell *et al.*, 2011), which requires considerations of trade-offs that are not purely related to the consequences of ecosystem change but have wider social, economic and often political dimensions. Given that human activities are both causing biodiversity loss and changing environmental conditions (Millennium Ecosystem Assessment, 2005), it is essential that these consequences are better understood to better inform societal responses and actions.

1.3 Policy and legislative context

In September 2000, world leaders came together at the headquarters of the United Nations and adopted the UN Millennium Declaration (United Nations, 2000). This set a series of time-limited targets, that became known as the Millennium Development Goals (MDGs) (Table 1.1a), to be achieved by 2015. Progress in meeting these goals has been a central tenet of UN development work and international policy ever since. While human health and security dominate these goals, two are tightly linked to ecological processes. MDG1 can be summarised as food security and MDG7 as sustainability and the targets for these goals (Table 1.1b) provide the international policy context for the development of much of the environmental agenda in the first part of the twenty-first century.

Perhaps the single most important international agreement driving the conservation of biodiversity pre-dates the establishment of these goals,

Table 1.1 *(a) UN Millennium Development Goals and (b) the targets associated with Goals 1 and 7, which have the most direct links to ecological processes.*

		(a)	
Goal 1		Eradicate extreme poverty and hunger	
Goal 2		Achieve universal primary education	
Goal 3		Promote gender equality and empower women	
Goal 4		Reduce child mortality	
Goal 5		Improve maternal health	
Goal 6		Combat HIV/AIDS, malaria and other diseases	
Goal 7		Ensure environmental stability	
Goal 8		Develop a global partnership for development	
		(b)	
Goal 1	Eradicate extreme poverty and hunger	Target 1	Halve, between 1990 and 2015, the proportion of people whose income is less than $1 per day
		Target 2	Halve, between 1990 and 2015, the proportion of people who suffer from hunger
Goal 7	Ensure environmental stability	Target 9	Integrate the principles of sustainable development into country policies and programmes and reverse the loss of environmental resources
		Target 10	Halve, by 2015, the proportion of people without sustainable access to safe drinking water and basic sanitation
		Target 11	Have achieved by 2020 a significant improvement in the lives of at least 100 million slum dwellers

however. The Convention on Biological Diversity (CBD) was agreed in Rio de Janeiro in 1992 and has been ratified by the majority of the world's nations (United Nations, 1992; Convention on Biological Diversity, 2013). In the run up to the 20th anniversary of the CBD, the international community met in Nagoya, Japan, and agreed five Strategic Goals for global biodiversity and backed these with 20 quantified and time-limited targets (UNEP, 2010; Table 1.2). In both agreements, the fundamental dependence of human life (and lifestyles) on functioning ecosystems through ecosystem services and benefits provides the underlying rationale for conserving biodiversity and natural ecosystems. This has been a key factor in the propagation of that rationale in conservation

Table 1.2 *Strategic goals of the Convention on Biological Diversity's strategy for 2011–20 (UNEP, 2010). Each Strategic Goal has a set of associated Aichi Targets (20 in total) which provide a more detailed set of aspirations, such as: at least halve and, where feasible, bring close to zero the rate of loss of natural habitats; establish a conservation target of 17% of terrestrial and inland water areas and 10% of marine and coastal areas; restore at least 15% of degraded areas through conservation and restoration activities; make special efforts to reduce the pressures faced by coral reefs.*

Strategic Goal A	Address the underlying causes of biodiversity loss by mainstreaming biodiversity across government and society
Strategic Goal B	Reduce the direct pressures on biodiversity and promote sustainable use
Strategic Goal C	To improve the status of biodiversity by safeguarding ecosystems, species and genetic diversity
Strategic Goal D	Enhance the benefits to all from biodiversity and ecosystem services
Strategic Goal E	Enhance implementation through participatory planning, knowledge management and capacity building

policy and practice to many parts of the world, driving the establishment of a range of legislative instruments at regional and national scales.

At a national level, most governments have a developed system of planning and environmental management and regulation for activities occurring within their terrestrial territories. The extent to which these extend across the intertidal and into the coastal seas is highly variable. Many countries have or are developing approaches, broadly grouped as 'coastal zone management', that seek to provide integrated planning and environmental protection for the coastal zone. These generally focus on development, infrastructure and sea defences with water quality and fisheries managed by separate processes. The development of the UN Convention on the Law of the Sea (UNCLOS) provided a framework for the development of legislative frameworks covering the high seas (beyond 12 nautical miles from the coast).

UNCLOS allows nations with a coastline to establish Exclusive Economic Zones (EEZ) out to 200 nautical miles from their coasts (or to the median line with another state). Within its EEZ, each nation controls the exploitation of living and non-living resources. The UNCLOS was first published in the 1970s (having been in negotiation since the early 1950s) but the value of the potentially rich mineral resources on and under the seafloor have ensured that claims and counter claims for parts of the ocean floor continue to be made to this day. Even if agreement was reached on the geographically defined units that constitute each nation's EEZ, biological resources and ecological processes extend across these boundaries.

A variety of international treaties have been agreed to address the transboundary nature of, for example, pollution dispersion and fish stock migration (e.g. United Nations Conference on Straddling Fish Stocks and Highly Migratory Fish Stocks; United Nations, 1995). The CBD ushered in a more holistic approach to environmental management, requiring protection of functioning ecosystems (United Nations, 1992). This has created an impetus to develop environmental management of the seas and oceans around ecological boundaries. A decade-long study initiated by the UN Environment Programme (UNEP) and driven by the US National Oceans and Atmosphere Authority (NOAA), has delimited and characterised the world's Large Marine Ecosystems (NOAA, 2009; Sherman and Hempel, 2009). These large units are being used, to some extent, as management units when they fall entirely within the jurisdiction of a single authority (EU, Australia, Canada) but, to date, there has been little international collaboration to develop management around such ecologically based areas.

In 1997, Canada became the first nation to introduce a legislative framework for integrated management for its EEZ. The Canadian Oceans Act has three guiding principles; sustainable development, integrated management and a precautionary approach. Five spatial management units, known as Large Ocean Management Areas (LOMAs) were designated and a commitment made to develop management plans for each one. The smaller of the LOMAs (i.e. the Beaufort Sea and St Lawrence Seaway) might be regarded as ecological units. In 2002, a strategy document was published and this in turn led to an action plan for management within the LOMAs that was published in 2005. Since then the focus has been on establishing the current health of the ecosystems and identifying the main threats to maintaining healthy seas.

Other holistic marine ecosystem management initiatives include the EU Marine Strategy Framework Directive and the US and Australian 'aspirational' ocean policy documents. For example, the US government has produced a framework document published in 2004 entitled *An Ocean Blueprint for the 21st Century* (US Commission on Ocean Policy, 2004). It is not a statutory instrument but seeks to facilitate federal, regional and local initiatives and foster the development of a common framework for management to improve decision making, promote effective co-ordination and 'move towards' an ecosystem-based approach. In 2008, Australia published an Oceans Policy (Commonwealth of Australia, 1998), that explicitly sought to promote the development of Regional Marine Plans, based on Large Marine Ecosystems boundaries. The policy aimed to deliver integrated and ecosystem-based planning and regulation.

In Europe, the *Marine Strategy Framework Directive* (MSFD) was enacted in 2008, building on a range of previous directives and initiatives, including the Habitats Directive and the Water Framework Directive. The MSFD can be seen as the result of an evolution in the European Union's approach to environmental management. In the 1970s, directives focused on protection of human health and activities, the Dangerous Substances Directive and its daughter directives, restricted levels of the most toxic substances in the environment, including marine and freshwaters; the Bathing Waters Directive sought to limit the exposure of water users to pathogens and toxins; and the Shellfish Directive did the same for waters used to raise shellfish for human consumption. Application of these directives delivered benefits for the 'health' of the wider ecosystem. The Water Framework Directive sought to provide a more holistic approach to protecting fresh, coastal and transitional waters and established the principle of using biological indices as a measure of environmental health; the MSFD goes further. The overarching aim of the MSFD is to move all of Europe's marine waters into a condition of 'Good Environmental Status'. While directives are legally binding on member states of the EU, the MSFD provides a framework within which actions need to be taken rather than comprising a prescriptive set of regulations. Those actions include assessing the current status of each member state's seas, establishing targets and indicators of good status, formulating measures to achieve it and undertaking monitoring to evaluate success. Each nation can independently decide what criteria to use to assess 'good environmental status' for its waters. To provide some structure to this, the EU is focusing on 11 'descriptors' of the marine environment (Table 1.3), these being a mix of environmental/ecosystem properties and measures of human pressures on the environment.

In general, the emerging marine policy framework aspires towards more holistic management regimes, taking an ecosystem-based approach, taking account of ecosystem services and managing multiple sectors in a unified framework (Kidd *et al.*, 2011). The emerging field of *marine spatial planning* is akin to established terrestrial planning processes and is expected to provide a key tool for delivery of this aspiration through

Table 1.3 *The 11 descriptors of the marine environment used under the EU Marine Strategy Framework Directive as a basis for making an assessment of whether a system achieves 'Good Environmental Status'.*

	Environmental Aspect
Descriptor 1:	Biological diversity
Descriptor 2:	Non-indigenous species
Descriptor 3:	Population of commercial fish/shellfish
Descriptor 4:	Elements of marine food webs
Descriptor 5:	Eutrophication
Descriptor 6:	Sea-floor integrity
Descriptor 7:	Alteration of hydrographical conditions
Descriptor 8:	Contaminants
Descriptor 9:	Contaminants in fish and seafood for human consumption
Descriptor 10:	Marine litter
Descriptor 11:	Introduction of energy, including underwater noise

Adapted from Annex 1 of MSFD (European Union, 2008).

the simultaneous consideration of pressures from multiple sectors and the ability to assess and manage trade-offs between benefits derived from a proposed activity and any costs in terms of impairment of other ecosystem services and societal benefits.

1.4 Scientific knowledge required to underpin implementation of legislative frameworks

In the preceding sections, we have established that marine ecosystems are intrinsically valuable and that human activities are affecting them in complex ways. Society is recognising that action must be taken to prevent their degradation and has enacted legislation to drive that action. To take effective action with limited resources requires decisions to be made about which management interventions to deploy to maximise the benefits to society. Making good decisions requires a good understanding of what the consequences are likely to be of choosing different options. In this context, decision makers need to understand how the range of human activities that do or could take place within their jurisdiction affect the capacity of ecosystems to sustainably deliver services to society. Armed with such knowledge, decision makers should be able to determine which activities should be restricted as a matter of highest priority in order to conserve, restore or ensure sustainable delivery of the particular services that are most highly valued that arise from within their management area. In this section, we examine the extent to which relevant knowledge is available and accessible to environmental decision makers.

Extensive research has been completed on impacts of human activities on biodiversity and environmental conditions. The most wide-reaching review of this work was undertaken in the Millennium Ecosystem Assessment (2005), but a wealth of individual articles and reviews is available (e.g. Polunin, 2008, and references therein and see Chapters 6–10). A lot of work focuses on the effects of individual stressors on individual taxa and is dominated by laboratory-based research, e.g. as ecotoxicological studies. Field-based studies on assemblages of organisms are rarer, but those that are well designed and executed provide particularly valuable information in this context (see review by Mayer-Pinto *et al.*, 2010). Data from Environmental Impact Assessments and other forms of environmental assessment and monitoring are also of value, but are not always widely accessible or included in syntheses.

As mentioned above, since the early '90s, there has also been a considerable research effort to understand how changes in biodiversity can affect ecosystem functioning (see also Chapter 5). Most of this work, however, has been geared towards testing theoretical models and has not been done at the appropriate scale or on the appropriate communities or processes for maximal relevance to environmental decision makers (Kremen, 2005; Raffaelli, 2006; Naeem, 2006; Duffy, 2009; Crowe *et al.*, 2012). Many of the findings to date suggest that consequences of biodiversity loss can be context dependent and strongly influenced by the identities of the species lost and of those remaining. Research to date has focused primarily on a limited number of tractable model systems and research localities and findings may not be directly applicable to new contexts and ecosystems.

More recently, researchers (and decision makers) have begun to consider the links between ecosystem functioning and ecosystem services (UK National Ecosystem Assessment, 2011; Barbier *et al.*, 2011). Efforts have focused on the easier to study services such as food provision, regulatory services, such as gas and climate regulation and flood and storm defence, and intermediate (supporting) services, such as nutrient cycling (Liquete *et al.*, 2013). Often they focus on easy to study habitats, such as sea grasses, mangroves and coral reefs (Liquete *et al.*, 2013); or locations where biodiversity and functions that underpin regulatory services might be comparatively simple, such as seagrass (e.g. Duarte *et al.*, 2013), saltmarsh and coastal wetlands (e.g. Beaumont *et al.*, 2014) and mangroves (e.g. Donato *et al.*, 2011) for carbon regulation (Pendleton *et al.*, 2012); or the links between cultural services or food provision to habitat are very apparent, such as coral reefs. Much of the research has been piecemeal with a focus on developing and implementing methodology for practical assessments and an emphasis on using existing data which was not collected for this purpose. Such data have been shown to be inadequate for comprehensive assessment (e.g. Austen *et al.*, 2011) and, in addition, are rarely collected at the appropriate scale for management and policy purposes. In contrast to the research on how biodiversity can affect ecosystem functioning, the marine ecosystem functioning-ecosystem service research is largely without reference to theory. Marine ecosystem service research has been confounded by lack of agreement on the typology and conceptual frameworks (Chapter 2), although there is a generally agreed dissatisfaction with the direct transfer of terrestrial frameworks to the marine environment.

Regardless of the shortcomings of these three fields of research (on impacts, biodiversity–ecosystem function, or BEF, relationships and ecosystem services), the evidence they have yielded is what is required as a basis for decisions geared towards maximising societal benefits from marine ecosystems while minimising impacts on them. To a considerable degree, however, these three fields have remained distinct from each other. There is therefore an urgent need to integrate them, such that the consequences of a decision to restrict or permit a particular activity can be traced through changes to biodiversity and ecosystem functioning to changes in ecosystem services and benefits. Such integration requires a concerted effort in terms of collating and synthesising existing information, as well as targeted research to redress the shortcomings identified above. Models are needed to integrate data with mechanistic understanding to provide predictions of changes to ecosystems and services under different scenarios of environmental change and management intervention. Decision-support tools must be developed to make these outcomes accessible to decision makers in a useable form.

A key data need is for maps of environmental conditions, ecosystems and habitats to develop an understanding of how these will respond to different levels of pressure from human activities as well as similar spatial information concerning the current and future demand for ecosystem services. Planners and environmental managers can, armed with this information, engage with society about managing the priorities and trade-offs that will always arise in holistic, multi-sector approaches.

To improve decision making into the future, it is necessary to test how effective the actions taken have been in achieving their objectives. This is reflected in much of the legislation, which requires monitoring to determine the current status of ecosystems and then to assess whether status improves after management action has been taken (see above). Doing this effectively requires carefully designed monitoring programmes focused on appropriate response variables. Logical designs for such programmes have been developed but are not always adopted (Underwood, 1991, 1995; Spellerberg, 2005; Lindenmayer and Likens, 2010). Thus, some of the approaches in use do not lead to clear-cut interpretations of changes and their causes. Considerable effort has also gone into and continues to go into selecting appropriate response variables and developing cost-effective metrics of ecological integrity (e.g. Borja and Dauer, 2008). Most indicators and metrics developed to date focus essentially on biodiversity. New tools will be needed which more explicitly capture status in terms of ecosystem functioning and service provision.

1.5 Outline and scope of book

Many marine ecosystems are extremely productive and highly valued by society, yet are threatened by human activities (Beaumont *et al.*, 2007; Halpern *et al.*, 2008; Barbier *et al.*, 2011). In many parts of the world, they are now legally protected and in some cases subject to detailed spatial planning (Kidd *et al.*, 2011), but managers often lack ready access to research findings to underpin effective management decisions. This book aims to provide that information in an accessible format, emphasising findings that can underpin environmental decision making and intervention. We will provide a detailed review of impacts of a range of specific human activities and pressures on biodiversity and ecosystem functioning. Uniquely, this book will also seek to follow these changes through to impacts on ecosystem services, such that managers can make informed decisions on how to prioritise management interventions to maximise benefits to society, as required under emerging legislation. It will also summarise key concepts and critically review recent developments in relevant research areas. As such we hope it will also be valuable to researchers and advanced students in marine ecology and environmental management.

In the introductory chapters, we present critical summaries of the key concepts that constitute the scientific basis for environmental decision making in relation to marine ecosystems and the goods and services they provide. Chapter 2 will consider the range of ecosystem services that society derives from marine ecosystems. We then undertake a broad examination of the ways in which human activities that are undertaken to derive benefits from marine ecosystems affect marine biodiversity and ecosystems, the mechanisms by which they do so and the kinds of scientific approaches that are used to reveal their effects (Chapter 3). In Chapter 4, we describe how the ways in which anthropogenic stressors affect ecosystems can be strongly influenced by the ways in which they arrive in the system (so-called disturbance regimes). In recognition of the fact that many ecosystems are affected by many different kinds of stressor, we also argue that it is vital to understand how multiple stressors interact to influence each other. We will review this knowledge to some degree, but will argue that it is largely incomplete. In its absence, the primary focus of the book is on impacts of individual stressors.

The fifth chapter reviews current understanding of relationships between biodiversity and ecosystem functioning in general terms that relate to the themes of this book. It will ask whether and how changes in different measures of biodiversity could be used to predict changes in functioning in a given context.

The core chapters, 6 to 10, each explore the impacts on biodiversity, ecosystem functioning and services of a particular human activity or of pressures derived from several activities (see Chapter 3 for full details of coverage). These chapters are arranged in part to reflect traditional divisions in environmental decision making and management. There are generally separate departments responsible for fisheries, spatial planning and the control of contaminants, nutrients and invasive species. We have arranged the chapters to maximise the accessibility of information relevant to individuals working in those separate departments. However, we also present strong arguments in favour of greater integration between departments in risk assessment and priority setting.

In this book, climate change and other consequences of raised CO_2, such as ocean acidification, will be discussed as potential modifiers of the impact of local stressors (in Chapter 4) rather than being treated in detail in their own right. This is first because these topics have recently been reviewed in considerable depth and detail (see Chapter 3 for references), and second because over the next few decades at least, they will continue to occur regardless of any decisions made at a local or regional level. Thus, aside from efforts for mitigation and adaptation, the important thing is for managers to be able to assess whether the effects of a particular local stressor are going to be exacerbated by climate change in years to come and thus to be able to identify which local stressors to prioritise for control for benefits now and in the future.

In Chapter 11, we synthesise the information presented in Chapters 6–10 and summarise the resultant knowledge of how particular activities and pressures influence ecosystem services. Chapter 11 then discusses how this knowledge can be applied to the decision-making process as trade-offs between the costs and benefits of a range of possible management interventions. In Chapter 12, we summarise the key insights identified in this book, reflect on the contribution of ecological insight towards the achievement of the relevant Millennium Ecosystem Goals (see Section 1.3) and consider the value of the ecosystem services concept in further improving the relationship between human society and the sea.

Throughout the book, examples will be primarily drawn from temperate systems in Europe, North America and Australasia and will generally emphasise coastal habitats.

References

Angel, M. V. (1993). Biodiversity of the pelagic ocean. *Conservation Biology*, 7, 760–772.
Austen, M. C., Malcolm S. J., Frost, M. *et al.* (2011). Marine. In *The UK National Ecosystem Assessment Technical Report*. Cambridge: UNEP-WCMC, pp. 459–499.
Balvanera, P., Pfisterer, A. B., Buchmann, N. *et al.* (2006). Quantifying the evidence for biodiversity effects on ecosystem functioning and services. *Ecology Letters*, 9, 1146–1156.
Barbier, E. B., Hacker, S. D., Kennedy, C. *et al.* (2011). The value of estuarine and coastal ecosystem services. *Ecological Monographs*, 81, 169–193.
Beaumont, N. J., Austen, M. C., Atkins, J. P. *et al.* (2007). Identification, definition and quantification of goods and services provided by marine biodiversity: Implications for the ecosystem approach. *Marine Pollution Bulletin*, 54, 253–265.
Beaumont N. J., Jones, L., Garbutt, A. *et al.* (2014). The value of blue carbon sequestration and storage in coastal habitats. *Estuarine and Coastal Shelf Science*, 137, 32–40.
Borja, A. and Dauer, D. M. (2008). Assessing the environmental quality status in estuarine and coastal systems: Comparing methodologies and indices. *Ecological Indicators*, 8, 331–337.
CBD (2013). Convention on Biological Diversity. Available at: http://www.cbd.int/convention/, accessed December 2013.
Census of Marine Life (2010). Available at: http://www.coml.org, accessed December 2013.
Commonwealth of Australia 1998. *Australia's Oceans Policy: caring, understanding, using wisely*. Commonwealth Government of Australia, Canberra, Australia.

Costanza, R., d'Arge, R., de Groot, R. et al. (1997). The value of the world's ecosystem services and natural capital. *Nature*, 387, 253–260.

Crowe, T. P., Bracken, M. E. and O'Connor, N. E. (2012). Reality check: issues of scale and abstraction in biodiversity research, and potential solutions. In *Marine Biodiversity Futures and Ecosystem Functioning: Frameworks, Methodologies and Integration*, ed. M. Solan, R. J. A. Aspden and D. M. Paterson. Oxford: Oxford University Press, pp. 185–199.

de Groot, R. S., Fisher, B., Christie, M. et al. (2010). Integrating the ecological and economic dimensions in biodiversity and ecosystem service valuation. In *The Economics of Ecosystems and Biodiversity: Ecological and Economic Foundations*, ed. P. Kumar. London and Washington DC: Earthscan, pp. 9–40.

Desse, J. and Desse-Berset, N. (1993). Pêche et surpêche en Mèditerranée: le temoinage des os. In *Exploration des Animaux Sauvages à Travers le Temps*, ed. J. Desse and F. Audoin-Rouzeau. Juan-les-Pins, France: Editions APDCA, pp. 327–339.

Donato, D. C., Kauffman, J. B., Murdiyarso, D. et al. (2011). Mangroves among the most carbon-rich forests in the tropics. *Nature Geoscience*, 4, 293–297.

Duarte, C. M., Sintes, T. and Marba, N. (2013). Assessing the CO_2 capture potential of seagrass restoration projects. *Journal of Applied Ecology*, 50, 1341–1349.

Duffy, J. E. (2009). Why biodiversity is important to the functioning of real-world ecosystems. *Frontiers in Ecology and the Environment*, 7, 437–444.

European Union (2008). Directive 2008/56/EC Of The European Parliament and of the Council of 17 June 2008 establishing a framework for community action in the field of marine environmental policy (Marine Strategy Framework Directive). *Official Journal of the European Union*, L164, 19–40.

FAO (2012). *The State of World Fisheries and Aquaculture*. Rome: Food and Agriculture Organization.

Field, C. B., Behrenfeld, M. J., Randerson, J. T. and Falkowski, P. (1998). Primary production of the biosphere: integrating terrestrial and oceanic components. *Science*, 281, 237–240.

Frid, C. L. J. and Paramor, O. A. L. (2012). Feeding the world: what role for fisheries? *ICES Journal of Marine Science*, 69, 145–150.

Halpern, B. S., Walbridge, S., Selkoe, K. A. et al. (2008). A global map of human impact on marine ecosystems. *Science*, 319, 948–952.

Isbell, F., Calcagno, V., Hector, A. et al. (2011). High plant diversity is needed to maintain ecosystem services. *Nature*, 477, U199–U196.

Jackson, J. B. C. (2001). What was natural in the coastal oceans? *Proceedings of the National Academy of Sciences of the United States of America*, 98, 5411–5418.

Jackson, J. B. C., Kirby, M. X., Berger, W. H. et al. (2001). Historical overfishing and the recent collapse of coastal ecosystems. *Science*, 293, 629–638.

Kidd, S., Plater, A., and Frid, C. L. J. (2011). *The Ecosystem Approach to Marine Planning and Management*. Oxford: Earthscan, Taylor and Francis Group.

Kremen, C. (2005). Managing ecosystem services: what do we need to know about their ecology? *Ecology Letters*, 8, 468–479.

Lawton, J. H. and Brown, V. K. (1993). Functional redundancy. In *Biodiversity and Ecosystem Function*, ed. E.-D. Schulze and H. A. Mooney. Berlin: Springer-Verlag, pp. 255–270.

Lindenmayer, D. B. and Likens, G. E. (2010). *Effective Ecological Monitoring*. London: Earthscan.

Liquete, C., Piroddi, C., Drakou, E. G. et al. (2013). Current status and future prospects for the assessment of marine and coastal ecosystem services: as systematic review, 8(7), e67737.

Martinez, N. D. (1996). Defining and measuring functional aspects of biodiversity. In *Biodiversity: A Biology of Numbers and Difference*, ed. K. J. Gaston. Oxford: Blackwell Science, pp. 114–148.

Mayer-Pinto, M., Underwood, A. J., Tolhurst, T. et al. (2010). Effects of metals on aquatic assemblages: What do we really know? *Journal of Experimental Marine Biology and Ecology*, 391, 1–9.

Millennium Ecosystem Assessment (2005). *Ecosystems and Human Well-being: Biodiversity Synthesis*. Washington DC: World Resources Institute.

Naeem, S. (2006). Expanding scales in biodiversity-based research: challenges and solutions for marine systems. *Marine Ecology Progress Series*, 311, 273–283.

Naeem, S., Bunker, D. E., Hector, A. et al. (2009). *Biodiversity, Ecosystem Functioning, and Human Wellbeing: An Ecological Perspective*. Oxford: Oxford University Press.

NOAA 2009. NOAA Large Marine Ecosystems of the World. Available at: http://www.lme.noaa.gov/index.php?option=com_content&view=article&id =47&Itemid=41.NOAA, accessed December 2014.

Pacala, S. and Kinzig, A. P (2002). Introduction to theory and the common ecosystem model. In *Functional Consequences of Biodiversity: Empirical Progress and Theoretical Extensions*, ed. A. P. Kinzig, S. W. Pacala, and D. Tilman. Princeton, NJ: Princeton University Press, pp. 169–174.

Paterson, D. M., Defew, E. and Jabour, J. (2012). Ecosystem function and co-evolution of terminology in marine science and management. In *Marine Biodiversity and Ecosystem Functioning: Frameworks, Methodologies and Integration*, ed. M. Solan, R. Aspden and D. M. Paterson. Oxford: Oxford University Press, pp. 24–33.

Pendleton, L., Donato, D. C., Murray, B. C., et al. (2012). Estimating global 'blue carbon' emissions from conversion and degradation of vegetated coastal ecosystems. *PLoS ONE*, 7, e43542.

Polunin, N. V. C. (2008). *Aquatic Ecosystems: Trends and Global Prospects*. Cambridge: Cambridge University Press.

Raffaelli, D. G. (2006). Biodiversity and ecosystem functioning: issues of scale and trophic complexity. *Marine Ecology Progress Series*, 311, 285–294.

Sherman, K. and Hempel, G. (2009). The UNEP Large Marine Ecosystems report: a perspective on changing conditions in the LMEs of the world's regional seas. Nairobi, Kenya: UNEP, p. 851.

Speight, M. and Henderson, P. (2010). *Marine Ecology: Concepts and Applications*. Oxford: John Wiley and Sons.

Spellerberg, I. F. (2005). *Monitoring Ecological Change*. Cambridge: Cambridge University Press.

Srivastava, D. S. and Vellend, M. (2005). Biodiversity–ecosystem function research: is it relevant to conservation? *Annual Review of Ecology Evolution and Systematics*, 36, 267–294.

Stachowicz, J. J., Bruno, J. F. and Duffy, J. E. (2007). Understanding the effects of marine biodiversity on communities and ecosystems. *Annual Review of Ecology and Systematics*, 38, 739–766.

Steele, J. H. (1985). A comparison of terrestrial and marine ecological systems. *Nature*, 313, 355–358.

Steele, J. H. (1991). Can ecological theory cross the land-sea boundary? *Journal of Theoretical Biology*, 153, 425–436.

Sumaila U. R. and Cisneros-Montemayor, A. (2010). A global estimate of benefits from ecosystem-based marine recreation: potential impacts and implications for management. *Journal of Bioeconomics*, 12(3), 245–268.

TEEB (2010). *The Economics of Ecosystems and Biodiversity: Mainstreaming the Economics of Nature: A Synthesis of the Approach, Conclusions and Recommendations of TEEB*. Paris: UNEP.

UK National Ecosystem Assessment (2011). The UK National Ecosystem Assessment: synthesis of key findings. UNEP-WCMC, Cambridge.

Underwood, A. J. (1991). Stupidity, myths and guess-work in the detection of marine environmental impacts: dead bird-watching for hydrographers. *Second Australasian Hydrographic Symposium: The Hydrographic Society Special Publication*, 27, 43–53.

Underwood, A. J. (1995). Ecological research and (and research into) environmental management. *Ecological Applications*, 5, 232–247.

UNEP (2010). Aichi biodiversity targets. Available at: https://www.cbd.int/sp/targets/, accessed December 2014.

United Nations (1992). *Convention on Biological Diversity*. Rome: United Nations.

United Nations (1995). *Agreement for the Implementation of the Provisions of the United Nations Convention on the Law of the Sea of 10 December 1982 relating to the Conservation and Management of Straddling Fish Stocks and Highly Migratory Fish Stocks*. Rome: United Nations.

United Nations (2000). *United Nations Millennium Declaration*. New York: UN General Assembly.

US Commission on Ocean Policy (2004). *An Ocean Blueprint for the 21st Century*. Washington DC: US Commission on Ocean Policy.

Webb, T. J. (2012). Marine and terrestrial ecology: unifying concepts, revealing differences. *Trends in Ecology and Evolution*, 27, 535–541.

Whittaker, R. H. and Likens, G. E. (1975). The biosphere and man. In *Primary Productivity of the Biosphere*, ed. H. Leith and R. H. Whittaker. New York: Springer-Verlag, p. 306.

Wilson, E. O. and Peters, F. M. (1988). *Biodiversity*. Washington DC: National Academy Press.

Worm, B., Barbier, E. B., Beaumont, N. *et al.* (2006). Impacts of biodiversity loss on ocean ecosystem services. *Science*, 314, 787–790.

2 · Ecosystem services and benefits from marine ecosystems

MELANIE AUSTEN, CAROLINE HATTAM
AND TOBIAS BÖRGER

2.1 Introduction

For some time environmentally minded citizens, nongovernmental organisations (NGOs) and other pressure groups have advocated the need for protection and conservation of species and habitats, in both terrestrial and marine ecosystems. There has been frustration that this advocacy has not been fully taken up by environmental policy makers and managers. Similarly, the same policy makers and managers are aware that significant changes to ecosystems have been taking place that require policies and management actions to halt, slow or reverse these changes but that these are costly to implement. In times of competing needs for public funds, it has always been difficult to justify such measures. Yet there has been a growing awareness that changes in ecosystems impact on humans both directly and indirectly, for example through global reductions in fish stocks for food and polluted waters that are unfit for bathing. Humans have recognised their connections and dependence on ecosystems for millennia and Mooney and Ehrlich (1997) assert that concern for our impacts on these connections was formally raised by Marsh in 1864.

In the 1970s, the terms environmental services and then ecosystem services were coined to indicate the positive benefits society gained from the functioning and properties of ecosystems. Mooney and Ehrlich (1997) and Gómez-Baggethun et al. (2009) provide a modern history of the increasing interest in ecosystem services. For marine scientists, a pivotal paper was Costanza et al. (1997) in which the value of the world's ecosystems to people was estimated in monetary terms with marine ecosystems having the highest values of all. This presentation of valuation of ecosystem services was published in *Nature*, a prestigious and widely read journal for natural scientists. It alerted researchers and those who had been concerned about the state of the environment and its ongoing degradation to the opportunity to attempt to quantify and disseminate the importance of ecosystems and their biodiversity from an anthropocentric viewpoint. In addition, it indicated that economists could quantify this importance in an easily understood metric: monetary value. It was hoped that such measures would have considerably greater impact on environmental policy and management than ecologists' metrics, which include diversity and abundance of species and flows of nutrients, or the moral argument put forward by many pressure groups.

Since then, the Millennium Assessment (MA, 2005), The Economics of Ecosystems and Biodiversity (TEEB, 2010) and, more recently, the UK National Ecosystem Assessment (Mace *et al.*, 2011) have attempted to conduct global and national assessments of ecosystem services and all have tackled marine ecosystem services to some extent. These latter assessments reflect an increasing drive by environmental policy makers and managers to consider ecosystems from an anthropocentric perspective in terms of their goods and services or benefits. They have fostered interdisciplinary collaboration between natural scientists and social scientists (particularly economists) to a previously unprecedented extent, working towards the aims and objectives of policy makers.

As this science of ecosystem services has developed rapidly, so have the definitions and typologies (classifications). The whole area is still in a state of flux as will be reflected on in this chapter. Three papers were published recently redefining and re-describing marine ecosystem services (Liquete *et al.*, 2013, Böhnke-Henrichs *et al.*, 2013; Hattam *et al.*, 2015).

In this chapter, a definition of ecosystem services for this book will be presented along with a consideration of some of the different conceptual frameworks. A working typology of marine ecosystem services will be outlined. This is followed by a brief discussion of the characteristics of ecosystem services from spatial and temporal perspectives by considering where and to whom ecosystem services are finally delivered and the dynamic nature of ecosystems. The chapter will conclude with a reflection on the uses of the ecosystem service concepts for science, policy, management and communication.

2.2 What are ecosystem services and benefits?

Ecosystem services are made up of tangible goods (e.g. food and raw materials) and intangible services (e.g. the regulation of our climate and the remediation of waste). There are several definitions of ecosystems services, but here we use the definition from TEEB: 'The direct and indirect contributions of ecosystems to human well-being. The concept of "ecosystem goods and services" is synonymous with ecosystem services' (TEEB, 2010). They are ecological in nature and would continue to exist irrespective of the presence of humans (Fisher et al., 2009).

For the purposes of economic valuation, it is important to distinguish between ecosystem services and the benefits they generate (Boyd and Banzhaf, 2007). The UK's National Ecosystem Assessment proposed that benefits are derived from ecosystem service outcomes (Mace et al., 2011) and goes on to describe the value of welfare improvements as 'benefits'. Thus there is a clear distinction between ecosystem services and the benefits they provide. In this book, we emphasise the contribution of ecosystems to service outcomes, which should be considered from an ecological perspective, and the benefits, which must be considered and measured from social, economic and health perspectives. It is the benefits that can be measured in terms of monetary value. There is also growing interest in supplementing monetary valuation with non-monetary measures such as contributions to happiness, subjective well-being and health improvements (e.g. Vemuri and Costanza, 2006; MacKerron and Mourato, 2013).

Ecosystem services are commonly divided into three major categories (e.g. Fisher et al., 2009; Atkins et al., 2011; Balmford et al., 2011; Mace et al., 2011): provisioning, regulating and cultural. A fourth category of habitat services was added by TEEB (de Groot et al., 2010), although these services are elsewhere included as regulating services (e.g. in the Common International Classification of Ecosystem Services (CICES), EEA, 2013). Descriptions of these categories help understanding of the concept of ecosystem services.

Provisioning services are the products obtained from ecosystems including food, fibres, fuels, genetic resources, medicines and pharmaceuticals, ornamental resources and fresh water.

Regulating services refer to the contribution of ecosystems to the regulation of ecosystems processes including, for example, the regulation of climate, water and some human diseases.

Cultural services generate the non-material benefits that people obtain from ecosystems through spiritual enrichment, cognitive development, reflection, recreation and aesthetic experience, including knowledge systems, social relations and aesthetic values.

Habitat services represent the importance of ecosystems in the provision of living space for resident and migratory species (thus maintaining the gene pool and nursery service).

2.3 Ecosystem service assessment frameworks

Although classifications for ecosystem services were under development well before the UN-supported MA, the MA approach is perhaps still the best known. It defined ecosystem services as 'the benefits people obtain from ecosystems' (MA, 2005), and classified them by functional group, dividing them into supporting, provisioning, regulating and cultural services. Subsequently, it has been widely applied in many different contexts and has formed the basis for a number of alternative classifications.

The loose definition of ecosystem services by the MA (in which services are defined as benefits) undermines the application of accounting systems to ecosystem services (Boyd and Banzhaf, 2007). It is also difficult to apply the MA classification in a decision-making context because supporting services were considered equal to other services, despite the fact that the value of supporting services is considered to be inherent in the value (which is not necessarily reflected in the market price) of all other services (Fisher *et al.*, 2009). The absence of hierarchy within the MA classification therefore makes it inappropriate for use with ecosystem service valuation (Wallace, 2007; Fisher and Turner, 2008) as it may lead to considerable double counting (Boyd and Banzhaf, 2007).

Boyd and Banzhaf (2007) emphasised the need to distinguish ecosystem services from the benefits they generate and the processes and functions from which they are derived. They made a distinction between final ecosystem services and intermediate products. Final ecosystem services were defined as the 'components of nature, directly enjoyed, consumed, or used to yield human well-being', while intermediate products were the components and functions of the ecosystem that input into final ecosystem services. Fisher *et al.* (2009) refined this further by redefining ecosystem services as 'the aspects of ecosystems utilised (actively or passively) to produce human well-being'. They emphasised that ecosystem services are ecological in nature but that they do not have to be used directly (as Boyd and Banzhaf suggest), recognising the indirect use of many ecosystem services (e.g. many regulating services such as climate regulation). To operationalise this approach, Balmford et al. (2011) positioned the Fisher et al. (2009) ecosystem service classification within a general impact assessment framework.

Johnston and Russell (2011) made further clarifications, recommending that the identification of final and intermediate ecosystem services should follow a set of rules, where a biophysical outcome (i.e. the output from an ecosystem process and/or function) from an ecosystem can only be considered a service for a particular beneficiary if:

1. changes in that outcome influence the welfare of that beneficiary;
2. the biophysical outcome is an output of an ecological system prior to any combination with other forms of capital (e.g. labour, finance or technology);
3. the beneficiary is willing to pay for an increase in these outputs (all other aspects of the ecosystem being held constant, i.e. the influence on welfare is not contingent on the change or existence of any other service).

They also recognised that ecosystem services can be beneficiary specific and state that, while the output from an ecosystem process and/or function can represent a final ecosystem service to one beneficiary, they might be only an underlying function to another. It is important to stress that the impact of any ecosystem service on human well-being or welfare, as measured through valuation such as willingness to pay for an increase in that service, is not an indicator of the service itself. Rules 1 and 3 instead state that a specific biophysical outcome has to be perceived by a beneficiary to be classified as an ecosystem service. The extent of this service can only be expressed by certain indicators. Valuation, such as willingness to pay, is just one expression of the benefit that any specific beneficiary gets from the service. It is possible, however, that an ecosystem service can exist and affect the life conditions of a human-being but it is not recognised and hence valued by that person, especially if non-use values are taken into account.

Amidst the ongoing discussion about an appropriate ecosystem service classification, the UN supported a second ecosystem service initiative: TEEB project. TEEB built on the assessment framework developed by Balmford et al. (2008, 2011), emphasising both the ecological and economic aspects important to ecosystem service valuation, biodiversity loss and ecosystem degradation. As indicated earlier, the TEEB project defined ecosystem services as 'the direct and indirect contributions of ecosystems to human well-being', highlighting that individual services secured from the natural environment can have multiple benefits and that services are actually conceptualisations ('labels') of the 'useful things' ecosystems 'do' for people, directly and indirectly (de Groot et al., 2010). Using the 'cascade' model, adapted from Haines-Young and Potschin (2010), TEEB distinguished biophysical structures and processes from ecological functions that may provide services to humans. The benefits derived from these services can be quantified (subject to available enumeration/valuation techniques) and included in economic analyses and decision making. While it is useful to distinguish between ecosystem processes and functions, ecosystem services and the benefits they generate to avoid double counting, it is acknowledged that it is not always possible to do so.

Further developments in classification systems include the incorporation of abiotic components of ecosystems. Often their position in a classification system is not mentioned (e.g. Fisher et al., 2009). Others suggested that their inclusion dilutes the arguments when ecosystem services are used to promote the conservation of biodiversity (Haines-Young and Potschin, 2009). Nevertheless, many naturally occurring abiotic materials and processes are used actively and passively by humans, generating substantial benefits to society. For example, naturally occurring aggregates provide the raw materials for the construction industry, wave and tidal flows can be used to generate electricity, and geological structures may act as tourist attractions (e.g. the Jurassic coast, Dorset, UK) providing inspiration for a number of cultural goods and services. Atkins et al. (2011) therefore explicitly incorporated raw materials, such as aggregates, and energy within their classification system, while Saunders et al. (2010) also included geological processes, space and waterways, and physical barriers. CICES suggests an alternative approach whereby the services generated by abiotic resources are classified in an annex to the ecosystem service classification to ensure they are not overlooked (EEA, 2013).

Once classified, the economics literature provides further approaches to group ecosystem services by how they impact human well-being and by the type of service they constitute within the traditional goods classification of economics. According to the total economic value (TEV) framework ecosystem services benefit humans through direct use, indirect use and non-use (e.g. Pearce and Moran, 1994). This classification has consequences for the choice of appropriate valuation techniques. The economic goods classification groups ecosystem services according to the criteria of rivalry in consumption and excludability of use (e.g. Brown

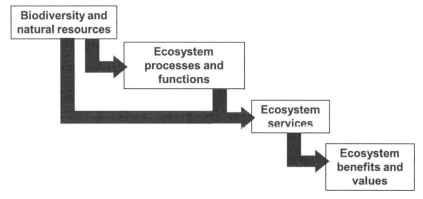

Figure 2.1 Ecosystem service conceptual framework.

et al., 2007; Fisher *et al.*, 2009). Again, the crucial point here is that these economic classification frameworks play a role in the valuation of benefits from ecosystem services rather than in the quantification of services and are thus not further discussed here.

Each of the classification systems has been developed and applied for different contexts and each approach has its merits; however, as de Groot *et al.* (2010) suggested 'perhaps we should accept that no final classification can capture the myriad of ways in which ecosystems support human life and contribute to human well-being and that no fundamental categories or completely unambiguous definitions exist for such complex systems'. Classification systems need to be suitable to the policy and management problem at hand, and therefore different interpretations may be needed depending on the context (Fisher *et al.*, 2009; de Groot *et al.*, 2010). There is much to be gained, however, from building on existing classification systems and the wealth of information already in the ecosystem services literature.

This is the approach that has been taken for marine ecosystem services by Böhnke-Henrichs *et al.* (2013), Liquete *et al.* (2013) and Hattam *et al.* (2015) who, along with others (e.g. Beaumont *et al.*, 2007; Atkins *et al.*, 2011), have tailored ecosystem service classifications specifically for use in the marine environment.

2.4 Ecosystem services typology

In this chapter we use a framework (Figure 2.1) that clearly separates marine ecosystem services from the benefits that accrue from them. The framework also separates the ecosystem services (and benefits) from the underpinning biodiversity, natural resources, ecosystem processes and functions that provide the services. This framework and the Ecosystem Service Typology (Table 2.1) that will be used in this book were developed in the EU FP7 project VECTORS by Hattam et al. (2015). The typology builds upon those developed by Böhnke-Henrichs et al. (2013), Beaumont et al. (2008), Fisher et al. (2009) and Atkins et al. (2011). In this framework, marine ecosystem processes and functions, and biodiversity and natural resources can all be described in ecological terms whilst the benefits and their values are described in the context of human well-being and measured in human value systems. Ecosystem services are described in a combination of ecological and human well-being terms but are measured from an ecological perspective. For example, the ecosystem service wild-capture seafood provision is described as the supply of marine organisms for human consumption. It is measured in terms of populations and type of organisms that are available to be captured for food. The benefits are food for human consumption and can be measured in terms of weight and value of fish landings for consumption.

Although it has been proposed that abiotic natural resources, such as aggregates (sand and gravel for building) and marine energy (wind, wave, tidal), should be included in ecosystem services by Atkins et al. (2011) and Saunders et al. (2010), they are not explicitly included in the typology here. They are referred to in several chapters as their exploitation can have impacts on biodiversity and ecosystems and therefore need to be taken into account in spatial planning and environmental management. The typology for ecosystem services, however, is constrained to those services that derive from the living components of an ecosystem and are influenced by this living component. The quantity and quality of abiotic resources are not typically affected by the living part of an ecosystem. Where they are (e.g. water quality), this is already captured by other ecosystem services (e.g. waste treatment and assimilation).

2.5 Ecosystem service characteristics

2.5.1 Interlinkages of services and benefits

As is evident from the typology, marine ecosystems provide an extensive number of services. The same services tend to be delivered by different habitat types (i.e. sediment, rock or pelagic) regardless of where they are (i.e. intertidal, coastal shelf, transitional waters, deep sea). Although

Table 2.1 *Marine ecosystem service typology. Developed in the EU FP7 project VECTORS by Hattam et al. (2015) the typology builds upon those developed by Beaumont et al. (2008), Fisher et al. (2009), Atkins et al. (2011) and Böhnke-Henrichs et al. (2013).*

	Ecosystem service	Description	Examples
		PROVISIONING SERVICES	
1	Food provision:		
	(a) Wild capture seafood	All available marine flora and fauna extracted from unmanaged marine environments for consumption by humans	Fish, shellfish, seaweed
	(b) Farmed seafood	Food from aquaculture for consumption by humans	
2	Biotic raw materials (non-food):		
	(a) Genetic resources	The provision/extraction of genetic material from marine flora and fauna for use in non-medicinal contexts	Genetic enhancement of biofuel-producing microalgae
	(b) Medicinal resources	Any material that is extracted from or used in the marine environment for its ability to provide medicinal benefits	Marine-derived pharmaceuticals; neutraceuticals
	(c) Ornamental resources	Any material that is extracted for use in decoration, fashion, handicrafts, souvenirs, etc.	Shells, aquarium fish, pearls, coral
	(d) Other biotic raw materials	Extraction of all other renewable biotic resources	Extraction of marine products for cosmetics, the chemical industry, biofuels, fish for fish feed etc.
		REGULATING SERVICES	
3	Air purification	Influence of a marine ecosystem on concentration of pollutants from the atmosphere	The removal from the air of pollutants such as fine dust and particular matter, sulfur dioxide, carbon dioxide, etc.

(cont.)

Table 2.1 (cont.)

	Ecosystem service	Description	Examples
4	Climate regulation	The contribution of a marine ecosystem to the maintenance of a favourable climate through impacts on the hydrological cycle and the contribution to climate-influencing substances in the atmosphere	The production, consumption and use by marine organisms of gases such as carbon dioxide, water vapour, nitrous oxides, methane and dimethyl sulfide
5	Disturbance prevention or moderation	The contribution of marine ecosystem structures to the dampening of the intensity of environmental disturbances such as storm floods, tsunamis and hurricanes	The reduction in the intensity of and/or damage caused by high energy waves by salt marshes, sea grass beds, reefs and mangroves; absorption of excess flood water by saltmarshes
6	Regulation of water flows	The contribution of marine ecosystems to the maintenance of localised coastal current structures	The effect of reefs and macroalgae on localised current intensity
7	Waste treatment and assimilation	The removal of contaminant and organic nutrient inputs to marine environments from humans	The breakdown of chemical contaminants and organic nutrients by marine microorganisms; filtering of coastal water by shellfish; burial of radioactive materials by burrowing seabed animals
8	Coastal erosion prevention	The contribution of marine ecosystems to coastal erosion prevention	The maintenance of saltmarshes by coastal vegetation and mudflat ecosystems; reduction in scouring potential that results from nearshore macroalgae forests and seagrass beds

Table 2.1 (cont.)

	Ecosystem service	Description	Examples
9	Biological control	The contribution of coastal ecosystems to the maintenance of population dynamics, resilience through food web dynamics, disease and pest control	The support of reef ecosystems by herbivorous fish that keep algae populations in check; the role that top predators play in limiting the population sizes of opportunistic species like jellyfish; the biological control of bacterial and viral diseases that impact humans directly or via the seafood they eat
		HABITAT SERVICES	
10	Migratory and nursery habitat	The contribution of a particular marine habitat to migratory and resident species' populations through the provision of critical temporary habitat for feeding, or reproduction and juvenile maturation	The importance of reefs (e.g. cold water corals, mussel and oyster beds), seagrass beds, macroalgae stands and estuaries as reproduction and nursery habitat of commercially valuable species that are harvested elsewhere
11	Gene pool protection	The contribution of marine habitats to the maintenance of viable gene pools through natural selection/evolutionary processes which enhances adaptability of species to environmental changes, and the resilience of the ecosystem	Importance of all marine habitats (e.g. coral reefs, gravel seabeds) to maintain inter- and intraspecific genetic diversity of species with potential benefits and/or commercial use
		CULTURAL SERVICES	
12	Leisure, recreation and tourism	The provision of opportunities for tourism, recreation and leisure that depend on a particular state of marine ecosystems	Bird/whale watching, beachcombing, sailing, recreational fishing, SCUBA diving, etc.

(cont.)

Table 2.1 *(cont.)*

	Ecosystem service	Description	Examples
13	Aesthetic experience	The contribution that a marine ecosystem makes to the existence of a surface or subsurface landscape that generates a noticeable emotional response within the individual observer. This includes informal spiritual individual experiences but excludes that which is covered by services 17	The particular visual facets of a 'seascape' (like open 'blue' water), a 'reef-scape' (with abundant and colourful marine life), a 'beach-scape' (with open sand), but also geological features, etc. that emotionally resonate with individual observers
14	Inspiration for culture, art and design	The contribution that a marine ecosystem makes to the existence of environmental features that inspire elements of culture, art, and/or design. This excludes that which is covered by services 2c, 13 and 16	The use of a marine landscape or species in pictures, films, paintings, architecture and design (like waves in jewellery)
15	Cultural heritage	The contribution of marine ecosystems to the maintenance of cultural heritage, and providing a 'sense of place'	Appreciation of coastal/marine ecosystems as a source of cultural identity and heritage (e.g. UNESCO World Heritage sites)
16	Cultural diversity	The contribution of marine ecosystems to social and cultural values and adaptations that pertain to living at coasts and exploiting marine resources	Social and cultural structures and relations of fishing/coastal/island communities
17	Spiritual experience	The contribution that a marine ecosystem makes to formal and informal collective religious experiences. This excludes that which is covered by services 13–14	e.g. Several Greek and Roman gods were connected to the sea; fish as a Christian symbol; whales and salmon play important roles in indigenous communities' religions

Table 2.1 (cont.)

	Ecosystem service	Description	Examples
18	Information for cognitive development	The contribution that a marine ecosystem makes to education, research, and individual and collective cognitive development	Scientific research (natural and social sciences) and applied scientific research (e.g. use of marine genetic materials/biological information, blue biotechnology and bionics/biomimetics); local ecological knowledge of coastal communities (e.g. for exploiting resources); environmental education of children and adults through field excursions

the organisms and their biological activity and functions differ between these habitats and locations, the ecosystem processes and functions that underpin the benefits are similar for provisioning, regulating and cultural services (Austen et al., 2011). However, the amount of service, and hence the benefit derived, varies according to the habitat/location.

In some cases, these services can deliver multiple benefits. For example, climate regulation delivers a habitable climate for people, but also for food provision, for biodiversity in general and the services it supports, and a pleasant climate for recreation. In contrast, many services may contribute to the same benefit. Often it may be the ecosystem as a whole that contributes to many cultural benefits, such as inspiration for culture, art and design. Furthermore, the uses made of these services may not be compatible in space and time. Demand for one or other service (e.g. food provision) may influence the provision of all other services. This highlights the complex and interlinked nature of ecosystem services (Fisher et al., 2009). Many of these linkages are nonlinear and poorly understood, and use of one ecosystem service may influence others in unexpected ways (Barbier et al., 2008).

2.5.2 Spatial and temporal dimensions

Ecosystem services are used and provide benefits over a range of scales, both temporal and spatial. Some services are experienced by beneficiaries very locally to the ecosystems that provided them. For example, food may be captured and consumed locally; recreational benefits and other cultural services may be experienced by those living at the coast; and coastal communities may benefit from flood and storm dampening and reduction in coastal erosion provided locally by salt marshes and mud and sand flats. In many instances, the place at which the benefits from a service are finally experienced can be very distant, often inland, and appear to be spatially and temporally very disconnected from the ecosystem habitat(s) and organisms that enabled the service to be provided (Table 2.2). For example, waste may enter the sea from inland conurbations via rivers discharging into estuaries and hence the coast where the service of waste treatment and assimilation is undertaken over a period of time; wild capture or aquaculture-produced fish and shellfish may be consumed locally, regionally or air freighted and shipped globally, and once processed, consumed months or even years after their catch; coastal tourism is a global industry; and climate regulation facilitated locally by sea grass beds, or globally by plankton is of global benefit.

Changes in ecosystem services and the processes and functions that generate them also occur over a number of different, often co-existing, spatial and temporal scales. Global scale services (e.g. climate regulation) may change slowly over long time frames, with a lag between the initial shock that initiated change and the manifestation of that change. In contrast, small-scale local services such as local food provision from mobile fish stocks may respond quickly to changing conditions. The response of ecosystems and humans to change may be variable. In some cases, ecosystem change may occur much faster than humans can adapt, while in other cases ecosystem services many not be able to respond or replenish fast enough to support the human demand for them (MA, 2005).

What is considered an ecosystem service may also change over time, shifting with societal preferences, irrespective of the state of the ecosystem. For example, whales were once seen as a resource for food and other raw materials and as such were caught in their thousands by people from many countries; now they are the subject of the multi-billion dollar whale-watching business for tourists (Collet, 2007). The closure of marine areas to fishing activities also suggests a shift in societal preferences, with greater emphasis being placed on habitat

Table 2.2 *Beneficiaries of ecosystem service provision.*

Ecosystem service	Beneficiaries of service		
	Local	Regional	Global

PROVISIONING SERVICES
Food provision:
  (a) Wild capture seafood
  (b) Farmed seafood
Biotic raw materials (non-food):
  (a) Genetic resources
  (b) Medicinal resources
  (c) Ornamental resources
  (d) Other biotic raw materials
	▓	▓	▓

REGULATING SERVICES Air purification Climate regulation Disturbance prevention or moderation Regulation of water flows Waste treatment and assimilation Coastal erosion prevention Biological control	▓	▓	▓
HABITAT SERVICES Migratory and nursery habitat Gene pool protection	▓	▓	
CULTURAL SERVICES Leisure, recreation and tourism Aesthetic experience Inspiration for culture, art and design Cultural heritage Cultural diversity Spiritual experience Information for cognitive development	▓	▓	▓

services and the regenerative capacity of the ecosystem as opposed to provisioning.

2.5.3 Ecosystem service accessibility

The nature of ecosystem services, and the benefits they generate, also affects the ways that they are accessed and consequently the ways that they can be managed, assessed and potentially valued. Ecosystem services and benefits can be characterised according to their degree of rivalry and excludability (Costanza, 2008; Fisher et al., 2009). A good or service is rival if its use by one person means there is less of that good or service for others to use, while it is excludable if once a unit of it has been used by an individual, others can be kept from having the same unit. Services that are both rival and excludable are easily traded through markets and are known as 'private goods.' This is true of many benefits that result from extractive resources, such as fish for consumption.

The degree to which ecosystem services are rival and excludable varies, but many services can be considered completely non-rival and non-excludable and are called pure public goods. The capacity of the ecosystem to assimilate waste is one such public good. It is difficult to prevent people putting waste products into the environment, and the use of the environment for waste assimilation by one person does not exclude the same use by another. Overuse of public goods can occur, however, resulting in a process called congestion. The waste assimilation service provides a good example of this. In some places so much waste has entered the marine environment so quickly that marine ecosystems are unable to break it down (e.g. dead zones). In this situation, the service of waste assimilation has become increasingly rival, to a point where no one can use it for further waste breakdown. The use of fish stocks for food provision shows a similar fate. The same could also be increasingly argued for the service of climate regulation, and in particular for the ability of the environment to store and sequester carbon (Fisher et al., 2009).

Management of public goods is often challenging (Ostrom et al., 1999). New institutions are being developed in attempts to support their management, such as markets for ecosystem services and payments for ecosystem services (Kumar, 2005; Jack et al., 2008; Kinzig et al., 2011). This should provide greater incentive for the protection and sustainable use of ecosystem services. Making an ecosystem service increasingly rival and excludable, however, will have implications for access to ecosystem services by some users.

2.6 Assessing ecosystem services

To assess changes in ecosystem services that might arise from human impacts, a large amount of information is potentially required. Various approaches can be taken. This book takes as its initial starting point impacts on habitat and/or organisms, their functioning, ecosystem processes affected and the implications in terms of services and beneficiaries. A different approach could be taken from the perspective of what ecosystem services are required, where and by whom, and then to consider where these services can be best provided and the management requirements to ensure sustainable provision. In reality, currently neither approach is fully adopted in the marine environment although there may be an implicit assumption that marine planning will ultimately take a more holistic approach that embraces ecosystem services (Börger et al., 2014).

Thus, when considering the quantification of service provision and benefits in support of management and policy, a considerable amount of spatially disaggregated data and understanding is required. This includes data on different habitats and their living components, their spatial extent and their functioning. Additional data are also required on the benefits that are actually realised from these services and where the benefits accrue, as well as the different values of these benefits, bearing in mind that value is likely to change spatially and according to supply and demand. In the UK National Ecosystem Assessment it was apparent that most of the ecosystem service and benefit data is not available at the disaggregated level of marine habitat/location type in the UK (Austen et al., 2011).

An additional consideration is that supply of ecosystem services as well as demand for them varies temporally. Food provision may be seasonal; carbon fixation, storage and sequestration to facilitate climate regulation varies seasonally with phytoplankton growth; protective regulatory services are more likely to be required during winter; cultural services, such as recreation, are predominantly exploited during warmer months in temperate climates. The spatial and temporal disconnects between service provision and beneficiaries are therefore key points for consideration in any ecosystem assessment.

2.7 Why consider ecosystem services – for management, for communication

Understanding of the changes caused by human impacts on ecosystem services and the value of the benefits they provide can be used in several ways. People often relate more readily to things that affect their daily lives. They are more accustomed to think in terms of relative values of goods and services, particularly, but not exclusively, if the valuation term is expressed in monetary units. Hence the expression of ecosystem change in terms of ecosystem services and benefits as well as their values provides a useful tool for communication.

The principal use, however, is to provide the justification for the development and implementation of policy and management measures to reduce harmful impacts and/or to mitigate them. This requires an understanding of the underlying causes of change due to impacts upon biodiversity and ecosystem functioning. The need for such policy and management measures is more obvious if consequent feedbacks of inaction, positive and negative, to society are understood and communicated. Valuation of ecosystem services is then a tool that can be used to explore in more depth the economics effect of ecosystem change. By understanding the value that society places on ecosystems, and how this value is changing, the welfare effects of ecosystem change can be identified. The drive towards understanding and valuing marine ecosystem services in monetary terms is further elaborated by the TEEB study (2012). TEEB (2010) suggests that the assessment and valuation of ecosystem services helps society to rethink its relationship with ecosystems, highlighting the consequences of its actions. It also helps to explicitly demonstrate the cost of management decisions and identify where the costs of action outweigh the benefits or vice versa.

Through understanding and enhanced communication, policy makers, managers and society can more fully consider the trade-offs between activities that impact on different ecosystem services. These trade-offs are often linked in both time and space and often revolve around provisioning services, for example, excessive nutrients coming from agricultural land affect the ability of some coastal areas to provide fish. When decisions are made to maximise one service at the expense of others, losses may result from the inherent trade-offs made (Rodríguez et al., 2006). Negative impacts from such decisions often occur due to the lack of knowledge of how these services interact in both space and time (Carpenter et al., 2009). Spatial and temporal issues therefore need to be taken into account in the management of ecosystem services, as well as the linkages to the social system. Failure to do so typically leads to environmental degradation and policy failure (Cash et al., 2006). These trade-offs will be discussed further in Chapter 11.

References

Atkins, J. P., Burdon, D., Elliott, M. and Gregory, A. J. (2011). Management of the marine environment: integrating ecosystem services and societal benefits with the DPSIR framework in a systems approach. *Marine Pollution Bulletin*, 62, 215–226.

Austen, M. C., Malcolm, S. J., Frost, M. et al. (2011). Marine. In *The UK National Ecosystem Assessment Technical Report*. Cambridge: UNEP-WCMC, pp. 459–499.

Balmford A., Rodrigues A. S. L., Walpole M. et al. (2008). *The Economics of Ecosystems and Biodiversity: Scoping the Science.* Cambridge: European Commission.

Balmford, A., Fisher, B., Naidoo, R. et al. (2011). Bringing ecosystem services into the real world: an operational framework for assessing the economic consequences of losing wild nature. *Environmental and Resource Economics*, 48, 161–175.

Barbier, E. B., Koch, E. W., Silliman, B. R. et al. (2008). Coastal ecosystem-based management with nonlinear ecological functions and values. *Science*, 319(5861), 321–323.

Beaumont, N. J., Austen, M. C., Atkins, J. P. et al. (2007). Identification, definition and quantification of goods and services provided by marine biodiversity: implications for the ecosystem approach. *Marine Pollution Bulletin*, 54, 253–265.

Beaumont, N. J., Austen, M. C., Mangi, S. C. and Townsend, M. (2008). Economic valuation for the conservation of marine biodiversity. *Marine Pollution Bulletin*, 56, 386–396.

Böhnke-Henrichs, A., de Groot, R., Baulcomb, C. et al. (2013). Typology and indicators of ecosystem services for marine spatial planning and management. *Journal of Environmental Management*, 130C, 135–145.

Börger, T., Beaumont, N. J., Pendleton, L. et al. (2014). Incorporating ecosystem services in marine planning: the role of valuation. *Marine Policy*, 46, 161–170.

Boyd, J. and Banzhaf, S. (2007). What are ecosystem services? The need for standardized environmental accounting units. *Ecological Economics*, 63(2–3), 616–626.

Brown, T. C., Bergstrom, J. C. and Loomis, J. B. (2007). Defining, valuing and providing ecosystem goods and services. *Natural Resources Journal* 47, 329–376.

Carpenter, S. R., Mooney, H. A., Agard, J. et al. (2009). Science for managing ecosystem services: Beyond the Millennium Ecosystem Assessment. *Proceedings of the National Academy of Science of the United Sates of America*, 106(5), 1305–1312.

Cash, D. W., Adger, W. N., Berkes, F. et al. (2006). Scale and cross-scale dynamics: governance and information in a multilevel world. *Ecology and Society*, 11(2), 181–192.

Collet, S. (2007). Values at sea, value of the sea: mapping issues and divides. *Social Science Information*, 46(1), 35–66.

Costanza, R. (2008). Ecosystem services: multiple classification systems are needed. *Biological Conservation*, 141(2), 350–352.

Costanza, R., d'Arge, R., de Groot, R. et al. (1997). The value of the world's ecosystem services and natural capital. *Nature*, 387, 253–260.

de Groot, R. S., Fisher, B., Christie, M. et al. (2010). Integrating the ecological and economic dimensions in biodiversity and ecosystem service valuation. In *The Economics of Ecosystems and Biodiversity (TEEB): Ecological and Economic Foundations*, ed. P. Kumar. London: Earthscan, pp. 9–40.

EEA (2013). Towards a Common International Classification of Ecosystem Services. CICES Version 4.3.

Fisher, B. and Turner, K. R. (2008). Ecosystem services: classification for valuation. *Biological Conservation*, 141, 1167–1169.

Fisher, B., Turner, R. K. and Morling, P. (2009). Defining and classifying ecosystem services for decision making. *Ecological Economics*, 68(3), 643–653.

Gómez-Baggethun, E., de Groot, R., Lomas, P. and Montes, C. (2009). The history of ecosystem services in economic theory and practice: from early notions to markets and payment schemes. *Ecological Economics*, 69(6), 1209–1218.

Haines-Young, R. and Potschin, M. (2009). Methodologies for defining and assessing ecosystem services. Final Report, JNCC, Project Code C08–0170–0069.

Haines-Young, R. and Potschin, M. (2010). The links between biodiversity, ecosystem services and human well-being. In *Ecosystem Ecology: A New Synthesis*, ed. Raffaelli, D and Frid. Cambridge: Cambridge University Press, pp. 110–139.

Hattam C., Atkins J. P., Beaumont N. J. et al. (2015). Marine ecosystem services: linking indicators to their classification. *Ecological Indicators*, 49, 61–75.

Jack, B. K., Kousky, C. and Sims, K. R. E. (2008). Designing payments for ecosystem services: Lessons from previous experience with incentive-based mechanisms. *Proceedings of the National Academy of Science of the United Sates of America PNAS*, 105(28), 9465–9470.

Johnston, R. J. and Russell, M. (2011). An operational structure for clarity in ecosystem service values. *Ecological Economics*, 70, 2243–2249.

Kinzig, A. P., Perrings, C., Chapin, I. F. S. et al. (2011). Paying for ecosystem services: promise and peril. *Science*, 334, 603–604.

Kumar, P. (2005). *Market for Ecosystem Services*. Winnipeg, Canada: International Institute for Sustainable Development.

Liquete, C., Piroddi, C., Drakou, E. G. et al. (2013). Current status and future prospects for the assessment of marine and coastal ecosystem services: as systematic review. *PLoS ONE*, 8(7), e67737.

Mace, G. M., Bateman, I., Albon, S. et al. (2011). *The UK National Ecosystem Assessment Technical Report, UK National Ecosystem Assessment*. Cambridge: UNEP-WCMC.

MacKerron, G. and Mourato, S. (2013). Happiness is greater in natural environments. *Global Environmental Change*, 23(5), 992–1000.

Millennium Ecosystem Assessment (MA) (2005). *Ecosystems and Human Well-being: A Framework for Assessment*. Washington DC: Island Press.

Mooney, H., Ehrlich, P. (1997). Ecosystem services: a fragmentary history. In *Nature's Services. Societal Dependence on Natural Ecosystems*, ed. G. Daily. Washington DC: Island Press.

Ostrom, E., Burger, J., Field, C. B., Norgaard, R. B. and Policansky, D. (1999). Revisiting the commons: local lessons, global challenges. *Science*, 284, 278–282.

Pearce, D. W. and Moran, D. (1994). *The Economic Value of Biodiversity*. London: Earthscan.

Rodríguez, J. P., Beard Jr., T. D., Bennett, E. M. et al. (2006). Trade-offs across space, time, and ecosystem services. *Ecology and Society*, 11(1), 28.

Saunders, J., Tinch, R. and Hull, S. (2010). *Valuing the Marine Estate and UK Seas: An Ecosystem Services Framework*. London: The Crown Estate.

TEEB (2010). *The Economics of Ecosystems and Biodiversity: Ecological and Economic Foundations*, ed. P. Kumar. London and Washington: Earthscan.

TEEB (2012). Why value the oceans. Discussion paper, TEEB.

UK National Ecosystem Assessment (2011). *The UK National Ecosystem Assessment: Synthesis of the Key Findings*. Cambridge: UNEP-WCMC.

Vemuri, A. W. and Costanza, R. (2006). The role of human, social, built, and natural capital in explaining life satisfaction at the country level: toward a National Well-Being Index (NWI). *Ecological Economics*, 58(1), 119–133.

Wallace, K. J. (2007). Classification of ecosystem services: problems and solutions. *Biological Conservation*, 139, 235–246.

3 · *Assessing human impacts on marine ecosystems*

CHRISTOPHER FRID AND
TASMAN CROWE

3.1 Introduction

Human society derives considerable benefit from marine ecosystems, as described in Chapters 1 and 2. It does so through a wide range of activities. In deriving these benefits humans exploit a range of ecological and environmental resources and services (Chapter 2). The majority of human uses of the marine environment have some measurable impact on the supporting ecosystem (Figure 3.1). Our treatment of these impacts starts from a consideration of the benefits human society is deriving from the system but recognises that the process of obtaining many of these benefits is carried out by what can be termed 'economic sectors': fishing, construction, agriculture, tourism, etc. (Table 3.1). Impacts are direct or indirect and may be sustainable, in which case the system will continue to provide the service indefinitely and will recover upon cessation of the impacting activity, or they may be unsustainable (Frid and Dobson, 2013). Human activities may extract living components of the ecosystem (for example, as food or for the aquarium trade), non-living materials (salt, sediment, mineral deposits), and both sustainable (wind, tides) and unsustainable (oil, gas) energy resources. In each of these cases the system is altered by the removal (a direct effect) and often also by the removal process which may have direct and indirect effects on the ecosystem. Humans discharge their waste directly or indirectly (via rivers and agricultural runoff) into the seas and in doing so they exploit regulating services to, for example, dilute, transform, detoxify and sequester wastes (Peterson and Lubchenko, 1997). The cultural importance of natural systems is receiving increasing recognition for its support of the very profitable tourism, leisure and recreation sectors, as well as benefiting health and well-being. These sectoral activities can bring their own

Marine Ecosystems: Human Impacts on Biodiversity, Functioning and Services, eds T. P. Crowe and C. L. J. Frid. Published by Cambridge University Press. © Cambridge University Press 2015.

Assessing human impacts on marine ecosystems · 43

Figure 3.1 Some examples of human activities and pressures potentially impacting on marine ecosystems: (a) shipping, (b) aquaculture, (c) fisheries, (d) marine litter, (e) power generation. Images by C. Frid (a, c) and T. Crowe (b, d, e). A black and white version of this figure will appear in some formats. For the colour version, please refer to the plate section.

Table 3.1a *Matrix of pressures associated with sectoral activities (P, physical; C, chemical; B, biological). Pressures and sectors are derived from Robinson et al. (2008). Numbers in brackets indicate chapters in which sectors/pressures are covered. * indicates pressures which are reviewed in detail in books/papers on impacts of climate change and are covered in this book only as potential modifiers of impacts of other stressors (Chapter 4). Explanatory notes for the different sectors are provided below (Table 3.1b).*

☐ – no known association between sector and pressure ▨ – potential association between sector and pressure

Ecosystem service:	Benefits from provisioning services		Benefits from regulating services			Benefits from cultural services	Benefits from abiotic ecosystem components			
	Provision of food and biotic raw materials		Waste treatment and assimilation			Leisure, recreation, tourism and other cultural				
Sector:	Fisheries (6) Active / Passive	Aquaculture (6)	Sewage disposal (8,9)	Agriculture (runoff) (8, 9)	Industry (discharge) (8, 9)	Leisure and tourism (7, 8)	Salt, sediment and mineral extraction (3)	Non-renewable energy (9)	Renewable energy (7)	Construction/development (7)	Shipping (7, 9, 10)

Pressure↓

- P Habitat loss to land (7)
- P Habitat change to another marine habitat (7)
- P Physical disturbance (6)
- P Siltation rate changes (7)
- P Temperature change (*)
- P Salinity change (*)
- P Water flow (7)
- P Tidal emergence regime (7)
- P Wave exposure changes – local (7)
- P Litter (9)
- C Non-synthetic compounds (9)
- C Synthetic compounds (6, 9)
- C De-oxygenation (8)
- C Inorganic nutrients (8)
- C Organic enrichment (8)
- B Introduction of microbial pathogens (6, 10)
- B Introduction/spread of non-indigenous species (10)
- B Removal of target and non-target species (6)

Table 3.1b *Explanatory notes for the sectors and sub-sectors in Table 3.1a. The explanations given in this table are examples and are not intended to be exhaustive for the sectors described.*

Sector	Description of sector and clarification of pressures
Fisheries – active	Biomass is removed with the use of mobile gear through trawling and/or dredging
Fisheries – passive	Biomass is removed with the use of stationary gear such as potting, staked nets and lines
Aquaculture	The cultivation of fish, invertebrates or algae
Sewage discharge	Includes the discharge of raw, primary and secondary treated effluent and of storm water runoff from roads
Agricultural discharge	Includes diffuse inputs of nutrients from land, often via freshwater systems
Industrial discharge	Includes effluent (not sewage) resulting from industrial activities such as brewing, pharmaceuticals, metal works and food processing
Construction development	Construction of coastal infrastructure and activities related to this; including navigational dredging, aggregate extraction, sea defences, barrages, weirs, marinas and harbours and beach replenishment
Salt, sediment and mineral extraction	Self-explanatory
Shipping	Includes shipping in industrial sectors such as oil and gas and container shipping
Leisure and tourism	Activities include angling, bait collection and the use of small motor craft for pleasure
Energy	Non-renewable includes extraction and processing of oil and gas, and the operation of power stations where cooling water maybe produced. Renewable covers the construction of marine-based renewable energy structures such as wind, tidal and wave turbines

impacts. For example, living at the coast requires the building of infrastructure and generates wastes, but living at the coast may be a cultural choice or it may be a decision made as result of economic pressure, i.e. employment. The provision of infrastructure is a requirement of many human activities that occur at the coast and so emphasises that we need to consider sectors of human activity as having two dimensions: (1) the reason for the activity, which we could think of in terms of the economic or socio-cultural driver, and (2) the pressure that the activity exerts on the receiving system (Rogers and Greenaway, 2005; EEA, 2007). The first is important because policy takes account of drivers and the resultant laws and regulations have, historically, been developed to control activities/sectors. However, from the perspective of the affected ecosystem, it is the pressure they exert that is important not the economic driver. The 'ecological pressure' of anoxia is produced by multiple different sectors, e.g. sewage disposal, organic enrichment from a fish farm, or as a result of eutrophication via agricultural runoff, but the environmental impacts of this pressure do not differ. So ultimately it is the societal drivers that dictate which benefits we seek and hence what pressures we exert on the ecosystem and it is the pressures that cause impact on the ecosystem.

In this chapter, we begin by briefly considering the range of ecological pressures exerted on marine ecosystems by human activities. We will then consider how these impacts impinge on different levels of ecological organisation, and how the system responds to changing patterns of pressures. In particular, we will explore the scope for recovery following cessation or reduction of an impacting activity. Chapters 6–10, are structured to link the activities of a sector to the resulting pressures and review their impacts comprehensively and without overlap (based on the classification provided in Table 3.1). Chapters 6 and 7 are structured around activities that either impose specific pressures not covered in other chapters (fisheries and construction of physical structures) or impose a range of pressures, all separately covered in other chapters (e.g. aquaculture). Chapters 8 and 9 describe the impacts of nutrients and chemical contaminants, respectively, which are pressures imposed by a range of activities, and include links with the activities that drive them. Chapter 10 focuses primarily on the impacts of invasive species introductions, but it also comments on how sectoral activities can underpin introductions of non-indigenous species and their spread as this may guide management interventions.

3.2 Human impacts on marine ecosystems: a framework and overview

Human uses of the marine environment to provide benefits are many and most have a long history. Most activities have some effect on the ecosystem, causing physical change, altering levels of chemicals and their cycling, and affecting biological components of the system (Figure 3.1; Table 3.1). The activities and the pressures they cause are categorised here in relation to the typology of ecosystem services outlined in Chapter 2 to facilitate consideration of the potential impacts in relation to the potential benefits of the activities. Not all benefits from the marine environment are underpinned by ecosystem processes based on biological activity, some derive solely from physical properties of the environment. Abiotic benefits such as energy, aggregate extraction and shipping therefore fall outside the typology of ecosystem services presented in Chapter 2. Activities to extract such abiotic services can nevertheless affect biodiversity and ecosystem processes and thus impact on the delivery of biotic benefits, so are included in Table 3.1 and covered by chapters in this book. Conversely, not all benefits based on ecosystem services cause impacts when being derived by society, although they can themselves be affected by human activity. For example, the category 'habitat services' refers to the ability of a habitat to support the characteristic biological assemblage that may in turn support ecotourism. The use of the habitat by those organisms does not degrade the habitat. Similarly, 'climate regulation' services may well be compromised by human activities but the regulation of climate that we take a benefit from does not degrade the environment. The same logic applies to all regulating services except waste treatment and arguably all cultural services except leisure, recreation and tourism. Our focus here and in the subsequent chapters is on those activities that have potential to impact directly or indirectly on the biological components of the system (Table 2.1). In this volume we use the term 'stressor' when referring to the combination of human activities and the pressures they place on the ecosystem.

Impacts of different stressors can vary considerably in spatial extent. Fishing occurs over extremely large parts of the sea and so the impacts are equally spatially extensive. Discharges of urban wastewater occur from pipes and represent point sources. If one maps the area impacted by each discharge it is small at the scale of the coastal sea, and even when summed the total area may be far less than the area impacted by fishing, but at a particular locality the impact of the discharge may be much greater on the local ecosystem than the more widespread activity. Point-source discharges also tend to be better regulated and controlled than widespread impacts; contrast, for example, nutrients which enter the sea via wastewater discharge from sewage treatment works and those entering via rivers draining diffuse inputs from agricultural activity. Coastlines and coastal embayments may therefore suffer considerable human pressure from the high density of activities occurring and their combined impacts even if the individual impacts are small. These effects may be further confounded when pressures interact, often synergistically (Chapter 4). Recognition of these challenges for managers has, in part, fuelled the development of explicitly spatial approaches to managing pressure-generating activities, such as marine spatial planning (Chapter 1).

3.2.1 Provisioning services

Food provision

The oldest form of human use of marine ecosystems is probably the exploitation of marine animals and plants. Archaeological evidence suggests that early hominids living at coastal sites collected food from the shore (e.g. Craig et al., 2013) and where their ranges overlapped, hunted seals, otters and other fur-bearing animals. It is impossible to exploit living resources without having some effect – even killing a single organism is a source of additional mortality. However, all populations can withstand a level of elevated mortality without significant impact on the population dynamics. The threshold between non-significant and significant impact varies between species; it is much lower for long-lived taxa such as whales than for short-lived taxa such as sandeels (Hall, 1999). Given that global food security is one of the United Nation's Millennium Development Goals (http://www.un.org/millenniumgoals/; Chapter 1) the policy aim is actually to exploit fish (including shellfish and marine plants) stocks at the *maximum* sustainable level. This level of exploitation is higher than that giving 'no significant ecological change' and will result in widespread alterations to ecosystems. Effectively, humanity is an additional, voracious, predator in the system. This predator is size selective (usually) and species selective, so altering population dynamics of the 'predated' stocks and consequently the competitive interactions and food web dynamics of the remaining community (Chapter 6).

In addition to the direct effects on the exploited populations and effects on the food web caused by these changes the harvesting of marine living resources can have a number of other impacts (Table 3.1). These include physical impacts on seabed habitats including resuspension of sediments (Chapter 6), the discarding of materials such as fishing nets that act as litter (Chapters 6 and 9), and, for static structures, changes in the flow regime, for example by the building of fish weirs and dams (Chapter 7).

While some harvesting techniques are extremely efficient at selecting the target species (hunting seals and whales, spear fishing) most commercial fisheries are not selective. Many fishing techniques lead to mortality of non-target organisms that are not suitable for human consumption. Examples include, sea birds caught on the baited hooks of long lines, porpoises and seals drowned after entanglement in nets, benthic organisms caught in trawls and dredges. The non-target organisms are often thrown back dead, which further alters marine food webs by providing a food subsidy to scavengers (Chapter 6). Mortality of non-target species also occurs as a result of the impact of bottom-fishing gears, particularly those that are towed, on the physical habitat. These gears are often designed to dig into the sediment to catch shellfish or to scare up flat fish, and in so doing resuspend and move sediments, smother the benthos, reduce turbidity, produce scour and also potentially alter geochemical conditions (Chapter 6).

Recent decades have seen a rapid expansion in the production by aquaculture, that to some extent mirrors the shift from hunter-gathering to agriculture on land. Aquaculture now accounts for almost 50% of total fisheries production (FAO, 2012). Aquaculture, like agriculture, can have profound impacts on the environment in which it is practiced (Table 3.1; Chapter 6). Onshore intensive aquaculture systems draw in resources including energy and, often, fisheries resources in the form of processed food pellets, while producing organic-rich effluent and releasing other potential toxicants such as pharmaceuticals (Lucas and Southgate, 2012). Offshore and coastal aquaculture may also, depending on the taxa farmed, produce large quantities of waste and consume fisheries production. The stock may also consume a large proportion of the available oxygen, while the waste may also place an oxygen demand on the receiving environment, in some cases leading to deoxygenation. While the waste from onshore facilities may be passed through a treatment plant, offshore sites generally release waste directly into the environment often raising concerns about environmental impacts (Gowan and Bradbury, 1987).

Biotic raw materials (non-food)
Some biological material is also sourced to provide non-food benefits. The oldest example of this is the extraction of sponges which, in the Mediterranean and Japan, has a documented history measured in thousands of years. Other examples include seaweed harvesting and cultivation for conversion to fertiliser, animal feed or for non-food uses such as agar for microbiological gels, harvesting of shells/coral to form ornaments and the exploitation of sponges, bryozoans, algae and other taxa for biologically active compounds for medicinal or cosmetic applications (Chapter 2).

While the scale of this, in tonnes extracted, is small compared to fisheries, the effects can be locally significant. For example the collection of 'black coral' for jewellery has caused the species to decline to such an extent that it is on the IUCN Red List of endangered species (Grigg, 1984).

3.2.2 Regulating services

Many of the benefits derived from the regulating services described in Chapter 2, such as air purification and climate regulation, are passively derived from marine ecosystems without impacting them. The exception is waste treatment and assimilation, the impacts of which are outlined below (Table 3.1).

Waste treatment and assimilation

Human activities result in a wide range of substances entering the marine environment, some directly and some indirectly either from the atmosphere or via river systems that drain into coastal seas. These substances are termed pollutants when their presence results in a risk to human or ecosystem health (Chapter 9). For the vast majority of substances, any polluting effect is limited to areas of elevated concentrations near to source. Given the vast volume of the sea and the continual mixing and dispersion of water by the wind and tides material is diluted, mixed and transported away from the discharge site.

Seawater is a complex chemical solution that supports a wide diversity of life. Most substances that enter the marine environment undergo some degree of transformation by chemical or biological processes, or a combination of the two. The simplest example of such a transformation is carbohydrate waste discharged from a food processing plant. This waste would be subject to biological degradation that could proceed as follows:

1. Break down of organic wastes by physical and biological processes releases carbon dioxide, water and various inorganic nutrients.
2. During biological breakdown of the waste microbes utilise oxygen.
3. High levels of waste stimulate this microbial activity to the extent that the system becomes oxygen depleted (Díaz and Rosenberg, 1995).
4. The liberated nutrients could stimulate plant growth and so fuel eutrophication (Chapter 8).

Some substances do not undergo marked transformations. For example, some organic molecules degrade incompletely, or only very slowly, whereas metals may remain as metal ions or become weakly bound to other ions or organic material (and may sometimes become more toxic as a result).

Ecological processes can, on the one hand, contribute to the 'removal' of wastes, but, on the other hand, the wastes may impact the biodiversity and compromise short-term functioning of the ecosystem and hence a wide variety of functions/services (Chapter 9). Toxicity effects from heavy metals, radio-nucleotides and complex organic molecules tend to be localised near to source. However, as toxicity is species specific and as toxins can bioaccummulate up the food chain these effects can alter community composition, food-web structure and functioning.

There is considerable interest at present in the ability of biogeochemical processes to sequester substances, removing them from the active elemental cycle and locking them into geological reservoirs. This interest has focused on radio-nucleotides, but it is now focused on anthropogenic carbon. It is estimated that the oceans have sequestered ~30% of the excess carbon emitted from the burning of fossil fuels in the last 200–300 years (Takahashi et al., 2001). So, without the oceans we can expect that the effects of climate change would be more pronounced. This raises the questions: can we stimulate this process to offset current anthropogenic inputs of carbon, and is there a limit to how much carbon can be sequestered in this way? Carbon is primarily sequestered by conversion of carbon dioxide into carbonate skeletal materials, particularly coral reefs, coccolithoporid plates and mollusc shells. Upon the death of an organism, a large proportion of this skeletal material sinks rapidly to the seafloor where it can become buried in the sediment, or overgrown by new reef in the case of corals, and so is removed from the short-term carbon cycle. Some organic matter may also sequester carbon, oil and coal are in fact the remains of partly decomposed organic matter that became locked into ancient sedimentary rocks. While sequestration, by definition, removes the material from the ecological system the process of sequestration may have impacts on ecosystem dynamics. Radio-nucleotides may exert a toxicity/mutagenic effect of organism they come into contact with while carbon drawdown increases seawater acidity (see below).

3.2.3 Cultural services

The coastline and features of the marine environment have cultural importance in many societies. These range from natural geological features of cultural importance to ship wrecks which act as memorials for lives lost. In recent decades the growth of leisure time has also led to increased importance of marine environments as venues for recreation (Chapter 2). In many cases, these benefits can be obtained with little or no impact, for example, just observing the coastal landscape and habitats, enjoying the benefits by looking at artworks (photographs, paintings, TV programmes or films) of or inspired by the natural environment. This concept extends to the idea that without necessarily ever seeing a whale or a coral reef, we feel our lives are enriched by knowing that they do exist. Economists capture this in the 'concept of existence value' and the 'willingness to pay' to know that these aspects of the environment are protected (Chapter 2).

Recreation and tourism involving larger numbers of individuals can place a range of pressures on the environment (Table 3.1). While impacts from physical visitors to the marine environment are often minor, the movement between different localities by recreational craft, including surfboards, and the use of angling tackle in multiple water bodies can act as vectors for pathogens and non-native species (Chapter 10). Disturbance to wildlife by visitors is the most obvious stressor arising from 'ecotourism' type activities such as whale/seal/bird watching. However visitors accessing the shore for bathing, diving on coral reefs, trampling on rocky shores and walking across mudflats all exert a physical impact pressure on the environment (Chandrasekara and Frid, 1996; Fletcher and Frid, 1997). Visitors also often generate chemical contamination/wastes (Chapter 9). For example, tourists vising the Red Sea and other tropical holiday resorts use a variety of sunscreen preparations. These wash off the bodies of swimmers, snorkelers and divers giving rise to a potential threat to the coral organisms (Danovaro et al., 2008).

Construction of infrastructure at the coast in order to facilitate cultural and recreational activities is widespread. This can lead to loss of habitat, for example, when building occurs on the foreshore, and modification of coastal dynamics, by coastal protection works and harbours (Chapter 7). Over 60% of the global population lives within 100 km of the coast and this proportion is increasing (McGranahan et al., 2007). So the demand on coastal land will grow even faster than suggested by the 50% increase in global population predicted between 2000 and 2050. This will set up a conflict with the demand for coastal space to develop aquaculture to feed the growing population (Chapter 6) and also increase the demands in coastal ecosystems for waste treatment/assimilation. The impact of global warming on sea level will also mean that coastal infrastructure will increasingly be defended and this will result in further widespread habitat modification. The detailed outcome of these changes is difficult to predict at a macro-scale except that it is clear that the growing population in the coastal zone will drive large-scale habitat modification and this will result in changes in ecosystem functioning and service delivery (Chapter 7).

3.2.4 Benefits from abiotic raw materials

Non-living resources: salt, sediment, fossil fuels, other minerals

Non-living resources from the marine environment provide a wide range of benefits to society and the diversity of resources exploited and the range of techniques used to gain these benefits exert a wide range of pressures and impacts on the environment (Table 3.1). These include impacts on the physical environment as the resources are physically removed – impacts include habitat, water flow, oxygen, salinity, temperature (from cooling water, or mine water discharge) and water quality changes. Processing of the material may then add further impacts from water quality changes, the release of chemicals used in the processing and litter items.

Salt has been extracted from seawater by evaporation in tropical and temperate areas since prehistoric times (e.g. Prakash, 2004) and there is archaeological evidence of extensive salt-pans in many regions. The construction of these salt-pans will have involved the loss of the coastal habitat (usually wetlands) they replaced and, in operation, mortality of the biota impounded in the salt-pan as water evaporates. While the latter is probably a trivial effect, the scale of salt production at some sites is of local and regional significance in terms of impacts on the extent of wetland habitats.

The exploitation of minerals, such as sand, gravel, stone and coral rubble for building has a similarly long history, and while salt or mineral production is not inherently dependent on ecological processes, the availability of some building materials is (e.g. coral stone; Clark and Edwards, 1994). The physical ecosystem processes that sort these sediments underpin their ease of exploitation. The impact of marine-derived mineral extraction for construction materials vary considerably. For example, the impact of limited sand removal from a dynamic, low diversity beach may be undetectable, whereas the use of dynamite to extract building material from a coral reef can cause system-wide collapse including a shift to low diversity, unstable, sediment systems with a much lowered fisheries production (Clark and Edwards, 1999). In some situations, the impacts may arise at some distance from the site of the activity, for example, the removal of sand from a beach may cause the erosion of a shore down current that previously depended on it for sediment supply. Given concerns regarding the potential impact of shoreline extractions of aggregates, most commercial scale extraction now occurs offshore (de Groot, 1996). The impacts at the extraction site and nearby can be severe and the time scales of recovery can take several decades (Boyd et al., 2004).

During the late nineteenth century early attempts to sample the deep ocean revealed the presence of metal rich (particularly manganese) nodules in localised regions of the oceans. To date, the technical challenges of raising minerals from 4 km below the sea surface, and the legal uncertainties surrounding the UN Convention on the Law of the Sea (http://www.un.org/depts/los/convention_agreements/texts/unclos/UNCLOS-TOC.htm) have limited commercial exploitation. However, a number of nations, including China, Russia and the United States, are likely to begin operations in the near future. There has been considerable concern raised over the possible impacts that the extraction of these nodules would have on these physically stable and biologically diverse ecosystems (Thiel, 1992).

Non-renewable energy: fossil fuels and power stations

The highest profile minerals sourced from under the sea are hydrocarbons – oil and gas. Extensive offshore reserves are exploited in the Gulf of Arabia/Red Sea, North Sea, Gulf of Mexico, and off Alaska with exploration or smaller scale developments occurring off west Africa, in the Indo-Pacific region and the Arctic. Development follows a pattern of exploration and then exploitation. Exploration often involves seismic surveys which, while short term, cause impacts on marine mammals, birds, fish and other organisms that have sound receptors or gas-filled spaces. Exploratory drilling at likely sites follows this. Drilling of oil/gas wells whether for exploration or subsequent production, produces rock waste covered in drill lubricants (known as 'mud' in the industry) that are toxic (Chapter 9). So piles of contaminated spoil are deposited on the seafloor and over time this material leaches toxins and fine particles creating a footprint of impact extending down current. The seafloor under the drilling platform can become abiotic with a classic succession down current of a low diversity community of opportunists/tolerant taxa, that gives way to a more species rich transitional one before a return to 'normal'. This pattern may extend up to 10 km at some sites (Kingston, 1992; Pearson and Mannvik, 1998).

Energy installations have often been sited at the coast as thermal power stations need access to cooling water to operate efficiently. Irrespective of the fuel used, coal, peat, timber/biomass, oil, gas or nuclear, large volumes of water are drawn in, entraining marine organisms that suffer mortality, and then released, typically 10 °C warmer (Clark et al., 2001). The warm water has less oxygen-carrying potential, and may contain residues of chemical added to the inflow to prevent biofouling/corrosion of the pipelines. The volume of water may be sufficient to influence the ecology of the receiving estuary/embayment with the significantly warmer environments often promoting conditions that allow invasive species to become established (Hibbert, 1977). Some power stations use engineered structures or sound or bubble barriers to reduce entrainment of large fish, but in doing so cause other impacts on the system.

Renewable energy: wind, waves and tides
With increasing concerns over carbon emissions from fossil fuel power stations and the continuing safety issues surrounding nuclear sites, the development of renewable energy is a major policy objective. The marine environment offers various options for exploitation of 'green energy' resources, these are often referred to collectively as the 'wet renewables' and comprise wind, wave and tidal power and their construction and operation exerts a range of pressures on the marine ecosystem (Table 3.1).

Offshore wind farms are technically the same as their onshore counterparts, but being offshore means they exploit stronger and more consistent winds that predominate over the sea. Offshore turbines are also less likely to cause 'planning' issues with local residents. Offshore wind farms do present the additional challenge of transporting the power generated ashore to the consumers and the electric fields generated by long electrical cable runs have prompted some environmental concerns (Gill, 2005). Cables may be laid on the seabed, may be on the seabed and then covered in rock armour (forming an artificial reef structure) or buried in the seafloor. The latter involves trench excavation, cable laying and then covering. These have local impacts on the benthos and also generate sediment plumes that may impact benthic systems and water column processes remote from the site of operation. The main ecological impacts during operation arise from the noise produced during construction and the physical positioning of structures (new habitat) in the environment. These structures may in turn alter currents and associated patterns of sediment movement (Gill, 2005).

Wave energy remains an experimental technology but, like offshore wind, the most significant ecological effects are likely to arise from the structures, construction activity and the cables (Chapter 7).

Tidal energy may be exploited in two ways: barrages/dams and submerged turbines. The latter is an experimental technology and the structures, their constructions and power distribution will probably be the main ecological effects (Frid et al., 2012). Barrages are a well-established technology and the ecological effects can be highly significant, altering the ecological functioning of the entire area behind the dam. Typically, the tide is held back, keeping mud flats covered until such point that they drain rapidly. The net effect of this is to reduce feeding times for wading shore birds such as turnstones, oystercatchers, curlew and dunlin (Goss-Custard et al., 1991).

Construction of physical infrastructure
The derivation of many of the above benefits requires the construction of physical infrastructure in seas and oceans. This can have a range of impacts, ranging from modifying hydrodynamics and sedimentation, to altering light regimes and providing habitat for invasive species (Table 3.1). These impacts are discussed in detail in Chapter 7.

Shipping
After food gathering, the use of the seas as a means of transport is probably the longest standing human activity in marine ecosystems (Table 3.1). The globalisation of world trade has stimulated increases in the level of maritime transport and driven a move to large ocean-going vessels. In addition to the construction of ports and harbours (associated impacts considered above and in Chapter 7), these changes have led to increases in noise in the marine environment, pollution from ships (both waste discharges, i.e. sewage, food wastes and litter and accidental spills of cargo and fuel), transport of organisms in ballast and on hulls (including non-native invasive species, Chapter 10) and the release of toxins from antifouling paints (Chapter 9).

3.2.5 Global changes associated with emissions of carbon dioxide and other gases

In 2013, the atmospheric concentration of CO_2 reached 400 ppm (Tans and Keeling, 2013), a level greater than at any other time in human history and considered most likely to have been caused by human activity

(IPCC, 2013). Associated with this increase and that of other gases has been a rapid increase in global temperature (IPCC, 2013). The effects of global warming include increasing sea temperatures, changing circulation patterns, increased storminess and changing sea levels (IPCC, 2013). Each of these changes can potentially have profound influences on marine ecosystems in their own right, changing physiological performance of individuals, dynamics of populations, patterns of distribution of species, community interactions and ecosystem processes (Harley et al., 2006). What's more, temporal climatic regimes are also changing, with greater frequency of extreme temperature, rainfall and storm events, potentially increasing runoff from terrestrial systems, exacerbating contamination by nutrients and other chemicals and modifying their influence (Chapter 4).

The other major impact of rising CO_2 on marine ecosystems are changes to ocean carbonate chemistry and pH. It is now recognised that the pH of the oceans has fallen by 0.1 pH units, equivalent to an increase in acidity of around 30% since the beginning of the Industrial Revolution (Doney et al., 2009). If this trend continues, and as atmospheric CO_2 levels increase it will, then the increased acidity of seawater will have profound impacts on organisms secreting carbonate skeletons and shells, including coral reefs, planktonic foraminifera and molluscs, such as the economically important clams (Fabry et al., 2008; Doney et al., 2009).

Detailed consideration of the impacts of global climate change and ocean acidification on marine ecosystems is beyond the scope of this book, which focuses on stressors which are, in principle, directly manageable. Recent treatments of the subject include: Royal Society (2005), Harley et al. (2006), Hoegh-Guldberg and Bruno (2010), Gattuso and Hansson (2011), Wernberg et al. (2012). Effects of global changes on disturbance regimes and on impacts of local stressors are considered in Chapter 4.

3.3 Impacts at different levels of biological organisation

Human activities and anthropogenic pressures can affect biota at different levels of biological organisation, ranging from intra-individual effects on, for example, metabolism, cellular structure and integrity, immunology or physiology, individual-level effects such as changes to behaviour, morphology, growth and reproduction, changes at the population level, for example, through impacts on recruitment, mortality, migration, demography or genetic structure, or at the community level, through changes in the identity and relative abundance of taxa or functional groups or at the

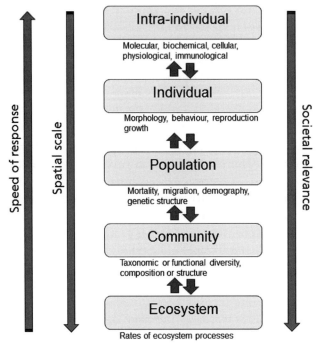

Figure 3.2 Levels of biological organisation at which impacts may occur and indications of variations speed and scale of responses and degree of societal relevance at different levels of the hierarchy. Arrows between levels indicate that effects at one level of organisation can have implications for others, both below and above them in the hierarchy. Adapted from Martínez-Crego et al. (2010).

ecosystem level, affecting structural components or functional processes (Figure 3.2).

At the ecosystem level, stressors can influence ecosystem functioning via their effects on the biota, but also by modifying the abiotic conditions, altering the context in which ecological processes occur and also by directly affecting the rates of chemical processes, for example, through differences in water temperature, light availability, nutrient supply and cycling.

Effects at one level of biological organisation can clearly underpin changes at other higher levels. Effects on communities, for example, are driven by changes in their constituent populations. Intra-individual effects may be sublethal, but if they influence the behaviour or reproductive output of individuals, they can ultimately reduce the long-term viability of populations and thereby affect community and ecosystem-level processes. Some ecological frameworks suggest that the likelihood that stressors will have effects at progressively higher levels of organisation depends primarily on their duration and intensity (Martínez-Crego et al., 2010). However, the impacts at one level of organisation do not necessarily translate predictably into impacts at other levels and, perhaps counter-intuitively, the influence of some stressors can be greater at higher levels of organisation than at lower levels. For example, Kozlov and Zvereva (2011) reported cases in which the abundance of herbivorous arthropods increased in polluted areas despite negative effects on individual performance and in which effects on diversity of plants were much greater than effects on performance or abundance of individual species.). It is therefore unwise to anticipate the effects of stressors at higher levels of organisation from existing knowledge of impacts at lower levels (Attrill and Depledge, 1997; Kozlov and Zvereva, 2011). It is important to recognise that changes at higher levels of organisation can also cascade down through the hierarchy, such that there are complex feedbacks between the different levels of biological organisation. For example, changes in the productivity of an ecosystem caused by increased nutrient availability can influence the suitability of the environment for different species and thus affect population and community structures (e.g. Rosenzweig and Abramsky, 1993); changes to the abundance or identity of predators in a community can modify the behaviour of individuals within prey populations (Forrester, 1994), which may also affect their ecological interactions with other species and their contribution to ecosystem processes.

The rates of biological response to stressors vary at different levels of organisation. Sublethal effects of low levels of stress tend to arise rapidly, but detectable effects at population, community and ecosystem levels can be slower to manifest (Figure 3.2).

3.4 How are 'impacts' determined?

Natural systems are variable at a variety of temporal and spatial scales and this can make it difficult to assess the impact of human activities on ecosystem components. Observed changes coincident with human activities may simply be the result of natural spatial or temporal variation. Some impacts are obvious, for example: (1) the destruction of an area of natural habitat for the development or construction of sea defences, (2) the development of hypoxic conditions from organic enrichment below fish farm cages, or (3) the physical destruction of deep-water corals by towed fishing gear. The majority of human pressures on marine environments do not have such clearly defined impacts, so environmental regulators and managers require robust scientific evidence to support them in targeting management actions effectively.

Best scientific practice requires that evidence, and indeed the only unequivocal evidence of causality, is derived from well-designed and conducted experimental studies that are appropriately analysed (Underwood, 1991a, b). However, there are a number of challenges which constrain the application of experimental approaches to impact assessment. For example, to assess the potential impact of a point discharge from a factory one might assess the toxicity of the substances in the effluent. There is a well-established series of protocols for such toxicity tests (for example, EPA, 2002a, b; OECD 2013; see also http://toxicitylab.com/content/toxicity_testing/epa_documents.php). These have the potential to demonstrate what biological/ecological effects might occur on exposure to the effluent, and these well-defined protocols permit comparison of different substances. However, it is extremely difficult to translate the results of laboratory toxicity tests, on one test organism or battery of organisms, to the 'real' field situation. In the field, the organisms at risk may differ genetically or be in different condition from the organisms in the laboratory tests (e.g. in terms of health or reproductive state) and so have different susceptibility. Additionally organisms in their natural habitat interact with other species, which may further modify their susceptibility. Furthermore, the toxicity of the stressors themselves may be modified by the environmental conditions and the presence of other stressors (Chapter 4). For some stressors, such as large-scale installations such as tidal turbines, laboratory investigation is impracticable or extremely difficult or costly. For these reasons, laboratory tests need to be supplemented by field trials of some kind. Field trials can involve comparisons of putatively impacted sites with carefully selected unaffected control sites (see below), transplantation of organisms into potentially impacted areas to test their responses (e.g. Honkoop et al., 2003) or direct experimental manipulation of the stressor in question, usually in small-scale experimental plots (e.g. McGuinness, 1990; Fitch and Crowe, 2012). Hybrid approaches can also be informative, in which organisms or whole assemblages are collected from putatively impacted and control sites in the field and assessed in the laboratory. A combined approach is often best, using laboratory experiments to establish possible responses to individual stressors in detail under controlled conditions and some field sampling or experimentation to test whether the same responses are observed in the field, in natural conditions with a range of other sources of variation (Crowe et al., 2012).

Field studies of anthropogenic impacts must address all of the challenges faced by ecological field experiments (Hurlbert, 1984; Underwood, 1992, 1994; Raffaelli and Moller, 1999). But it is often the case that a field experiment cannot be undertaken. For example, is it morally justified to release quantities of a known toxin or radioactive nucleotide into the environment at multiple locations in order to examine its effect? Investigations may therefore involve comparisons of sites at which impacts are suspected with control sites which are considered unimpacted. This approach is inherently weak as, at best, one is obtaining a correlation between the activity and changing environmental or biological response variables and it is further constrained when just one impacted and one control site are used, as there is no logical basis to attribute any difference between individual sites to any one cause (Hurlbert, 1984). The most powerful design for these field observational studies is a Beyond BACI (Before – After – Control – Impact) approach (Underwood, 1991a). In this case, multiple sites, matched as closely as possible in terms of salinity, grain size, wave exposure, initial community composition, etc. are sampled a number of times both before and after any putative impact occurs at one or more of the sites. Thus, a comparison of the control versus impact data can be used to check that prior to the impact the sites were comparable, and a comparison of the control sites from before and after the potentially impacting activity identifies the effects of any natural temporal change. In this way, any changes at the impacted site or sites, due to the anthropogenic activities, can be logically and statistically isolated from those due to natural processes (Underwood, 1991a, 1992).

Laboratory experiments, field experiments and even most comparative field studies are carried out over relatively short time scales (e.g. 1–2 years) and often on very localised spatial scales. Environmental policy and management is conducted on much larger spatial scales and needs to be employed over longer time frames. Therefore, there is a challenge in scaling the results from scientific studies up to management units (Underwood, 1995; Thrush et al., 1999). The sublethal impacts of toxins, population scale impacts, and the indirect effects on food web dynamics may only become apparent over long periods (for example, the impact of whaling in krill stocks and populations of penguins and seals that also consume krill; Laws, 1985; Fraser et al., 1992), whilst the large natural variability of most natural systems makes it difficult for large-scale comparisons to have sufficient power to detect effects (Thrush *et al.*, 1999). Long-term (e.g. multi-decadal) studies of marine ecosystems have proved to be powerful tools for understanding long-term human impacts (e.g. Hawkins *et al.*, 2002; Southward *et al.*, 2005), but they frequently lack the capacity to establish causality. In recognition of the data challenges in working with such variable and data-poor systems, marine pollution regulation and fisheries management in Europe and some other parts of the world have formally adopted 'precautionary' approaches, in which management action is triggered even if there is a lack of 'formal proof' of effects (Pikitch *et al.*, 2004).

3.5 Stability

The degree to which a particular ecosystem is impacted by a particular pressure varies depending on the ecosystem and the pressure involved. In other words, different ecosystems have different degrees of *resistance* to pressures. Biological systems also vary inherently in their ability to recover following perturbations, i.e. their *resilience*. Resistance and resilience, are thus distinct aspects of stability (Grimm and Wissel, 1997).

When considering the impacts of human stressors on the environment, it is therefore helpful to think in terms of a system's inherent ability to resist change and its ability to recover following damage. Sensitivity to stressors can be seen as a combination of these properties. Sensitivity then is the combination of exposure to damage and the ability to recover from that damage, such that, for example, a system would be considered more sensitive to an activity causing an equivalent amount of 'damage' if it recovers slowly than if it recovers more rapidly. Stability, in terms of resistance and resilience, has been the subject of considerable debate in ecology (e.g. Pimm, 1984; Grimm and Wissel, 1997; Ives and Carpenter, 2007). In practical terms, some schemes have been proposed to classify resistance, resilience and sensitivity to different stressors and these are outlined below.

Susceptibility to impact: resistance stability

The resistance categories used in Table 3.2 are based on Odum's (1989) definition of resistance, that is 'the ability of an ecosystem to withstand disturbance without undergoing a phase shift or losing structure or function'. Hall *et al.* (2008) developed a formal set of resistance categories based on this approach for the assessment of habitat sensitivity

Table 3.2 *The resistance categories used in Hall et al. (2008).*

Resistance category	Description
None	Removal of habitat, e.g. change in habitat type
Low	Effects on physical and biological structure of the habitat and causes widespread mortality
Medium	Some mortality at significant levels to species and some damage to the physical structure
High	No ecologically significant effect to the physical structure, no effect on viable population but may affect feeding, respiration and reproduction rates

to fishing impacts adopted by the Countryside Council for Wales in the management of fishing in Welsh waters and was subsequently used by the Oslo–Paris Commission (OSPAR).

Many habitats and organisms possess an inherent resistance to natural and anthropogenic pressures. However, the resistance of a habitat to loss of extent in response to a given pressure does not imply its resistance to a loss of quality in the functioning of the system. Natural variability and resistance to natural disturbance can sometimes make it difficult to detect the effects of human activities on marine ecosystems.

Capacity to recover: resilience stability

Resilience has been defined as 'the ability of a system to recover from disturbance or change while maintaining its function and service' (Carpenter *et al.*, 2001). It has been proposed that not all forms of impacting activity will result in the same types of changes in the ecosystem and that there are differences in the recovery rates between locations due, for example, to differences in hydrography and larval supply. Hall *et al.* (2008) showed that in most published and peer-reviewed studies, no relationship existed between the degree of damage to the habitat (expressed as % mortality) and the rate of the subsequent recovery, implying that in most marine habitats the supply of new recruits or adult immigrants was of greater importance than the response of any survivors in the impacted area. In developing a framework for multi-sectoral management in the North-east Atlantic, OSPAR used a categorical approach to assessing resilience for marine habitats (Robinson *et al.*, 2008; Table 3.3).

Table 3.3 *Definition of the four habitat resilience categories used by Robinson et al. (2008) for the purpose of comparing the sensitivity of different marine habitats in the North-east Atlantic.*

Resilience category	Description	Example
None	No recovery > 100 years	*Lophelia* reefs
Low	Recovery 10–100 years	Maerl beds, circalittoral communities of sponges, bryozoans and hydroids, infaunal communities dominated by long-lived bivalves, e.g. *Arctica*
Medium	Recovery 2–10 years	Moderately exposed rocky shores dominated by macroalgae
High	Recovery < 2 years	Wave-exposed intertidal communities dominated by barnacles and green algae

Table 3.4 *Marine habitat sensitivity matrix showing how expert judgements were used as the basis of the habitat sensitivity assessments for individual fishing gear–habitat interactions in Hall et al. (2008). Habitat sensitivity is classified as being High Sensitivity, Medium Sensitivity or Low Sensitivity, depending on the combination of resistance and resilience displayed by that habitat to each gear considered.*

		Resistance			
	Category	None	Low	Medium	High
Resilience	None	HIGH	HIGH	MEDIUM	LOW
	Low	HIGH	HIGH	MEDIUM	LOW
	Medium	HIGH	HIGH	MEDIUM	LOW
	High	MEDIUM	MEDIUM	LOW	LOW

Sensitivity: the combination of resistance and resilience

These resistance and resilience categories (Tables 3.2 and 3.3) can be combined to produce a matrix of marine habitat sensitivity (Table 3.4). There is an asymmetry to the matrix (Table 3.4). High resilience and medium resistance of a habitat suggests that it does not take that long to recover (<2 years) even if damage to the habitat is widespread. Medium resilience and medium resistance of a habitat equates to a system that takes a prolonged period to recover (2–10 years). As the system is perturbed for longer in the second case, and hence is, potentially, more vulnerable to further change this is seen as being of higher management concern and so warrants a higher (i.e. medium) sensitivity. This approach, supported by expert workshops, has been implemented by Natural Resources Wales for fisheries management in areas of nature conservation interest. For example, the habitat type 'sandy gravel with long-lived bivalves' is fished for scallops and on the basis of the presence of non-target species of long-lived, slow-growing, aperiodically recruiting bivalves was deemed to be of *low* resilience. This assessment was based on the extensive literature on the biology of the bivalves. Hence this assessment was deemed to be of *high* confidence. There are a number of studies that document the impact of (king) scallop dredges on the habitat features and non-target species on scallop grounds (Ramsay et al., 1998; Collie et al., 2000; Bradshaw et al., 2002). These studies show considerable levels of damage to the habitat and non-target species and the resistance to high-intensity fishing was deemed *none* and *low* for moderate and low intensities. As is usual, the peer-reviewed studies do not classify the habitats studies in the same terms as conservation and marine management organisations. In this case the experts considered there to be a number of high quality studies in this or similar habitats that could be used as the basis for the assessment. They therefore had *high* confidence in the assessment of resistance.

3.5.1 Stability and the balance of nature

Evidence is emerging (Chapter 4) that challenges the popular perception of a natural system's dynamics, which considers the ecological systems to be naturally balanced, with mechanisms existing that cause a return to a persistent state or equilibrium point. This notion has also informed the approach of policy makers. Ecologists would see this idea of the *balance of nature* as a simplification of the 'resilience' theory of ecological systems. The concept can be summarised as (1) ecosystems 'develop' over time through some kind of succession, and (2) that this development can reach an end-point, a 'climax community'. Environmental management has traditionally focused on this developmental trajectory, often seeking to deliver an end state deemed to be desirable because it produces economic benefits or fits the societal view of natural systems (Pimm, 1991).

Of course, resilience only operates within certain limits, and ecologists now recognise the existence of multiple stable states. For example, in the Antarctic whale fisheries, human activity has caused the system to flip into a new, different, stable configuration (Fraser et al., 1992). These tipping points (the point at which the existing stable state breaks down) are extremely difficult to predict in advance. May (1977) used simple population models and empirical data to demonstrate not just the existence of multiple stable states in natural systems (such as fish stocks and insect pest populations) but also the existence of such tipping points and of non-symmetrical onset and recovery trajectories. These non-symmetrical trajectories mean that reducing the impacting activities may not cause the system to recover via the same intermediate states as it did when the impact applied. This phenomenon is termed hysteresis and hysteresis can arise not just in terms of community composition but also of functional dynamics. Together these findings challenge the resilience/balance of nature view of natural ecosystems. A further complicating factor is the existence of 'slow' variables which change gradually over long time scales and thus are difficult to detect. Examples include the gradual accumulation of phosphorus in sediments in the Baltic (Reed *et al.*, 2011). Monitoring of the system shows all to be well until, in this example, the capacity of the system to absorb phosphorus is exceeded, at which point there are sudden and rapid changes. The system has reached a tipping point. Slow variables are difficult to track and detect because they have nonlinear dynamics, they may be under the control of multiple factors and their slow progress means that managers experience the shifting baseline phenomenon (Pauly, 1995), in which scientists and managers fail to correctly identify the true environmental baseline. For example, fisheries data rarely extend back beyond 100 years but fishing has been going on for thousands of years, and in many cases scientists use the earliest iteration of their current model as the baseline rather than even the potentially impacted populations sampled 100 years ago, let alone the truly unimpacted populations that existed before exploitation began. Scientists and managers may only become aware of the significance of slow acting variables after a threshold has been crossed and large-scale ecological and social changes have occurred.

3.6 Key points

- Human society derives considerable benefit from marine ecosystems, through a wide range of activities. The majority of human uses of the marine environment have some measurable impact on the supporting ecosystem. Impacts are direct or indirect and may be sustainable, in which case the system will continue to provide the service indefinitely

- and will recover upon cessation of the impacting activity, or they may be unsustainable.
- Most activities have some effect on the ecosystem, causing physical change, altering levels of chemicals and their cycling and/or affecting biological components of the system (Table 3.1). Human uses of the ecosystem can be categorised by the nature of the services being exploited: Provisioning Services, including living resources and abiotic (e.g. mineral) resources, Regulating Services such as nutrient cycling and carbon sequestration and Cultural Services such as tourism (Chapter 2). The impacts are best considered in terms of the pressures they exert on the system, for example, smothering by sediment, mortality of biota due to toxic effects.
- Human impacts can be observed at all levels of biological organisation and different pressures may influence different components of the system. However, ecological processes can cause such impacts to propagate through the ecosystem and in some cases these mitigate and in others magnify these impacts. There remain considerable gaps in our understanding of how human pressures propagate through complex, natural ecosystems.
- Natural systems are variable at a variety of temporal and spatial scales and this can make it difficult to assess the impact of human activities on ecosystem components. Best scientific practice requires that evidence is derived from well-designed and conducted studies that are appropriately analysed. However, there are a number of challenges which constrain the application of experimental approaches to impact assessment, including the complexity of natural ecosystems, the open nature of marine communities, and the dynamic and temporal and spatial heterogeneity of the marine environment.
- The degree to which a particular ecosystem is impacted by a particular stressor varies depending on the ecosystem and the stressor involved (see Chapter 4). In assessing the significance of the impacts of human stressors on the environment, it is important to consider a system's inherent ability to remain unchanged when subjected to stress (its resistance) and its ability to recover following damage (its resilience), which can be combined into measures of overall sensitivity.
- There is increasing evidence of the existence of multiple stable states, thresholds and tipping points and hysteresis in natural ecosystems. These phenomena make the prediction of human impacts for combinations of conditions not previously observed impossible and therefore constitute a major challenge for scientists and policy makers.

References

Attrill, M. J. and Depledge, M. H. (1997). Community and population indicators of ecosystem health: targeting links between levels of biological organisation. *Aquatic Toxicology*, 38, 183–197.

Boyd, S. E., Cooper, K. M., Limpenny, D. S. et al. (2004). Assessment of the rehabilitation of the seabed following marine aggregate dredging, Science Series Technical Report, CEFAS Lowestoft.

Bradshaw, C., Veale, L. O. and Brand, A. R. (2002). The role of scallop-dredge disturbance in long-term changes in Irish Sea benthic communities: a re-analysis of an historical dataset. *Journal of Sea Research*, 47, 161–184.

Carpenter, S., Walker, B., Anderies, J. and Abel, N. (2001). From metaphor to measurement: resilience of what to what? *Ecosystems*, 4, 765–781.

Chandrasekara, W. U. and Frid, C. L. J. (1996). Effects of human trampling on tidal flat infauna. *Aquatic Conservation: Marine and Freshwater Ecosystems*, 6, 299–311.

Clark, S. and Edwards, A. J. (1994). Use of artificial reef structures to rehabilitate reef flats degraded by coral mining in the Maldives. *Bulletin of Marine Science*, 55, 724–744.

Clark, S. and Edwards, A. J. (1999). An evaluation of artificial reef structures as tools for marine habitat rehabilitation in the Maldives. *Aquatic Conservation: Marine and Freshwater Ecosystems*, 9, 5–21.

Clark, R. B., Frid, C. L. J. and Attrill, M. (2001). *Marine Pollution*. Oxford: Oxford University Press.

Collie, J. S., Hall, S. J., Kaiser, M. J. and Poiner, I. R. (2000). A quantitative analysis of fishing impacts on shelf-sea benthos. *Journal of Animal Ecology*, 69, 785–798.

Craig, O. E., Saul, H., Lucquin, A. et al. (2013). Earliest evidence for the use of pottery. *Nature*, 496, 361–364.

Crowe, T. P., Bracken, M. E. and O'Connor, N. E. (2012). Reality check: issues of scale and abstraction in biodiversity research, and potential solutions. In *Marine Biodiversity Futures and Ecosystem Functioning: Frameworks, Methodologies and Integration*, ed. M. Solan, R. J. A. Aspden and D. M. Paterson. Oxford: Oxford University Press, pp. 185–199.

Danovaro, R., Bongiorni, L., Corinaldesi, C. et al. (2008). Sunscreens cause coral bleaching by promoting viral infections. *Environmental Health Perspectives*, 116, 441–447.

de Groot, S. J. (1996). The physical impact of marine aggregate extraction in the North Sea. *ICES Journal of Marine Science*, 53, 1051–1053.

Díaz, R. J. and Rosenberg, R. (1995). Marine benthic hypoxia: a review of its ecological effects and the behavioral responses of benthic macrofauna. *Oceanography and Marine Biology*, 33, 245–303.

Doney, S. C., Fabry, V. J., Feely, R. A. and Kleypas, J. A. (2009). Ocean acidification: the other CO_2 problem. *Annual Review of Marine Science*, 1, 169–192.

EEA (2007). The DPSIR framework used by the EEA. Available at: http://root-devel.ew.eea.europa.eu/ia2dec/knowledge_base/Frameworks/doc101182, accessed December 2013.

EPA (2002a). *EPA-821-R-02–012 Methods for Measuring Acute Toxicity of Effluents and Receiving Waters to Freshwater and Marine Organisms* (5th edn). Washington DC: US Environmental Protection Agency.

EPA (2002b). *EPA-821-R-02–014 Short-term Methods for Estimating the Chronic Toxicity of Effluents and Receiving Water to Marine and Estuarine Organisms*; (3rd edn). Washington DC: US Environmental Protection Agency.

Fabry, V. J., Seibel, B. A., Feely, R. A. and Orr, J. C. (2008). Impacts of ocean acidification on marine fauna and ecosystem processes. *Ices Journal of Marine Science*, 65, 414–432.

FAO (2012). *The State of World Fisheries and Aquaculture 2012*, Rome: Food and Agriculture Organisation.

Fitch, J. E. and Crowe, T. P. (2012). Combined effects of inorganic nutrients and organic enrichment on intertidal benthic macrofauna: an experimental approach. *Marine Ecology Progress Series*, 461, 59–70.

Fletcher, H. and Frid, C. L. J. (1997). Impact and management of visitor pressure on rocky intertidal algal communities. *Aquatic Conservation-Marine and Freshwater Ecosystems*, 7, 287–297.

Forrester, G. E. (1994). Influences of predatory fish on the drift density of stream insects. *Ecology*, 75, 1208–1218.

Fraser, W. R., Trivelpiece, W. Z., Ainley, D. G. and Trivelpiece, S. G. (1992). Increases in Antarctic penguin populations: reduced competition with whales or a loss of sea ice due to environmental warming? *Polar Biology*, 11, 525–531.

Frid, C. and Dobson, M. (2013). *The Ecology of Aquatic Management*. Oxford: Oxford University Press.

Frid, C. L. J., Andonegi, E., Depestele, J. et al. (2012). The Environmental Interactions of Tidal and Wave Energy Generation Devices. *Environmental Impact Assessment Review*, 32, 133–139.

Gattuso, J.-P. and Hansson, L. (2011). *Ocean Acidification*. Oxford: Oxford University Press.

Gill, A. B. (2005). Offshore renewable energy: ecological implications of generating electricity in the coastal zone. *Journal of Applied Ecology*, 42, 605–615.

Goss-Custard, J. D., Warwick, R. M., Kirby, R. et al. (1991). Towards predicting wading bird densities from predicted prey densities in a post-barrage Severn Estuary. *Journal of Applied Ecology*, 28, 1004–1026.

Gowan, R. J. and Bradbury, N. B. (1987). The ecological impact of salmonid farming in coastal waters: a review. *Oceanography and Marine Biology: Annual Review*, 25, 563–575.

Grigg, R. W. (1984). Resource management of precious corals: A review and application to shallow water reef-building corals. *Marine Ecology*, 5, 57–74.

Grimm, V. and Wissel, C. (1997). Babel, or the ecological stability discussions: an inventory and analysis of terminology and a guide for avoiding confusion. *Oecologia*, 109, 323–334.

Hall, K., Paramor, O. A. L., Robinson, L. A., Winrow-Giffin, A. and Frid, C. L. J. (2008). Mapping the sensitivity of benthic habitats to fishing in Welsh waters: development of a protocol. Countryside Council for Wales Policy Research Report No. 08/12.

Hall, S. J. (1999). *The Effects of Fishing on Marine Ecosystems and Communities*. Oxford: Blackwell.

Harley, C. D. G., Randall Hughes, A., Hultgren, K. M. *et al.* 2006. The impacts of climate change in coastal marine systems. *Ecology Letters*, 9, 228–241.

Hawkins, S. J., Gibbs, P. E., Pope, N. D. *et al.* (2002). Recovery of polluted ecosystems: the case for long-term studies. *Marine Environmental Research*, 54, 215–222.

Hibbert, C. J. (1977). Growth and survivorship in a tidal-flat population of the bivalve *Mercenaria mercenaria* from Southampton water. *Marine Biology*, 44, 71–76.

Hoegh-Guldberg, O. and Bruno, J. (2010). The impact of climate change on the world's marine ecosystems. *Science*, 328, 1523–1528.

Honkoop, P. J. C., Bayne, B. L. Underwood, A. J. and Sevensson, S. (2003). Appropriate experimental design for transplanting mussels (*Mytilus* sp.) in analyses of environmental stress: an example in Sydney Harbour (Australia). *Journal of Experimental Marine Biology and Ecology*, 297, 253–268.

Hurlbert, S. H. (1984). Pseudoreplication and the design of ecological field experiments. *Ecological Monographs*, 54, 187–211.

IPCC (2013). Climate Change 2013. The Physical Science Basis. Working Group I Contribution to the IPCC 5th Assessment Report: Changes to the Underlying Scientific/Technical Assessment. IPCC, Geneva.

Ives, A. R. and Carpenter, S. R. (2007). Stability and diversity of ecosystems. *Science*, 317, 58–62.

Kingston, P. F. (1992). Impact of offshore oil production installations on the benthos of the North Sea. *Ices Journal of Marine Science*, 49, 45–53.

Kozlov, M. V. and Zvereva, E. L. (2011). A second life for old data: global patterns in pollution ecology revealed from published observational studies. *Environmental Pollution*, 159, 1067–1075.

Laws, R. M. (1985). The ecology of the Southern Ocean. *American Scientist*, 73, 26–40.

Lucas, J. S. and Southgate, P. C. (2012). *Aquaculture: Farming Aquatic Animals and Plants*. Oxford: Wiley-Blackwell.

McGranahan, G., Balk, D. and Anderson, B. (2007). The rising tide: assessing the risks of climate change and human settlements in low elevation coastal zones. *Environment and Urbanization*, 19, 17–37.

McGuinness, K. A. (1990). Effects of oil spills on macro-invertebrates of saltmarshes and mangrove forests in Botany Bay, New South Wales, Australia. *Journal of Experimental Marine Biology and Ecology*, 142, 121–135.

Martínez-Crego, B., Alcoverro, T. and Romero, J. 2010. Biotic indices for assessing the status of coastal waters: A review of strengths and weaknesses. *Journal of Environmental Monitoring*, 12, 1013.

May, R. M. (1977). Thresholds and breakpoints in ecosystems with a multiplicity of stable states. *Nature*, 269, 471–477.

OECD (2013). *OECD Guidelines for the testing of Chemicals, Section 2: Effects on biotic systems*. Paris: OECD.

Pauly, D. (1995). Anecdotes and the shifting baseline syndrome of fisheries. *Trends in Ecology and Evolution*, 10, 430.

Pearson, T. H. and Mannvik, H. P. (1998). Long-term changes in the diversity and faunal structure of benthic communities in the northern North Sea: natural variability or induced instability? *Hydrobiologia*, 376, 317–329.

Peterson, C. H. and Lubchenko, J. (1997). Marine ecosystem services. In *Nature's Services: Societal dependence on natural ecosystems*, ed. G. C. Daily. Washington DC: Island Press, pp. 177–194.

Pikitch, E. K., Santora, E. A., Babcock, A. *et al.* (2004). Ecosystem-based fishery management. *Science*, 305, 346–347.

Pimm, S. L. (1984). The complexity and stability of ecosystems. *Nature*, 307, 321–326.

Pimm, S. L. (1991). *The Balance of Nature?* Chicago, IL: Chicago University Press.

Prakash, O. (2004). *Cultural History of India*. New Delhi, India: New Age International.

Raffaelli, D. and Moller, H. (1999). Manipulative field experiments in animal ecology: do they promise more than they can deliver? *Advances in Ecological Research*, 30, 299–338.

Ramsay, K., Kaiser, M. J. and Hughes, R. N. (1998). Responses of benthic scavengers to fishing disturbance by towed gears in different habitats. *Journal of Experimental Marine Biology and Ecology*, 224, 73–89.

Reed, D. C., Slomp, C. P. and Gustafsson, B. G. (2011). Sedimentary phosphorus dynamics and the evolution of bottom-water hypoxia: a coupled benthic-pelagic model of a coastal system. *Limnology and Oceanography*, 56, 1075–1092.

Robinson, L. A., Rogers, S. and Frid, C. L. J. (2008). A marine assessment and monitoring framework for application by UKMMAS and OSPAR – Assessment of pressures. A report to the JNCC from University of Liverpool, Liverpool, UK.

Rogers, S. I. and Greenaway, B. (2005). A UK perspective on the development of marine ecosystem indicators. *Marine Pollution Bulletin*, 50(1), 9–19.

Rosenzweig, M. L. and Abramsky, Z. (1993). How are diversity and productivity related? In *Species Diversity in Ecological Communities: Historical and Geographical Perspectives*, ed. R. E. Ricklefs and D. Schluter. Chicago, IL: Chicago University Press, pp. 52–53.

Royal Society (2005). *Ocean Acidification Due to Increasing Atmospheric Carbon Dioxide*. London: The Royal Society.

Southward, A. J., Langmead, O., Hardman-Mountford, N. J., *et al.* (2005). Long-term oceanographic and ecological research in the Western English Channel. *Advances in Marine Biology*, 47, 1–105.

Tans, P. and Keeling, R. (2013). Trends in atmospheric carbon dioxide. US National Oceanic and Atmospheric Administration. Available at: http://www.esrl.noaa.gov/gmd/ccgg/trends/#mlo_full, accessed December 2013.

Takahashi, T., Sutherland, S. C., Sweeney, C. *et al.* (2001). Global sea–air CO_2 flux based on climatological surface ocean pCO_2, and seasonal biological and temperature effects. *Deep Sea Research Part II: Topical Studies in Oceanography*, 49, 1601–1622.

Thiel, H. (1992). Deep-sea environmental disturbance and recovery potential. *Internationale Revue der Gesamten Hydrobiologie und Hydrographie*, 77, 331–339.

Thrush, S. F., Lawrie, S. M., Hewitt, J. E. and Cummings, V. J. (1999). The problem of scale: uncertainties and implications for soft-bottom marine communities and the assessment of human impacts. In *Biogeochemical Cycling and Sediment Ecology*, ed. J. S. Gray, W. Ambrose Jr. and A. Szaniawska. Dordrecht, the Netherlands: Kluwer Academic Publishers, pp. 195–210.

Underwood, A. J. (1991a). Beyond BACI: experimental designs for detecting human environmental impacts on temporal variations in natural populations. *Australian Journal of Marine and Freshwater Research*, 42, 569–587.

Underwood, A. J. (1991b). The logic of ecological experiments: a case history from studies of the distribution of macroalgae on rocky intertidal shores. *Journal of the Marine Biological Association of the UK*, 71, 841–866.

Underwood, A. J. (1992). Beyond BACI: the detection of environmental impacts on populations in the real, but variable, world. *Journal of Experimental Marine Biology and Ecology*, 161, 145–178.

Underwood, A. J. (1994). On Beyond BACI: sampling designs that might reliably detect environmental disturbances. *Ecological Applications*, 4, 3–15.

Underwood, A. J. (1995). Ecological research and (and research into) environmental management. *Ecological Applications*, 5, 232–247.

Wernberg, T., Smale, D. S. and Thomsen, M. S. (2012). A decade of climate change experiments on marine organisms: procedures, patterns and problems. *Global Change Biology*, 18, 1491–1498.

4 · Modifiers of impacts on marine ecosystems: disturbance regimes, multiple stressors and receiving environments

DEVIN LYONS, LISANDRO BENEDETTI-CECCHI, CHRISTOPHER FRID AND ROLF VINEBROOKE

4.1 Introduction

Effective management and the maintenance of marine ecosystem services rely on a capacity to predict the ecological consequences of environmental change and potential management interventions (Chapter 1). Making these predictions is difficult because anthropogenic stressors do not produce uniform or consistent impacts on biodiversity and ecosystem functioning. Rather, their effects can be modified by a variety of factors that cause them to vary among locations and different points in time. Thus, the effectiveness of actions taken to manage environmental problems is likely to vary in a similar way: interventions that are sufficient to mitigate a stressor's impacts in one situation might be inadequate or excessive in others. Both sound science and efficient management require us to recognise that spatial and temporal variability are inherent to natural systems, and that the ecosystem complexity places inherent limits on our ability to predict future ecological conditions. However, many of the causes of this variability have been identified. Careful consideration of these factors will enhance scientific understanding, improve ecological prediction and enhance our efforts to optimise marine policy and management by reducing the uncertainty associated with the effects of stressors.

Marine Ecosystems: Human Impacts on Biodiversity, Functioning and Services, eds T. P. Crowe and C. L. J. Frid. Published by Cambridge University Press. © Cambridge University Press 2015.

In this chapter, we examine three factors that cause anthropogenic activities to have inconsistent impacts on marine ecosystems. Both individual organisms and entire ecosystems tend to respond nonlinearly to gradients in stressor severity. Thus, we begin by discussing how differences in stressor intensity, as well as spatial and temporal characteristics of stressors' disturbance regimes can influence their impacts. Next, we discuss how characteristics of receiving ecosystems influence their sensitivity to stressors' impacts. These include biological characteristics such as the genetic structure of populations and community composition, as well as physical and chemical variables that may alter stressors' effects. Finally, we will discuss an issue of increasing concern among scientists and policy makers: the impacts of multiple stressors and the ways in which stressors can modify one another's effects.

4.2 Disturbance regimes

One of the primary reasons that the impacts of stressors vary from one situation to another is that the same stressor can operate under different disturbance regimes. Both natural disturbance and anthropogenic stressors vary in their intensity (or magnitude), spatial extent, frequency, duration, timing and temporal patterning of occurrence and this can affect the ultimate impact of a stressor (Miller, 1982; Pickett and White, 1985; Fraterrigo and Rusak, 2008). There has been substantial research addressing how changes in disturbance regimes affect biodiversity in marine systems (Sousa, 1984). There is also increased understanding of the consequences of biodiversity loss for marine ecosystem functioning (e.g. productivity and stability) and services (Solan et al., 2012). Here, we review the key findings of these studies to assess the potential for disturbance regimes to affect ecosystem functioning and services, with a focus on how their impact is mediated through their effects on biodiversity and interactions between organisms.

4.2.1 The intermediate disturbance hypothesis and its variants

The intermediate disturbance hypothesis (IDH) is one of the best-known conceptual models relating variation in disturbance to biodiversity (Connell, 1979). IDH postulates that peaks in species diversity should be observed when the intensity or frequency of disturbance is not too high to make the environment unfavourable to most species, and not too low to favour the spread of competitive dominants that would reduce diversity through competitive displacement – leading to a hump-shaped relationship. Despite the intuitive appeal of the IDH, both theoretical and empirical evidence suggest that the simple mechanisms and predictions of the IDH are not as well supported as previously thought (Fox, 2013). Instead, ecosystems display a much broader range of disturbance–diversity relationships, and hump-shaped relationships are no more common than flat, monotonically increasing or monotonically decreasing functional forms (Mackey and Currie 2001; Hughes *et al.*, 2007).

Variation in the form of disturbance–diversity relationships may be partially explained by interactions among different aspects of disturbance regimes (Miller *et al.*, 2011). Hall *et al.*, (2012) used microcosm experiments to show that the form of the disturbance–diversity relationship could be flat, monotonically increasing or hump-shaped, depending on the combination of disturbance frequency and intensity that was applied. Similarly, in examining how the rate of disturbance (area per unit time) affected the diversity of hard-bottom assemblages, Svensson *et al.*, (2009) found only weak support for the pattern predicted by the IDH at only one of three sites they examined. The effects they observed depended on whether equivalent disturbance rates were accomplished through larger infrequent disturbances or smaller frequent disturbances.

Although the IDH is not as well supported as once thought, the IDH and other diversity–disturbance relationships illustrate a number of key points. First, the effects of anthropogenic stressors are often likely to be nonlinear. The IDH predicts a nonlinear change in diversity, but other individuals, populations, and community metrics will also respond nonlinearly over gradients in disturbance. For example, the ability of algal turfs to become established within stands of canopy-forming algae is a nonlinear function of the size of disturbed gaps produced in these stands (Tamburello *et al.*, 2013). Second, effects at the community level emerge as the result of impacts at lower levels of biological organisation. Although diversity–disturbance relationships describe community-level patterns, they can result from reductions in the abundance of a single competitively dominant species. In this case, impacts on dominant species cascade up to generate community-wide patterns through direct and indirect effects on other organisms (see also the example of the *Exxon Valdez* oil spill, below). In other cases, impacts at lower levels of biological organisation may translate more directly to changes at the community or ecosystem level. Finally, although we will discuss disturbance intensity, extent, duration, etc., separately, these disturbance characteristics can interact.

4.2.2 Intensity of disturbance

Perhaps the most obvious aspect of a stressor's disturbance regime that influences its effects is its intensity. We generally expect the magnitude of a stressor's effects to increase with its intensity. Setting aside the sometimes hump-shaped relationships between disturbance severity and biodiversity discussed above, this expectation is born out by many types of biological responses at the individual, population, and community levels. For example, the productivity and calcification rates of individual coral colonies from a variety of species decrease with increasing concentrations of carbon dioxide (Anthony *et al.*, 2008). The growth of potentially harmful nuisance algae increases with increasing nutrient concentrations in eutrophic estuaries (Teichberg *et al.*, 2010). Predation by sea stars, fish and crustaceans declines over a gradient of declining oxygen concentration, allowing populations of hypoxia-tolerant bivalves to increase (Alteri, 2008). Unsurprisingly, populations and communities also tend to recover more slowly following severe disturbances than they do after a mild disturbance (Underwood, 1998; Speidel *et al.*, 2001; Dernie *et al.*, 2003).

4.2.3 Extent of disturbance

Spatial aspects of a stressor's disturbance regime affect whether mobile organisms are able to avoid or reduce exposure to the stressor, and influence recovery via immigration and recruitment. Disturbances that operate over large spatial extents may be more detrimental to species diversity and ecosystem functioning than smaller ones. In general, small areas have a large perimeter-to-area ratio, so lateral encroachment should be the primary mechanism of recovery of small disturbances in systems where vegetative growth is important, such as in algal-dominated assemblages. Recovery of large disturbed areas, in contrast, should be driven mostly by recruitment (Sousa, 1984). However, if the spatial extent of disturbance becomes too large, then recovery towards a pre-disturbed condition may be slow, or totally new species may appear, leading to an alternative community state (Petraitis and Latham, 1999).

We offer two examples of such shifts in species dominance and composition following human disturbance. The first is the *Exxon Valdez* oil spill, which occurred along the Alaskan coastal ecosystem in 1989. This dramatic event impacted more than 750 km of coastline, leading to profound changes in species composition and abundance of both soft- and hard-bottom assemblages (Paine et al., 1996). Over these large spatial scales, both recruitment limitation and the cascading indirect effects caused by alterations to species interactions contributed to delayed recovery (Peterson et al., 2003). The most evident of these cascades was triggered by loss of the dominant space occupier, the brown alga *Fucus gardneri*, with the subsequent loss of grazers (limpets and periwinkles) and predators (whelks). Open space was then colonised by filamentous algae and barnacles, with recovery to a canopy-dominated system taking place over a decade (Peterson et al., 2003).

The second example is provided by the highly destructive date mussel (*Lithophaga lithophaga*) fishery, as documented by studies along the Apulian coast in Italy (Fanelli et al., 1994). The date mussel, a boring bivalve living in calcareous rocks, is heavily exploited for commercial purposes. Collection requires drilling the rock surface, which causes the complete eradication of algal canopies (*Cystoseira* spp.) and associated understory assemblages in the shallow subtidal (0–10 m depth). This kind of impact extends for tens of kilometres along the Apulian coast and leads to an alternate, apparently stable state, in the structure of assemblages. The reduction in habitat complexity following the removal of algal canopies promoted the invasion of sea urchins, the colonisation of grazing resistant sponges and changes in the composition of fish assemblages (Fanelli et al., 1994; Guidetti et al., 2004). Sea urchins' ability to prevent algal recovery through grazing is considered the primary cause of stability of the highly depauperate, sponge-dominated benthic assemblage that developed in disturbed areas. This stability may have been prevented if harvesting had not occurred over such a large area. Macroalgae generally disperse over relatively short distances compared to fish and many marine invertebrates (Kinlan and Gaines, 2003). Thus, the large scale of *Cystoseira* removal may have contributed to the formation of this alternate state by preventing the alga from dispersing back into affected areas before sea urchins became established.

4.2.4 Duration of disturbance

The duration of disturbance is another important trait of disturbance regimes that determines the response of ecological communities and ecosystems to perturbations. It is important to distinguish between the duration of a disturbance and how long its effects last, as both brief and ongoing disturbances can have transient or long-lasting effects (Glasby and Underwood, 1996). However, we generally expect that ecosystems are more likely to recover from brief disturbances, whereas ongoing stressors are more likely to cause long-lasting or permanent changes to their structure and dynamics. Bender et al., (1984) referred to disturbances that cause relatively instantaneous change in species abundance and are followed by a return to the pre-disturbance conditions as pulse perturbations, and sustained alteration of species abundances that are maintained until the system shifts toward another 'equilibrium' or stable condition as press perturbations.

The temporal scale of disturbance events influences the severity of their effects, in large part because ecosystems react to environmental change through a series responses at the individual, population, and community levels, and these responses develop after different periods of time (Smith et al., 2009; Martinez-Crego et al., 2010). Initially, individual organisms make physiological adjustments that result in changes to community respiration and productivity to respiration ratios (Odum, 1985). Over time, community composition and ecosystem processes will change as stressors affect the interactions between species, as sensitive species are lost, and as new species immigrate into the community (Smith et al., 2009). Eventually, evolutionary processes that save populations from extinction, and allow them to rebuild, may come into play (Bell and Gonzales, 2011). Overall, these changes are likely to lead to nonlinear responses as the duration of a disturbance increases (Smith et al., 2009).

Long-lasting stressors do not need to be of large magnitude to be ecologically important because their effects can accumulate over time, and manifest themselves at higher levels of biological organisation. For example, exposure of mussels to sublethal concentrations of trace metals triggers detoxification mechanisms at the physiological level that may ultimately lead to population-level responses under chronic exposure (Regoli et al., 2011). Eventually, such chronic stressors can shape community composition and the functioning of ecosystems. In contrast, such sublethal stressors are unlikely to cause large population or community levels effects if they are of short duration. Even if short-term stressors have population and community-level consequences, these may often be transient, with the ecosystem recovering soon after the disturbance is over.

Although ecosystems may often recover from brief disturbances, transient stressors can have lasting ecological consequences if they are sufficiently severe, and occur over a large spatial scale. For example, large waves associated with hurricanes or strong storms may reshape marine coastal communities through the eradication of habitat-forming species such as bed-forming mussels and canopy-forming algae (Sousa, 1984; Schiel, 2011). Similarly, heat waves in the Mediterranean have triggered large-scale mass mortality of gorgonians and other macrobenthic species (Cerrano et al., 2000; Garrabou et al., 2009). Such short, intense stressors also have the potential to have lasting ecological effects when sudden removal of large numbers of individuals makes space and other resources available to stress-tolerant and rapidly colonising species (Glasby and Underwood, 1996). If these species are capable of pre-emptively excluding sensitive species, the ecosystem remains in an alternated state once the disturbance is over. Thus, the mechanisms involved are similar to those described in large-scale, long-lasting stressors such as the chemical pollution caused by oil spills and the continual harvesting of marine life discussed above.

4.2.5 Temporal pattern and timing of disturbance

Although examples of the effects of individual disturbances on marine coastal communities are numerous, much attention is now concentrated on the predicted increase in frequency of extreme events and variations in the timing and temporal pattern of disturbances, particularly in relation to climate change (Easterling et al., 2000; IPCC, 2007). Global warming will cause stronger temperature gradients between land and sea masses due to the larger heat capacity of the water. Increasing thermal gradients result in stronger winds and thus more frequent and more intense storms. Increasing atmospheric temperature also results in enhanced water evaporation, which leads to more intense and frequent extreme precipitation events (IPCC, 2007). Increasing frequency of these events results in a greater chance for disturbances to coincide with important population-level processes such as timing of reproduction and recruitment, or community-level processes such as trophic interactions. For example, disturbance experiments have shown that clearings produced in different periods of the year may undergo different trajectories of succession depending on seasonality in species reproduction or temporal variation in grazing intensity (Sousa, 1979; Hawkins, 1981; Benedetti-Cecchi, 2000). Thus, there is a great potential for increasing climate stochasticity to interact with population- and community-level processes, challenging species' ability to become adapted to new environmental conditions and generating new threats to ecosystem function and services (Bernhardt and Leslie, 2013).

Extreme climate events are also expected to become more concentrated within short periods of time, with potentially important, but

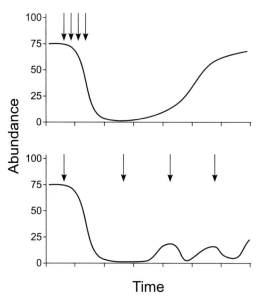

Figure 4.1 Diagram showing the hypothetical effect of clustered (upper panel) and more evenly distributed (lower panel) disturbances on population abundance over time (arbitrary units). Recovery is facilitated in the clustered scenario if disturbances are intense enough so that a single event can decimate the population.

largely unknown consequences for biodiversity and ecosystem functioning (Easterling et al., 2000; IPCC, 2007). Surprisingly, increased temporal clustering under the most intense disturbance regimes lessened the impact of simulated wave shock and desiccation stress on species abundance and diversity in experimental manipulations of intertidal ecosystems (Bertocci et al., 2005; Benedetti-Cecchi et al., 2006). These results provide an interesting parallel with a recent correlative analysis showing how the impact of cyclones on coral reefs is lower when extreme disturbances are clustered in time (Mumby et al., 2011). The explanation of these outcomes is that the temporal clustering of disturbance also involves prolonged interim periods allowing more time for assemblages to recover compared to situations where disturbances are more evenly distributed in time. If individual disturbances are strong enough so that one event already causes a large impact to assemblages, having one event or a series of events concentrated in a short period does not make much difference, but species will have more time available for recovery before a new disturbance occurs in the clustered scenario (Figure 4.1). This evidence suggests that the impact of extreme climate events may be less detrimental than expected, if climate change entails an increase in temporal clustering of pulse disturbances, as predicted by climate models (Easterling et al., 2000; Díaz and Murnane, 2008). This may also apply to clustered instances of other anthropogenic stressors.

On the other hand, repeated disturbances and long runs of unfavourable environmental conditions have been shown to enhance the risk of extinction in model ecosystems and laboratory experiments (Heino, 1998; Miramontes and Rohani, 1998; Pike et al., 2004). This increased risk may be due to reduced population asynchrony among species that respond similarly to stressors, which may inhibit the spatial and temporal partitioning of resources that allows species to coexist (Chesson, 2000). Whether more clustered disturbances increase or decrease extinction risk is likely to depend on the demographic traits of species that ultimately regulate growth rates, resource use, intra- and interspecific interactions and density dependence. Once again, understanding the complex interplay among life history, population growth and environmental variation is key to anticipate how changes in disturbance regimes will impinge on population dynamics and will affect ecosystem functions and services under a changing environment (Boyce et al., 2006).

4.3 Characteristics of the biota and receiving environment

If two ecosystems are subjected to the same anthropogenic stressor, operating under the same disturbance regime, they are unlikely to suffer the same ecological impacts. Similarly, the same ecosystem may respond differently to exactly the same disturbance, depending on when it occurs. Stressors' effects depend on their environmental context. There are at least three reasons that this is the case. First, environmental conditions can modify the actual stressor, or its disturbance regime, in ways that alter its effects. Second, environmentally and genetically induced differences in the sensitivity of the affected organisms can alter the effect that a stressor has. Third, differences in the structure and dynamics of the biological communities that a stressor affects can alter its effects. Although this suggests that there can be a high degree of variation in stressor's effects, understanding how and why these modifications occur makes it possible to discern some general patterns that can be used to predict the impacts of stressors in new situations. Subsequent chapters deal with specific classes of stressors, and include more detailed consideration of differences in their effects among different environments. In this section, we briefly examine some of the more general effects of the physical and biological environment on the action of anthropogenic stressors.

4.3.1 Direct modifications to stressors and disturbance regimes

Stressors will often have different effects in different contexts because they directly interact with local environmental conditions and are qualitatively changed as a result. For example, the bioavailability and toxicity of a range of toxic metals, organic pollutants and organometallic contaminants is altered by local physicochemical conditions. Environmental factors such as pH, temperature, salinity, redox potential and sediment resuspension influence the release of contaminants from sediments and can change their chemical structure or chelation (Eggelton and Thomas, 2004). Similarly, the effects of some invasive species on their native competitors can range from negative to positive over environmental gradients as the strength and relative importance of their competitive and facilitative effects changes (Rius and McQuaid, 2009; Branch et al., 2010).

Even if environmental conditions do not qualitatively modify a stressor, they can still modify its disturbance regime. For example, variation in wind and currents can alter the spatio-temporal distribution of stressors in the environment. Changes in the circulation of the North Pacific associated with the El Niño Southern Oscillation leads to changes in the distribution of marine litter in the region (Figure 4.2; Morishige et al., 2007). These changes cause an increased incidence of seal entanglement in plastic debris in El Niño years (Figure 4.3; Donohue and Foley, 2007). Local environmental conditions can also modify disturbance regimes at much smaller scales. For example, different habitat types within an ecosystem tend to accumulate different amounts of plastic debris, and may suffer different impacts as a result (Thompson et al., 2004). Burrowing activity by the crab *Chasmagnathus granulata* alters the accumulation of both plastic debris and organochlorine pesticides in South American mudflats, suggesting that biological differences between otherwise similar habitats can also alter the anthropogenic stress regimes (Iribarne et al., 2000; Menone et al., 2004).

In many cases it may be possible to predict the direct modification of stressors and their disturbance regimes. Many of the physical and chemical processes involved in these modifications are reasonably well understood,

Modifiers of impacts on marine ecosystems · 83

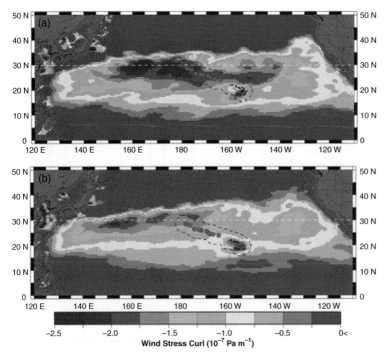

Figure 4.2 The 'Great Pacific Garbage Patch' is the name given to the accumulation of litter and debris in the area of convergent surface flows (blue and violet). The differential atmosphere–ocean conditions in El Niño years (a) result in a more intense and more extensive garbage patch and a greater spatial overlap between the convergent zone and the range of the Hawaiian monk seal (delimited by the black dashed line) than in non-El Niño years (b). The white dashed line at 30°N is for ease of comparison and the main islands of Hawaii are represented by red dots not to scale. Reprinted from Donohue and Foley (2007). A black and white version of this figure will appear in some formats. For the colour version, please refer to the plate section.

and this knowledge can be used in models to help predict and manage stressors. For example, models predicting the transport and accumulation of marine debris are being developed (Lebreton *et al.*, 2012; Maximenko *et al.*, 2012). Similarly, a large range of oil spill models capturing the key physical and chemical processed involved in the transport and degradation of oil have been developed, and are already used in both risk assessment and response to actual oil spills (Reed *et al.*, 1999; Price *et al.*, 2003).

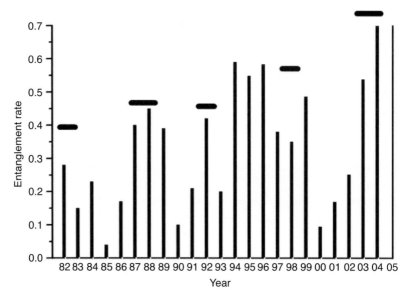

Figure 4.3 The proportion of the population of Hawaiian monk seals (*Monachus schauinslandi*) showing entanglement in plastic litter in the period 1982 to 2004, horizontal bars indicate El Niño periods. Reprinted from Donohue and Foley (2007).

4.3.2 Environmentally induced differences in sensitivity

The sensitivity of organisms to a stressor will vary in both time and space as a result of differences in their physiological and behavioural state, as well as genetic differences between and within populations. Organisms living in environments at the limits of their physiological tolerances may be less able to cope with an anthropogenic stressor. This could contribute to variation in the stressor effects at different sites along environmental gradients (salinity in estuaries, tidal emersion on a shore, etc.), and at different points in time. For example, the tidal cycle may provide a short-term pattern of changing sensitivity. At low tide, intertidal organisms experience a variety of potentially stressful environmental conditions including high light, thermal extremes, nutrient or food limitation, desiccation, and osmotic stress (Raffaelli and Hawkins, 1996). These natural stressors may make it more difficult for them to tolerate additional, anthropogenic stressors. Whether the environment varies spatially or temporally, the basic premise is that on each environmental axis there is an optimum condition for that organism. As conditions move away from that optimum the organism needs to invest progressively more of its 'energy' (or other resources) in basic maintenance under the progressively more challenging conditions. This model forms the basis of the scope for growth approach to pollution assessment, when an organism is stressed by pollution it has less 'energy' available to invest in growth as it is using more to counteract the effects of the pollutant (Van Haren and Kooijman, 1993).

Seasonal changes in the physiology and behaviour of individual organisms will contribute to temporal variation of sensitivity (Regoli et al., 2002). The clearest examples are those associated with seasonal breeding cycles. Seals will generally detect an oil spill at sea by olfactory means and avoid the area, but breeding (or moulting) seals are restricted to their breeding beaches and so are vulnerable to oil spills. Breeding periods also often involve the running down of lipid reserves, either at the time of breeding or during the maturation period of the gonads (Hagen and Schnack-Schiel, 1996). This causes major changes in the lipid physiology and can greatly increase vulnerability to lipid soluble toxins such as organohalides, many pesticides and heavy metals. Even extremely low concentrations (below routine detection limits) of substances that are hormone mimics/endocrine disruptors can be biologically significant and this sensitivity often varies with natural hormonal cycles associated with seasonal breeding (Porte et al., 2006).

In addition to altering organisms' sensitivity by altering their physiological state, local environmental conditions may drive genetic differences that influence stressors' effects. Genetic differences between populations living at different sites may alter their response to stressors, making some populations more vulnerable, while others have pre-adaptations that reduce their susceptibility. The broadcast breeding behaviour of most marine organisms often leads to a presumption of high gene flow and limited local adaptation. However, there is a growing body of data that show genetic differences between populations at even nearby locations (Sanford and Kelly, 2011). For example, populations of rag worms *Nereis* (= *Hediste*) *diversicolor* in metal-contaminated estuaries in Cornwall (UK) show a genetically based tolerance to the metals (Bryan and Hummerston, 1971). These populations might therefore be seen as pre-adapted to, or less vulnerable to stress from metal contamination by anthropogenic activities. This highlights that what is a stressor to one group of organisms may be the normal condition for other individuals of the same species living at a different site. Clearly, this only applies to stressors such as metal ions, temperature, noise, turbidity, sedimentation, etc. that occur naturally and pre-adaption to manufactured chemicals or other novel sources of stress will not exist. However, random genetic variation may still result in among-site variation in susceptibility to novel stressors.

Individuals within a population at one site will also differ in their genetic disposition and susceptibility to stressors. This genetic variation within populations may also have an important influence on their long-term response to anthropogenic stressors. Populations with greater genetic variability are more likely to contain individuals capable of coping with novel stressors and to evolve tolerance, so they may fare better in the long run (Bell, 2013). However, there is also some evidence to suggest that the small, peripheral populations at range edges/stressful environments may be more likely to produce the genetic variation required to deal with changing environments (Bell and Gonzalez, 2011). The traditional view has been that evolution is a slow process, but the evidence from changes in the biology of fish subjected to strong selection by fisheries (Law, 2000) is that it can occur rapidly. Cod in the North Sea now breed 2 years younger than 100 years ago (Olsen et al., 2004; Árnason et al., 2009). This hitherto unrecognised rapidity of evolutionary responses means that evolution may provide the means by which natural biodiversity survives assaults from anthropogenic environmental change (Gonzalez et al., 2013).

4.3.3 Differences in community structure and dynamics

Like individuals and populations, species differ in their sensitivity to different stressors, and communities composed of different species will differ in their sensitivity as a result. The environmental conditions that exist in different habitat types promote the development of different communities, composed of species with different traits, including differing sensitivity to particular stressors. Some habitats are more similar to one another than others. For example, an intertidal mudflat and a sandy beach share many physical constraints and so are similar in some respects, while each has fewer similarities to a coral reef (as reflected in classification schemes such as EUNIS (The European Nature Information System, http://eunis.eea.europa.eu/about.jsp)). Generally, we might expect that the responses of closely related habitat types may be similar for a given stressor. The ability to predict, in broad terms, the response of groups of habitats is important for the practical application of science to management. The practical (and financial) difficulties of having to provide robust scientific studies for every type of stressor in every type of habitat before management actions could be taken would be prohibitive. Thus we often focus on understanding the effects of common stressors on key habitats and species and try to extrapolate the findings to related cases.

Understanding differences between habitats can also be useful in understanding why the effects of stressors vary. Some of these differences will be strongly linked to the relative 'novelty' of the stressor affecting the ecosystem. For example, building a concrete structure on the seafloor will disturb the benthic community, regardless of whether the structure is built on a hard or soft bottom. However, the long-term effects of that structure may be very different. Whereas the new structure provides a relatively similar environment to the rocky substrate previously present in the hard-bottom habitat, it would be completely novel in a sandy area. Many of the species that formerly used the sand substrate may be lost, and new species may arrive to take advantage of the stable substrate. Differences between communities in their sensitivity to particular stressors may also be related to the natural disturbance regimes that they experience. For example, the effects of bottom trawling in shallow sandy habitats where sediments are continually disturbed by currents can be expected to be different from how they would be on rarely disturbed sponge reefs and deep-water coral reefs because the communities that inhabit each of these areas are accustomed to different regimes of natural disturbance and are characterised by species with different growth rates and life spans. Indeed, current evidence suggests that, although soft-bottom communities are surprisingly vulnerable to bottom fishing, biogenic habitats suffer the most severe impacts and that large, slow-growing organisms take longer to recover than those with short lifespans (Kaiser et al., 2006).

Physical and chemical differences between habitats may drive much of the spatial variation in stressors' effects. Climate fluctuations such as the El Niño Southern Oscillation and the North Atlantic Oscillation drive long-term variation in the structure and dynamics of marine ecosystems (Stenseth et al., 2002, 2003), and may thereby drive much of the long-term temporal variation in stressors' effects. However, less striking environmental differences, as well as stochastic variation in dispersal and recruitment also lead to continual spatial and temporal turnover of individuals and species that may have an influence on stressors' effects. There will also be site-specific differences in sensitivity and recovery that result from the supply of new recruits/individuals. This will vary depending on the local hydrography and the proximity and state of adjacent sources of potential colonists, as well as the dispersal ability of the species in the community. These issues are frequently considered together as 'connectivity' (Cowen et al., 2006) – how ecologically connected a site is.

Thus far we have primarily discussed how differences in the types or traits of species present in different communities can modify stressors' effects but community-level characteristics such as biodiversity and the interactions between species also play a role. Earlier we mentioned how the loss of canopy-forming algae following the *Exxon Valdez* spill led to cascading effects that caused other species to be lost as well. There are various mechanisms whereby this can occur, including the loss of an essential food resource (e.g. Lafferty and Kuris, 2009; Nichols *et al.*, 2009), competitive exclusion following the loss of a shared predator (Lubchenco, 1978; Bruno and O'Connor, 2005), and the loss of mutualistic or facilitative interactions (Memmott *et al.*, 2004; Kaiser-Bunbury *et al.*, 2010). The factors that determine the ecosystem's susceptibility to these 'extinction cascades' are still being studied and debated, but are expected to include the number and distribution of trophic (or mutualistic) links between species, the strength of species interactions, and the total number of species within the community (Ebenman and Jonsson, 2005). Extinction cascades demonstrate that interactions between species can extend the impacts of stressors to species that are not sensitive to the original stressors, or even exposed to them. Even if stressors do not result in extinctions, the number, type and strength of interactions between species within an ecosystem will influence their net effects.

According to the insurance hypothesis, greater biodiversity ensures ecosystems against declines in their productivity (Yachi and Loreau, 1999). This idea rests on the assumption that biodiversity imparts an ecosystem with a degree of functional redundancy. If several competing species have similar functional roles, this redundancy allows the loss of function caused by the removal or impairment of one species to be compensated for by one of its competitors (Walker, 1992; Naeem, 1998). Thus, the number of species present within a community may be important in determining its response to anthropogenic stressors because it influences the likelihood of species capable of both tolerating the stress and compensating for the loss of sensitive species, being present.

4.4 Multiple stressors

Ecosystems adjacent to centres of human population are subject to a wide variety of anthropogenic pressures, and even remote parts of the oceans face the impacts of fisheries' expanding footprint, plastic debris, synthetic chemicals and toxic metals (Davis, 1993; Lotze *et al.*, 2006; Halpern *et al.*, 2008; Rios *et al.*, 2010; Swartz *et al.*, 2010; Anderson *et al.*,

2011). All of these stressors are occurring alongside global warming and ocean acidification, which are expected to have their own direct effects on marine ecosystems, and to modify other natural and anthropogenic disturbance regimes (Harley *et al.*, 2006; Doney, 2010). Understanding how ecosystems respond to a multitude of anthropogenic stressors is one of the essential tasks facing those interested in preserving biodiversity, ecosystem functioning and the services that ecosystems provide to human society.

Scientists and policy makers have traditionally focused on investigating and managing anthropogenic pressures independently, but there is a growing recognition in both communities that a more integrative approach to science and management is required (Breitburg and Riedel, 2005; Rosenberg and McLeod, 2005; European Union, 2008; National Ocean Council, 2013). New approaches are required because the cumulative ecological impacts of several different anthropogenic stressors are not easily predicted from their individual effects. Rather, anthropogenic stressors frequently interact to produce unanticipated consequences. Often, the cumulative impacts of multiple stressors are simply larger or smaller than would have been expected because stressors modify each other's effects (Paine *et al.*, 1998; Crain *et al.*, 2008; Darling and Côté, 2008). In some instances, however, the collective action of several stressors has the potential to induce sudden, nonlinear regime shifts that drive ecosystems into novel, alternate states with fundamentally different structure, dynamics and capacity to deliver ecosystem services (Paine *et al.*, 1998; Scheffer *et al.*, 2001; Folke *et al.*, 2004). If we are to predict the consequences of ongoing environmental change and optimise the management of marine ecosystems in order to promote their resilience, it is important that we work to understand how the effects of different human activities combine to produce their cumulative effects. In this section we discuss different ways in which stressors interact to produce ecological surprises, the prevalence of these interactions and some examples of their effects.

4.4.1 Interactions between stressors: synergisms and antagonisms

We have previously used the term 'interaction' to refer to ecological processes such as competition, predation and facilitation. We have also used it to describe situations where the effects of a stressor are modified by the physical or chemical environment. We define interactions between stressors in a similar way. Specifically, we use the term (and

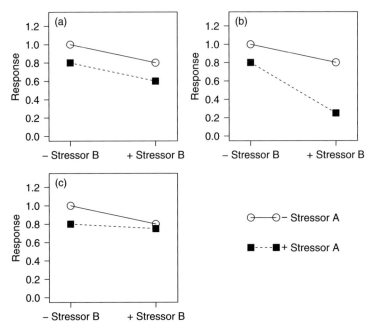

Figure 4.4 Plots showing the difference between non-interactive (a), synergistic (b), and antagonistic (c) effects of two stressors with negative effects on a response of interest. In this case, it is assumed that the two effects would combine additively in the absence of an interaction (a). If stressors interact synergistically, their total negative effect will be larger than if they did not interact (b), while it will be smaller if they interact antagonistically (c).

related terms) to refer to situations where the presence of one stressor modifies the effect of a different stressor through some chemical, physical or biological mechanism. The existence of interactions is typically inferred from differences between observed effects and our expectations of the stressors' independent effects. Efforts to understand the cumulative ecological impacts of anthropogenic activities typically begin with the assumption that their effects are additive in the absence of interactions between stressors (e.g. Ban and Alder, 2008; Crain *et al.*, 2008; Halpern *et al.*, 2008; Figure 4.4a). Throughout the rest of the chapter, we will use this 'additive model' to explain different types of interactions, and to illustrate particular examples. However, it is important to note that other models may provide a better description of some stressors' non-interactive effects. For example, the effects of some stressors combine multiplicatively (Folt *et al.*, 1999). Thus, although non-additive effects may be common when stressors interact, they are not necessarily indicative of mechanistic interactions between stressors.

Given the interdependence of biochemical, physiological and ecological systems, interactions between stressors are likely to be very common. Scientists and policy makers are concerned that stressors will often interact *synergistically* to produce net impacts that are larger than would have been predicted from their individual effects (Jackson *et al.*, 2001; Sala and Knowlton, 2006; European Union, 2008). Synergistic interactions occur when the effect of one stressor makes the organism, population or ecosystem of interest more sensitive to other stressors, or when stressors modify each other to become more potent (Figure 4.4b). If we do not anticipate such synergistic interactions between stressors, then we are more likely to implement management regimes that fail to protect sufficient levels of biodiversity and ecosystem functioning for ecosystems services to be maintained.

Although synergistic interactions are cause for concern, it is important to acknowledge that stressors can also interact *antagonistically*. Antagonistic interactions occur when stressors induce changes that make the affected system more resistant to subsequent stressors, or one stressor makes another less effective, thereby reducing their combined impacts (Figure 4.4c). Often, the net effects of antagonistically interacting stressors will still be larger than those of any of the individual stressors. However, in some cases one stressor may effectively cancel out the effect of another. For example, a climate-induced bleaching event on Kenyan reefs caused smaller losses of coral in areas exposed to fishing than in protected areas, resulting in cumulative impacts that were similar to that of bleaching alone (Darling *et al.*, 2010). Strong antagonisms can even result in cumulative effects that are actually smaller than those of individual stressors. For instance, a mesocosm experiment with natural plankton assemblages found that the joint effects of trace metal pollution and eutrophication on bacterial productivity and primary productivity are smaller than those of nutrients alone (Breitburg *et al.*, 1999). An understanding of such antagonisms may provide opportunities for management strategies that reduce overall environmental impacts, while satisfying the needs of different users of marine ecosystems, or that maximise the benefit of new management initiatives. For example, the data from the mesocosm experiment described above suggest that a manager with money to spend on reducing estuarine nutrient inputs might see a larger reduction in algal and bacterial blooms if they spend their funds in an estuary that is not affected by trace metal pollution.

Anthropogenic stressors can cause biodiversity, ecosystem functioning and other biological responses to increase or decrease. Thus, the terms synergism and antagonism refer to increases and decreases in the magnitude of stressors' effects, not to increases or decreases in the value of the response variable. When stressors have opposing effects they can still interact, but one cannot easily label their interactions as synergistic or antagonistic because it could be argued that an interaction has increased the magnitude of the net effect in one direction or decreased the magnitude of the effect in the opposite direction. Some have chosen to label negative interactions as synergisms and positive interactions as antagonisms (e.g. Crain et al., 2008), but this decision is arbitrary. A clear description of the direction and magnitude of the interaction effect is more informative.

4.4.2 Prevalence of interactions in marine ecosystems

Despite considerable interest in the topic, the prevalence of synergistic and antagonistic interaction in marine ecosystems remains unclear because of difficulties inherent in establishing causality from observational data and a lack of appropriate experimental data. Using observational data to investigate causal relationships and quantify the independent and interactive effects of spatially and temporally correlated stressors in complex ecosystems is very difficult (Adams, 2005). Replicated large-scale, long-term factorial experiments that would allow us to unambiguously detect synergisms and antagonism between stressors at the most relevant spatial and temporal scales are rare, perhaps due to the logistical and ethical difficulties associated with such experiments (Farnsworth and Rosovsky, 1993; Carpenter, 1998).

Despite the difficulties mentioned above, some patterns are beginning to emerge from shorter term, smaller scale experiments. Interactions among stressors appear to be very common, with clear evidence of synergisms or antagonisms in approximately three quarters of experiments investigating the impacts of multiple stressors (Darling and Côté, 2008; Crain et al., 2008). Although it is often feared that synergistic interactions between stressors are the norm (e.g. Jackson et al., 2001; Sala and Knowlton, 2006), recent meta-analyses suggest that antagonisms and synergisms occur with roughly equal frequency (Crain et al., 2008; Darling and Côté, 2008). These results give us an initial indication of how common synergisms and antagonisms may be, but they should be viewed with some caution. It is not clear how well the results of small-scale field and laboratory experiments represent the effects of the many stressors operating over larger spatial and temporal scales in the natural environment. Moreover, many knowledge gaps remain to be filled. One of these meta-analyses (Crain et al., 2008) was focused exclusively on marine ecosystems. It found that only a small proportion of the possible stressor combinations have been investigated, and relatively few factorial experiments have examined community- and ecosystem-level impacts. Rather, experimental research has been heavily weighted towards lab-based studies of effects on individual species. None of the studies included in the meta-analysis examined species richness or other measures of biodiversity, although at least two more recent studies have done so. Both failed to find strong evidence of interactions between stressors (Sundbäck et al., 2010; O'Gorman et al., 2012). Overall, studies examining the effects of multiple stressors on abundance, biomass, productivity, and growth rates revealed an overall antagonistic effect at the community level and an overall synergistic effect on individual species responses (Crain et al., 2008). However, the outcomes of individual studies were highly variable, with antagonisms, synergies and non-interactive effects commonly observed in both community- and species-level responses to multiple stressors. This suggests that a deeper understanding of interactions between stressors may be required to predict whether antagonistic or synergistic interactions are likely to occur in a particular situation.

4.4.3 Mechanisms of interaction

There appear to be no simple 'rules of thumb' that can be used to reliably predict the direction or magnitude of interactions between stressors. Rather, cumulative impacts by stressors depend on the identity, trophic position and life-history stage of the affected organisms, as well as the identity and number of stressors involved (Crain et al., 2008; Darling and Côté, 2008). Like the effects of individual stressors, interactions between stressors may also change with variation in disturbance regimes. Two stressors may have synergistic, antagonistic or non-interacting effects depending on their intensity, extent, duration or timing (e.g. Todgham et al., 2005; Reich et al., 2006; Wang et al., 2008). In the absence of simple generalisations, the best way to improve our knowledge and predictions of cumulative impacts may be to focus on the mechanisms whereby the stressors affecting an ecosystem potentially interact. Below, we discuss three classes of mechanism whereby stressors can modify one another's effects. Each recalls one of the three mechanisms we discussed above when describing why stressors' effects are shaped by the receiving environment.

Exogenous (direct) stressor interactions
The first way that some stressors interact is to directly modify one another through physical processes or chemical reactions. We refer to these direct abiotic interactions as *exogenous interactions* because they often occur outside of organisms, in the environment. Like the physical and chemical changes to stressors caused by their interaction with the environment, exogenous interactions between stressors change stressors' net effects because they alter quality or quantity of the stress experienced by the affected organisms. Certain types of exogenous interaction should be relatively predictable, based on our knowledge of chemical reactions and physical laws (i.e. low relative probability of an 'ecological surprise'). For example, the toxicity of some polycyclic aromatic hydrocarbons (PAHs) is enhanced in the presence of ultraviolet radiation (UV) because it induces them to undergo photochemical reactions that either transform them into more toxic compounds, or result in the formation of oxidising agents (Boese *et al.*, 1997; Swartz *et al.*, 1997; Aherns *et al.*, 2002). It is possible to predict which PAHs will undergo these photosensitisation reactions based on their structure (Diamond, 2003). Similarly, human activities that alter salinity, pH and sediment resuspension are likely to mediate organisms' exposure to toxins because, as we discussed earlier, these factors are known to influence the mobilisation and bioavailability of metals and other contaminants. Armed with such physical and chemical knowledge, we can improve the models used to predict stressors' effects. We can also enhance the effectiveness of monitoring and empirical research by tracking the physical and chemical processes that govern these exogenous interactions, and by focusing on the stressors that organisms are likely to experience, rather than their unmodified precursors.

Endogenous (indirect) stressor interactions
The second type of interaction between two or more stressors occurs indirectly, when the presence of one stressor alters the physiological or behavioural state of an affected organism, which causes it to respond differently to other stressors. These *endogenous interactions* recall the effect of environmental conditions on the sensitivity of organisms to the effects of individual stressors that we discussed above. There are numerous specific mechanisms whereby endogenous interactions can occur, but all

are mediated by interdependent biochemical and physiological processes within organisms. These processes may be linked because organisms use the same defence and repair mechanisms to deal with various stressors, because different stressors induce similar physiological responses within the organism or because all stressors affect an organism's energy balance through their effects on maintenance costs and energy acquisition.

Organisms respond to a diverse range of stressors through a general stress response that includes rapid synthesis of heat-shock proteins, release of stress hormones, increases in the activity antioxidant systems and a variety of other physiological adaptations controlled at the cellular and organismal levels (Davies, 2000; Kregel, 2002; Kultz, 2005; Charmandari et al., 2005). These changes help repair and prevent damage caused by the original stressor, but they can also help to protect an organism against subsequent stresses, a phenomenon known as cross-tolerance (Kultz, 2005; Young et al., 2009). Cross-tolerance may occur because organisms use the same machinery to deal directly with different environmental stressors or because different environmental stressors induce the same physiological stresses (e.g. reduced cellular and systemic oxygen levels, and increased levels of reactive oxygen species). Regardless of the situation, cross-tolerance reduces the cumulative damage caused by multiple stressful events and thus may provide the key to understanding antagonistic endogenous interactions.

Exposure to one stressor can also make organisms more sensitive to subsequent stressors, resulting in synergistic endogenous interactions. The downside to using the same defence and repair mechanisms to handle different stressors is that it can lead to synergistic interactions between stressors if several otherwise manageable stressors combine to overwhelm these mechanisms. If this occurs, the result may be a rapid, nonlinear decline in organismal performance. Endogenous synergisms will also result when one stressor impairs the physiological and biochemical means that an organism uses to cope with the subsequent stressor(s). For example, bisphenol A enhances the toxicity of cadmium by reducing the expression of metal-binding proteins in the liver through its effects on oestrogen receptors (Sogawa et al., 2001). Stress-induced behavioural changes that increase an organism's exposure or sensitivity to another stressor will also cause synergistic interactions between stressors. For instance, a variety of toxins cause marine bivalves to reduce production of byssal threads, which they use to anchor themselves to the substrate (Roberts, 1975; Karagiannis et al., 2011). This reduces the strength of their attachment to the substrate, and would make them more susceptible to trampling, predators, and other physical disturbances that could detach them from the substrate.

The net effects of multiple stressors vary because both the timing and intensity of stressors influence the total physiological stress that an organism experiences at any one time, and the degree to which cross-tolerance is possible. For cross-tolerance to occur, the stressors must not exceed an organism's defence and repair capacity. Moreover, sufficient time must have passed between stressors for the stress response to be induced, yet not so much time that the organism has recovered its pre-stress physiological state. For these reasons, it is just as important to consider the overall disturbance regime when assessing the threats posed by multiple stressors as when one is concerned with the effects of only a single stressor. Todgham *et al.* (2005) demonstrated the importance of considering disturbance regimes of multiple stressors in a series of experiments examining the survival of tidepool sculpins exposed to a sublethal heat shock and a severe hyperosmotic stress. In one experiment, ~45% of unstressed sculpins survived the hyperosmotic stress, but all of the fish that were consecutively exposed to both stressors were killed. When fish were allowed to recover between stressors, survival increased with the length of the recovery period, reaching 100% for fish given 24 hours and declining slightly for fish exposed to the salt stress 48 hours after the temperature shock. Levels of branchial heat-shock proteins, which depended on exposure to a prior stressor and the length of the recovery, were associated with increased survival, suggesting they were involved in cross-tolerance. In a second experiment, fish were exposed to non-lethal heat shocks of varying intensity, and then exposed to the same hyperosmotic shock after 8 hours of recovery. The cumulative impact of these stressors on fish mortality was additive at low intensity, antagonistic at moderate intensity and synergistic at high intensity.

The diversity of environmental stressors and the complexity of the systems organisms use to cope with them mean that a detailed mechanistic understanding of even a small proportion of the potential endogenous interactions between stressors would require a massive effort. Thus, there is a need for approaches that can provide general understanding of a broad range of endogenous interactions, and improve our ability to predict them. One promising approach is to study the bioenergetic dimensions of stress responses. There is considerable empirical evidence to suggest that an organism's energy balance plays a central role in determining its ability to tolerate stressors, and that stressors feed back to affect energy balance by determining the amount of energy that the organism must use to regain and preserve homoeostasis (reviewed by Sokolova et al., 2012). This influences the amount of energy available for growth, reproduction and other activities, including coping with additional environmental stressors. Dynamic energy budget (DEB) theory (Kooijman, 2010) provides a framework to investigate the bioenergetics aspects of endogenous interactions between stressors. This approach is only beginning to be used to investigate the effects of multiple stressors, but early results suggest that they can sometimes explain interactions between stressors observed at the organismal level as a consequence of their independent effects on metabolic processes (Jager, 2012). If the outputs of these models are be coupled to population, community and ecosystem models it may be possible to scale the physiological effects of stressors to other levels of biological organisation (Nisbet et al., 2008; Jager and Klok, 2010).

Ecologically mediated interactions
The third class of interactions between stressors comprises those that are contingent upon ecological interactions between organisms. These *ecologically mediated interactions* arise because, like spatially and temporally variable environments, individual stressors influence the composition of ecological communities and the strength of interactions between organisms. As a result, the presence of one stressor can alter the sensitivity of the ecosystem to other stressors and cause their effects to change (Figure 4.5). Although exogenous and endogenous interactions may manifest themselves in effects at the population, community and ecosystem level, we believe that most of the stressor interactions affecting biodiversity, ecosystem functioning, and the delivery of ecosystem services are likely to be mediated by ecological interactions between and within species.

Ecologically mediated interactions between stressors have the potential to produce large effects on marine biodiversity. In most cases, the net effects of multiple stressors are more likely to cause the loss of species than individual stressors acting alone. As a result, the extinction cascades we discussed earlier may also be more likely when an ecosystem is exposed to multiple stressors. Ecological regime shifts may be another example of ecological interactions causing the effects of multiple stressors to interact and cascade through ecosystems. As described in Chapter 2, regime shifts are abrupt, nonlinear changes in the structure and dynamics of an ecosystem. They occur when a change in the variables that define the ecosystem state (e.g. species abundances), or the parameters that define an ecosystem's stability (e.g. species interactions, environmental drivers),

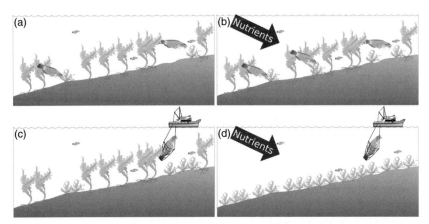

Figure 4.5 Illustration of ecologically mediated interactions between eutrophication and fishing in a hypothetical ecosystem. Under normal conditions, the community is composed of competing brown and green seaweed species, and a herbivorous fish that feeds on the green seaweed (a). Eutrophication favours increased production of the ephemeral green seaweed, but its abundance only increases slightly due to an increase in the intensity of grazing and abundance of the fish (b). Harvesting of large, adult fish releases the green seaweed from herbivory, but its abundance only increases slightly due to nutrient limitation and competition from the brown seaweed (c). When eutrophication and fishing are combined, the loss of herbivory leaves growth of the green seaweed unchecked, leaving it to outcompete the brown seaweed under nutrient-rich conditions. Created with symbols courtesy of the Integration and Application Network, University of Maryland Center for Environmental Science (ian.umces.edu/symbols/).

overcome the ecosystem's resilience and shift it into a different state (Beisner *et al.*, 2003). In the marine environment, many apparent regime shifts, including the change from coral- to macroalgal-dominated reefs in the Caribbean, the expansion of urchin 'barrens' at the expense of kelp beds in Tasmania, and the transformation of the Black Sea's benthic and pelagic ecosystems involved the effects of several anthropogenic pressures (Nyström *et al.*, 2000; Mee *et al.*, 2005; Ling *et al.*, 2009). Multiple anthropogenic stressors are not required for a regime shift to occur; any perturbation that overcomes an ecosystem's resilience can induce a regime shift. Nevertheless, regime shifts typically have multiple causes, often a gradual change in a driving variable that helps to define the stability of the ecosystem combined with an additional disturbance or with stochastic fluctuations in the system's state (Scheffer and Carpenter, 2003; Biggs *et al.*, 2009). Thus, anthropogenically induced regime shifts may be more likely where multiple stressors are present because prior stressors reduce ecological resilience, making it that much easier for subsequent stressors to drive the ecosystem into a different stable state.

The examples of extinction cascades and ecological regime shifts demonstrate that ecologically mediated synergisms are a major threat to marine ecosystems. However, these phenomena may be less common than ecologically mediated antagonisms. As mentioned previously, current experimental evidence suggest that stressors tend to have antagonistic effects on community-level productivity, growth, abundance and survival, while having synergistic effects on individual species (Crain et al., 2008). This suggests that interactions between species often moderate community-level responses to multiple stressors. There are at least two mechanisms whereby this moderating effect could occur. First, facilitative interactions between species (i.e. mutualism, commensalism) may buffer individual species against potentially limiting physical stresses, thereby dampening stressors' negative effects on community-level responses. The stress-gradient hypothesis predicts that the importance of facilitative interactions that ameliorate environmental conditions will increase with the intensity of environmental stress, and this prediction is supported by findings from a variety of different ecosystems (Bertness et al., 1999; Callaway et al., 2002; Bruno et al., 2003). Second, functional redundancy among competing species may buffer the community against stressors' negative effects. If an initial stressor favours species that are also tolerant of subsequent stressors over species that are sensitive to it (i.e. species tolerances for different stressors are positively correlated), the combined effects of the two stressors will be antagonistic.

In general, positive species interactions and functional redundancy may be more prevalent in ecosystems with many species, and ecosystems with high biodiversity may be less sensitive to effects of human activities as a result. However, ecosystem functioning will be eroded if the increasing number and intensity of stressors affecting ecosystems accelerates the loss of biodiversity, positive interactions and functional redundancy. This underscores the significance of the knowledge gap regarding stressors' net effects on marine biodiversity, and the importance of increased research in this area. Nevertheless, counting the number of species ecosystems gain or lose will not provide a complete picture of ecosystems' changing sensitivity to environmental change. An ecosystem's response to anthropogenic pressures will be determined by the functional capacity, stress tolerance and interactions of the species within it, rather than its biodiversity per se. Thus, a deeper understanding of the relationship between biodiversity and ecosystem functioning can be gained through an understanding of the functional traits of the species involved (Naeem and Wright, 2003; Hooper et al., 2005; Reiss et al., 2009). The utility of trait-based approaches is evident in several of the scenarios we discussed above. For instance, we described how both trade-offs in sensitivity to different stressors could lead to synergistic effects on ecosystem functioning, and positive associations between the traits governing species tolerances could lead to antagonistic effects. Similarly, Vinebrooke et al. (2004) showed how the relationship between species tolerances for different stressors could be used to predict whether they would have synergistic, antagonistic or additive effects on biodiversity. More generally, the relationship between ecological traits that determine species responses to environmental change ('response traits') and those that determine their effect on ecosystem functioning and services ('effect traits') can be used to gain additional insight into the effects of changing biodiversity on ecosystem functioning (Lavorel and Garnier, 2002; Suding et al., 2008). This is discussed further in Chapter 5.

4.5 Key points and recommendations

The impacts of anthropogenic stressors can vary greatly across time and space. As we have discussed in this chapter, this variation often occurs due to differences in stressors or disturbance regimes, differences in the physical, chemical and biological characteristics of affected ecosystems, or because several stressors interact with one another to produce different outcomes. Some key points to consider:

- The effects of stressors will change with their intensity, spatial extent, temporal duration, frequency and the temporal pattern of repeated disturbances.
- Generally, stressors' effects are likely to be smallest when intensity, extent, duration and frequency are low, though biodiversity may peak at intermediate levels of disturbance.
- Spatial characteristics of disturbances have an important effect on their impacts and how long it takes for ecosystems to recover because they influence which species are able to recolonise affected areas.
- Ongoing disturbances will normally have larger effects than brief stressors of equivalent severity because their effects are augmented over time, or because delayed recovery allows new kinds of effects to develop. The timing of disturbances and the temporal pattern of repeated disturbances also have complex influences on their effects

because they affect the sensitivity of individual species and interactions between species.
- Similar ecosystems are likely to respond similarly to anthropogenic stressors, permitting a degree of generalisation and pragmatism in developing management plans for a wide range of ecosystems, pending the accrual of more detailed ecosystem-specific understanding.
- The physical and chemical characteristics of an ecosystem can influence impacts in three ways. (1) by altering community composition, (2) by altering the physiological state and stress tolerance of individual organisms, (3) by directly interacting with stressors and altering their characteristics.
- There may be predictable differences in the sensitivities of different types of species. For example, the effects of anthropogenic stressors may affect higher trophic levels more severely.
- All marine ecosystems are affected by multiple human pressures. Science and management must recognise this and implement strategies that consider the net effects of different human activities, especially those that generate unexpected non-additive impacts or 'ecological surprises'.
- Both synergistic interactions, which enhance stressors' effects and antagonistic interactions, which mitigate stressors' effects are common. Synergisms create hazards for management, and antagonisms sometimes creates opportunities to reduce the net impact of human activities.
- Like the effects of individual stressors, the strength and direction of interactions between stressors will depend on their disturbance regime. The same two stressors may have antagonistic, synergistic or non-interactive effects, depending on their intensity, extent, duration and timing.
- Improving our understanding of the mechanisms whereby stressors interact will help us predict stressors' cumulative effects and choose the most effective means of managing the effects of different human activities.

References

Adams, S. M. (2005). Assessing cause and effect of multiple stressors on marine systems. *Marine Pollution Bulletin*, 51, 649–657.

Ahrens, M. J., Nieuwenhuis, R. and Hickey, C. W. (2002). Sensitivity of juvenile *Macomona liliana* (Bivalvia) to UV-photoactivated fluoranthene toxicity. *Environmental Toxicology*, 17, 567–577.

Altieri, A. H. (2008). Dead zones enhance key fisheries species by providing predation refuge. *Ecology*, 89, 2808–2818.

Anderson, S. C., Mills Flemming, J., Watson, R. and Lotze, H. K. (2011). Rapid global expansion of invertebrate fisheries: trends, drivers, and ecosystem effects. *PloS ONE*, 6, e14735.

Anthony, K., Kline, D., Díaz-Pulido, G., Dove, S. and Hoegh-Guldberg, O. (2008). Ocean acidification causes bleaching and productivity loss in coral reef builders. *Proceedings of the National Academy of Sciences*, 105, 17442–17446.

Árnason, E., Hernandez, U. B. and Kristinsson, K. (2009). Intense habitat-specific fisheries-induced selection at the molecular Pan I locus predicts imminent collapse of a major cod fishery. *PloS ONE*, 4, e5529.

Ban, N. and Alder, J. (2008). How wild is the ocean? Assessing the intensity of anthropogenic marine activities in British Columbia, Canada. *Aquatic Conservation: Marine and Freshwater Ecosystems*, 18, 55–85.

Beisner, B. E., Haydon, D. T. and Cuddington, K. (2003). Alternative stable states in ecology. *Frontiers in Ecology and the Environment*, 1, 376–382.

Bell, G. (2013). Evolutionary rescue and the limits of adaptation. *Philosophical Transactions of the Royal Society B: Biological Sciences*, 368, 20120080.

Bell, G. and Gonzalez, A. (2011). Adaptation and evolutionary rescue in metapopulations experiencing environmental deterioration. *Science*, 332, 1327–1330.

Bender, E. A., Case, T. J. and Gilpin, M. E. (1984). Perturbation experiments in community ecology: theory and practice. *Ecology*, 65, 1–13.

Benedetti-Cecchi, L. (2000). Predicting direct and indirect interactions during succession in a mid-littoral rocky shore assemblage. *Ecological Monographs*, 70, 45–72.

Benedetti-Cecchi, L., Bertocci, I., Vaselli, S. and Maggi, E. (2006). Temporal variance reverses the impact of high mean intensity of stress in climate change experiments. *Ecology*, 87, 2489–2499.

Bernhardt, J. R. and Leslie, H. M. (2013). Resilience to climate change in coastal marine ecosystems. *Annual Review of Marine Science*, 5. 371–392.

Bertness, M. D., Leonard, G. H., Levine, J. M., Schmidt, P. R. and Ingraham, A. O. (1999). Testing the relative contribution of positive and negative interactions in rocky intertidal communities. *Ecology*, 80, 2711–2726.

Bertocci, I., Maggi, E., Vaselli, S. and Benedetti-Cecchi, L. (2005). Contrasting effects of mean intensity and temporal variation of disturbance on a rocky seashore. *Ecology*, 86, 2061–2067.

Biggs, R., Carpenter, S. R. and Brock, W. A. (2009). Turning back from the brink: detecting an impending regime shift in time to avert it. *Proceedings of the National Academy of Sciences of the United States of America*, 106, 826–831.

Boese, B. L., Lamberson, J. O., Swartz, R. C. and Ozretich, R. J. (1997). Photoinduced toxicity of fluoranthene to seven marine benthic crustaceans. *Archives of Environmental Contamination and Toxicology*, 32, 389–393.

Boyce, M. S., Haridas, C. V. and Lee, C. T. (2006). Demography in an increasingly variable world. *Trends in Ecology and Evolution*, 21, 141–148.

Branch, G. M., Odendaal, F. and Robinson, T. B. (2010). Competition and facilitation between the alien mussel *Mytilus galloprovincialis* and indigenous species:

moderation by wave action. *Journal of Experimental Marine Biology and Ecology*, 383, 65–78.

Breitburg, D. L. and Riedel, G. F. (2005). Multiple stressors in marine systems. In *Marine Conservation Biology: The Science of Maintaining the Sea's Biodiversity*, ed. E. A. Norse and L. B. Crowder. Washington DC: Island Press, pp. 167–182.

Breitburg, D. L., Sanders, J. G., Gilmour, C. C. et al. (1999). Variability in responses to nutrients and trace elements, and transmission of stressor effects through an estuarine food web. *Limnology and Oceanography*, 44, 837–863.

Bruno, J. F. and O'Connor, M. I. (2005). Cascading effects of predator diversity and omnivory in a marine food web. *Ecology Letters*, 8, 1048–1056.

Bruno, J. F., Stachowicz, J. J. and Bertness, M. D. (2003). Inclusion of facilitation into ecological theory. *Trends in Ecology and Evolution*, 18, 119–125.

Bryan, G. W. and Hummerstone, L. G. (1971). Adaptation of the polychaete *Nereis diversicolor* to estuarine sediments containing high concentrations of heavy metals. I. General observations and adaptation to copper. *Journal of the Marine Biological Association of the United Kingdom*, 51, 845.

Callaway, R. M., Brooker, R., Choler, P. et al. (2002). Positive interactions among alpine plants increase with stress. *Nature*, 417, 844–848.

Carpenter, S. R. (1998). The need for large-scale experiments to assess and predict the response of ecosystems to perturbation. *Successes, Limitations, and Frontiers in Ecosystem Science*, 287–312.

Cerrano, C., Bavestrello, G., Bianchi, C. N. et al. (2000). A catastrophic massmortality episode of gorgonians and other organisms in the Ligurian Sea (northwestern Mediterranean), summer 1999. *Ecology Letters*, 3, 284–293.

Charmandari, E., Tsigos, C. and Chrousos, G. (2005). Endocrinology of the stress response. *Annual Review of Physiology*, 67, 259–284.

Chesson, P. (2000). Mechanisms of maintenance of species diversity. *Annual Review of Ecology and Systematics*, 31, 343–366.

Connell, J. H. (1979). Intermediate-disturbance hypothesis. *Science*, 204, 1345–1345.

Cowen, R. K. (2006). Scaling of connectivity in marine populations. *Science*, 311, 522–527.

Crain, C. M., Kroeker, K. and Halpern, B. S. (2008). Interactive and cumulative effects of multiple human stressors in marine systems. *Ecology Letters*, 11, 1304–1315.

Darling, E. S. and Côté, I. M. (2008). Quantifying the evidence for ecological synergies. *Ecology Letters*, 11, 1278–1286.

Darling, E. S., Mcclanahan, T. R. and Côté, I. M. (2010). Combined effects of two stressors on Kenyan coral reefs are additive or antagonistic, not synergistic. *Conservation Letters*, 3, 122–130.

Davies, K. (2000). Oxidative stress, antioxidant defenses, and damage removal, repair, and replacement systems. *IUBMB Life*, 50, 279–289.

Davis, W. J. (1993). Contamination of coastal versus open-ocean surface waters. *Marine Pollution Bulletin*, 26, 128–134.

Dernie, K. M., Kaiser, M. J., Richardson, E. A. and Warwick, R. M. (2003). Recovery of soft sediment communities and habitats following physical disturbance. *Journal of Experimental Marine Biology and Ecology*, 285–286, 415–434.

Diamond, S. A. (2003). Photoactivated toxicity in aquatic environments. In *UV Effects in Aquatic Organisms and Ecosystems,* ed. E. W. Z. Helbling. Cambridge: The Royal Society of Chemistry, pp. 219–250.

Díaz, H. F. and Murnane, R. J. (2008). *Climate Extremes and Society*. Cambridge: Cambridge University Press.

Doney, S. C. (2010). The growing human footprint on coastal and open-ocean biogeochemistry. *Science*, 328, 1512–1516.

Donohue, M. and Foley, D. G. (2007). Remote sensing reveals links among the endangered Hawaiian monk seal, marine debris, and El Niño. *Marine Mammal Science*, 23, 468–473.

Easterling, D. R., Meehl, G. A., Parmesan, C. *et al.* (2000). Climate extremes: observations, modeling, and impacts. *Science*, 289, 2068–2074.

Ebenman, B. and Jonsson, T. (2005). Using community viability analysis to identify fragile systems and keystone species. *Trends in Ecology and Evolution*, 20, 568–575.

Eggleton, J. and Thomas, K. V. (2004). A review of factors affecting the release and bioavailability of contaminants during sediment disturbance events. *Environment International*, 30, 973–980.

European Union (2008). Directive 2008/56/EC of the European Parliament and of the Council of 17 June 2008 Establishing a Frame- work for Community Action in the Field of Marine Environmental Policy (Marine Strategy Framework Directive). *Official Journal of the European Union*, l 164, 19–40.

Fanelli, G., Piraino, S., Belmonte, G., Geraci, S. and Boero, F. (1994). Human predation along Apulian rocky coasts (SE Italy): desertification caused by *Lithophaga lithophaga* (Mollusca) fisheries. *Marine Ecology Progress Series*, 110, 1–8.

Farnsworth, E. J. and Rosovsky, J. (1993). The ethics of ecological field experimentation. *Conservation Biology*, 7, 463–472.

Folke, C., Carpenter, S., Walker, B. *et al.* (2004). Regime shifts, resilience, and biodiversity in ecosystem management. *Annual Review of Ecology, Evolution, and Systematics*, 35, 557–581.

Folt, C., Chen, C., Moore, M. and Burnaford, J. (1999). Synergism and antagonism among multiple stressors. *Limnology and Oceanography*, 44, 864–877.

Fox, J. W. (2013). The intermediate disturbance hypothesis should be abandoned. *Trends in Ecology and Evolution*, 28, 86–92.

Fraterrigo, J. M. and Rusak, J. A. (2008). Disturbance-driven changes in the variability of ecological patterns and processes. *Ecology Letters*, 11, 756–770.

Garrabou, J., Coma, R., Bensoussan, N. *et al.* (2009). Mass mortality in northwestern Mediterranean rocky benthic communities: Effects of the 2003 heat wave. *Global Change Biology*, 15, 1090–1103.

Glasby, T. M. and Underwood, A. J. (1996). Sampling to differentiate between pulse and press perturbations. *Environmental Monitoring and Assessment*, 42, 241–252.

Gonzalez, A., Ronce, O., Ferriere, R. and Hochberg, M. E. (2013). Evolutionary rescue: an emerging focus at the intersection between ecology and evolution. *Philosophical Transactions of the Royal Society B: Biological Sciences*, 368, 20120404–20120404.

Guidetti, P., Fraschetti, S., Terlizzi, A. and Boero, F. (2004). Effects of desertification caused by *Lithophaga lithophaga* (Mollusca) fishery on littoral fish assemblages along rocky coasts of southeastern Italy. *Conservation Biology*, 18, 1417–1423.
Hagen, W. and Schnack-Schiel, S. B. 1996. Seasonal lipid dynamics in dominant Antarctic copepods: Energy for overwintering or reproduction? *Deep Sea Research Part I: Oceanographic Research Papers*, 43, 139–158.
Hall, A. R., Miller, A. D., Leggett, H. C. et al. (2012). Diversity-disturbance relationships, frequency and intensity interact. *Biology Letters*, 8, 768–771.
Halpern, B. S., Walbridge, S., Selkoe, K. A. et al. (2008). A global map of human impact on marine ecosystems. *Science*, 319, 948–952.
Harley, C. D. G., Randall Hughes, A., Hultgren, K. M. et al. (2006). The impacts of climate change in coastal marine systems. *Ecology Letters*, 9, 228–241.
Hawkins, S. J. (1981). The influence of season and barnacles on the algal colonization of *Patella vulgata* exclusion areas. *Journal of the Marine Biological Association of the United Kingdom*, 61, 1–15.
Heino, M. (1998). Noise colour, synchrony and extinctions in spatially structured populations. *Oikos*, 83, 368–375.
Hooper, D. U., Chapin, F. S., Ewel, J. J. et al. (2005). Effects of biodiversity on ecosystem functioning: a consensus of current knowledge. *Ecological Monographs*, 75, 3–35.
Hughes, A. R., Byrnes, J. E., Kimbro, D. L. and Stachowicz, J. J. (2007). Reciprocal relationships and potential feedbacks between biodiversity and disturbance. *Ecology Letters*, 10, 849–864.
IPCC (2007). *Climate Change 2007: The Physical Science Basis. Working Group I Contribution to the Fourth Assessment Report of the IPCC*. Cambridge: Cambridge University Press.
Iribarne, O., Botto, F., Martinetto, P. and Gutierrez, J. L. (2000). The role of burrows of the SW Atlantic intertidal crab Chasmagnathus granulata in trapping debris. *Marine Pollution Bulletin*, 40, 1057–1062.
Jackson, J. B., Kirby, M. X., Berger, W. H. et al. (2001). Historical overfishing and the recent collapse of coastal ecosystems. *Science*, 293, 629–637.
Jager, T. (2012). *Making sense of Chemical Stress: Applications of Dynamic Energy Budget Theory in Ecotoxicology and Stress Ecology*. Available at: http://www.Debtox.Info/book.Php.
Jager, T. and Klok, C. (2010). Extrapolating toxic effects on individuals to the population level: The role of dynamic energy budgets. *Philosophical Transactions of the Royal Society B: Biological Sciences*, 365, 3531–3540.
Kaiser, M. J., Clarke, K. R., Hinz, H. et al. (2006). Global analysis of response and recovery of benthic biota to fishing. *Marine Ecology Progress Series*, 311, 1–14.
Kaiser-Bunbury, C. N., Muff, S., Memmott, J., Müller, C. B. and Caflisch, A. (2010). The robustness of pollination networks to the loss of species and interactions: A quantitative approach incorporating pollinator behaviour. *Ecology Letters*, 13, 442–452.
Karagiannis, D., Vatsos, I. N. and Angelidis, P. (2011). Effects of atrazine on the viability and the formation of byssus of the mussel *Mytilus galloprovincialis*. *Aquaculture International*, 19, 103–110.

Kinlan, B. P. and Gaines, S. D. (2003). Propagule dispersal in marine and terrestrial environments: a community perspective. *Ecology*, 84, 2007–2020.

Kooijman, S. A. L. M. (2010). *Dynamic Energy Budget Theory for Metabolic Organisation.* Cambridge: Cambridge University Press.

Kregel, K. C. (2002). Invited review. Heat-shock proteins: modifying factors in physiological stress responses and acquired thermotolerance. *Journal of Applied Physiology*, 92, 2177–2186.

Kultz, D. (2005). Molecular and evolutionary basis of the cellular stress response. *Annual Review of Physiology*, 67, 225–257.

Lafferty, K. D. and Kuris, A. M. (2009). Parasites reduce food web robustness because they are sensitive to secondary extinction as illustrated by an invasive estuarine snail. *Philosophical Transactions of the Royal Society B: Biological Sciences*, 364, 1659–1663.

Lavorel, S. and Garnier, E. (2002). Predicting changes in community composition and ecosystem functioning from plant traits: revisiting the holy grail. *Functional Ecology*, 16, 545–556.

Law, R. (2000). Fishing, selection, and phenotypic evolution. *ICES Journal of Marine Science*, 57, 659–668.

Lebreton, L. C. M., Greer, S. D. and Borrero, J. C. (2012). Numerical modelling of floating debris in the world's oceans. *Marine Pollution Bulletin*, 64, 653–661.

Ling, S. D., Johnson, C. R., Frusher, S. D. and Ridgway, K. R. (2009). Overfishing reduces resilience of kelp beds to climate-driven catastrophic phase shift. *Proceedings of the National Academy of Sciences of the United States of America*, 106, 22341–22345.

Lotze, H. K. (2006). Depletion, degradation, and recovery potential of estuaries and coastal seas. *Science*, 312, 1806–1809.

Lubchenco, J. (1978). Plant species diversity in a marine intertidal community: importance of herbivore food preference and algal competitive abilities. *American Naturalist*, 23–39.

Mackey, R. L. and Currie, D. J. (2001). The diversity–disturbance relationship: is it generally strong and peaked? *Ecology*, 82, 3479–3492.

Martínez-Crego, B., Alcoverro, T. and Romero, J. (2010). Biotic indices for assessing the status of coastal waters: A review of strengths and weaknesses. *Journal of Environmental Monitoring*, 12, 1013.

Maximenko, N., Hafner, J. and Niiler, P. (2012). Pathways of marine debris derived from trajectories of lagrangian drifters. *Marine Pollution Bulletin*, 65, 51–62.

Mee, L., Friedrich, J. and Gomoiu, M. (2005). Restoring the Black Sea in times of uncertainty. *Oceanography*, 18, 100–111.

Memmott, J., Waser, N. M. and Price, M. V. (2004). Tolerance of pollination networks to species extinctions. *Proceedings of the Royal Society B: Biological Sciences*, 271, 2605–2611.

Menone, M. L., Miglioranza, K. S. B., Iribarne, O., Aizpún De Moreno, J. E. and Moreno, V. C. J. (2004). The role of burrowing beds and burrows of the SW Atlantic intertidal crab *Chasmagnathus granulata* in trapping organochlorine pesticides. *Marine Pollution Bulletin*, 48, 240–247.

Miller, A. D., Roxburgh, S. H. and Shea, K. (2011). How frequency and intensity shape diversity-disturbance relationships. *Proceedings of the National Academy of Sciences*, 108, 5643–5648.

Miller, T. E. (1982). Community diversity and interactions between the size and frequency of disturbance. *American Naturalist*, 120, 533–536.

Miramontes, O. and Rohani, P. (1998). Intrinsically generated coloured noise in laboratory insect populations. *Proceedings of the Royal Society B: Biological Sciences*, 265, 785–792.

Morishige, C., Donohue, M. J., Flint, E., Swenson, C. and Woolaway, C. (2007). Factors affecting marine debris deposition at French frigate shoals, Northwestern Hawaiian Islands Marine National Monument, 1990–2006. *Marine Pollution Bulletin*, 54, 1162–1169.

Mumby, P. J., Vitolo, R. and Stephenson, D. B. (2011). Temporal clustering of tropical cyclones and its ecosystem impacts. *Proceedings of the National Academy of Sciences of the United States of America*, 108, 17626–17630.

Naeem, S. (1998). Species redundancy and ecosystem reliability. *Conservation Biology*, 12, 39–45.

Naeem, S. and Wright, J. P. (2003). Disentangling biodiversity effects on ecosystem functioning: Deriving solutions to a seemingly insurmountable problem. *Ecology Letters*, 6, 567–579.

National Ocean Council (2013). National Ocean Policy Implementation Plan. National Ocean Council, Washington DC. Available at: http://www.whitehouse.gov/sites/default/files/national_ocean_policy_implementation_plan.pdf.

Nichols, E., Gardner, T. A., Peres, C. A. and Spector, S. (2009). Co-declining mammals and dung beetles: an impending ecological cascade. *Oikos*, 118, 481–487.

Nisbet, R. M., Muller, E. B., Lika, K. and Kooijman, S. A. L. M. (2008). From molecules to ecosystems through dynamic energy budget models. *Journal of Animal Ecology*, 69, 913–926.

Nyström, M., Folke, C. and Moberg, F. (2000). Coral reef disturbance and resilience in a human-dominated environment. *Trends in Ecology and Evolution*, 15, 413–417.

O'Gorman, E. J., Fitch, J. E. and Crowe, T. P. (2012). Multiple anthropogenic stressors and the structural properties of food webs. *Ecology*, 93, 441–448.

Odum, E. P. (1985). Trends expected in stressed ecosystems. *Bioscience*, 419–422.

Olsen, E. M., Heino, M., Lilly, G. R. *et al.* (2004). Maturation trends indicative of rapid evolution preceded the collapse of northern cod. *Nature*, 428, 932–935.

Paine, R. T., Ruesink, J. L., Sun, A. *et al.* (1996). Trouble on oiled waters: lessons from the Exxon Valdez oil spill. *Annual Review of Ecology and Systematics*, 27, 197–235.

Paine, R. T., Tegner, M. J. and Johnson, E. A. (1998). Compounded perturbations yield ecological surprises. *Ecosystems*, 1, 535–545.

Peterson, C. H., Rice, S. D., Short, J. W. *et al.* (2003). Long-term ecosystem response to the Exxon Valdez oil spill. *Science*, 302, 2082–2086.

Petraitis, P. S. and Latham, R. E. (1999). The importance of scale in testing the origins of alternative community states. *Ecology*, 80, 429–442.

Pickett, S. T. and White, P. S. (1985). *The Ecology of Natural Disturbance and Patch Dynamics*. San Diego, CA: Academic Press.

Pike, N., Tully, T., Haccou, P. and Ferriere, R. (2004). The effect of autocorrelation in environmental variability on the persistence of populations: an experimental test. *Proceedings of the Royal Society B-Biological Sciences*, 271, 2143–2148.

Porte, C., Janer, G., Lorusso, L. C. et al. (2006). Endocrine disruptors in marine organisms: Approaches and perspectives. *Comparative Biochemistry and Physiology Part C: Toxicology and Pharmacology*, 143, 303–315.

Price, J. M., Johnson, W. R., Marshall, C. F., Ji, Z.-G. and Rainey, G. B. (2003). Overview of the oil spill risk analysis (OSRA) model for environmental impact assessment. *Spill Science and Technology Bulletin*, 8, 529–533.

Raffaelli, D. H. and Hawkins, S. J. (1996). *Intertidal Ecology*. Dordrecht, The Netherlands: Kluwer Academic Publishers.

Reed, M., Johansen, Ø., Brandvik, P. J. et al. (1999). Oil spill modeling towards the close of the 20th century: overview of the state of the art. *Spill Science and Technology Bulletin*, 5, 3–16.

Regoli, F., Gorbi, S., Frenzilli, G. et al. (2002). Oxidative stress in ecotoxicology: from the analysis of individual antioxidants to a more integrated approach. *Marine Environmental Research*, 54, 419–423.

Regoli, F., Benedetti, M. and Giuliani, M. E. (2011). Antioxidant defenses and acquisition of tolerance to chemical stress. In *Tolerance to Environmental Contaminants*, ed. C. Amiard-Triquet, P. S. Raibow and M. Roméo. Boca Raton, FL: CRC Press, pp. 153–173.

Reich, P. B., Hobbie, S. E., Lee, T. et al. (2006). Nitrogen limitation constrains sustainability of ecosystem response to CO_2. *Nature*, 440, 922–925.

Reiss, J., Bridle, J. R., Montoya, J. M. and Woodward, G. (2009). Emerging horizons in biodiversity and ecosystem functioning research. *Trends in Ecology and Evolution*, 24, 505–514.

Rios, L. M., Jones, P. R., Moore, C. and Narayan, U. V. (2010). Quantitation of persistent organic pollutants adsorbed on plastic debris from the northern Pacific gyre's 'eastern garbage patch.' *Journal of Environmental Monitoring*, 12, 2226.

Rius, M. and McQuaid, C. D. (2009). Facilitation and competition between invasive and indigenous mussels over a gradient of physical stress. *Basic and Applied Ecology*, 10, 607–613.

Roberts, D. (1975). The effect of pesticides on byssus formation in the common mussel, *Mytilus edulis*. *Environmental Pollution (1970)*, 8, 241–254.

Rosenburg, A. A. and Mcleod, K. L. (2005). Implementing ecosystem-based approaches to management for the conservation of ecosystem services: Politics and socio-economics of ecosystem-based management of marine resources. *Marine Ecology Progress Series*, 300, 271–274.

Sala, E. and Knowlton, N. (2006). Global marine biodiversity trends. *Annual Review of Environment and Resources*, 31, 93–122.

Sanford, E. and Kelly, M. W. (2011). Local adaptation in marine invertebrates. *Annual Review of Marine Science*, 3, 509–535.

Scheffer, M., Carpenter, S., Foley, J. A., Folke, C. and Walker, B. (2001). Catastrophic shifts in ecosystems. *Nature*, 413, 591–596.

Scheffer, M. and Carpenter, S. R. (2003). Catastrophic regime shifts in ecosystems: linking theory to observation. *Trends in Ecology and Evolution*, 18, 648–656.

Schiel, D. R. (2011). Biogeographic patterns and long-term changes on New Zealand coastal reefs: non-trophic cascades from diffuse and local impacts. *Journal of Experimental Marine Biology and Ecology*, 400, 33–51.

Smith, M. D., Knapp, A. K. and Collins, S. L. (2009). A framework for assessing ecosystem dynamics in response to chronic resource alterations induced by global change. *Ecology*, 90, 3279–3289.

Sogawa, N., Onodera, K., Sogawa, C. A. *et al*. (2001). Bisphenol a enhances cadmium toxicity through estrogen receptor. *Methods and Findings in Experimental and Clinical Pharmacology*, 23, 395.

Sokolova, I. M., Frederich, M., Bagwe, R., Lannig, G. and Sukhotin, A. A. (2012). Energy homeostasis as an integrative tool for assessing limits of environmental stress tolerance in aquatic invertebrates. *Marine Environmental Research*, 79, 1–15.

Solan, M., Aspden, R. J. and Paterson, D. M. (2012). *Marine Biodiversity and Ecosystem Functioning: Frameworks, Methodologies, and Integration*. Oxford: Oxford University Press.

Sousa, W. P. (1979). Experimental investigations of disturbance and ecological succession in a rocky intertidal algal community. *Ecological Monographs*, 49, 227–254.

Sousa, W. P. (1984). The role of disturbance in natural communities. *Annual Review of Ecology and Systematics*, 15, 353–391.

Speidel, M., Harley, C. D. and Wonham, M. J. (2001). Recovery of the brown alga *Fucus gardneri* following a range of removal intensities. *Aquatic Botany*, 71, 273–280.

Stenseth, N. C. (2002). Ecological effects of climate fluctuations. *Science*, 297, 1292–1296.

Stenseth, N. C., Ottersen, G., Hurrell, J. W. *et al*. (2003). Review article. Studying climate effects on ecology through the use of climate indices: the North Atlantic oscillation, El Niño Southern Oscillation and beyond. *Proceedings of the Royal Society B: Biological Sciences*, 270, 2087–2096.

Suding, K. N., Lavorel, S., Chapin, F. S. *et al*. (2008). Scaling environmental change through the community level: a trait-based response-and-effect framework for plants. *Global Change Biology*, 14, 1125–1140.

Sundbäck, K., Alsterberg, C. and Larson, F. (2010). Effects of multiple stressors on marine shallow-water sediments: Response of microalgae and meiofauna to nutrient–toxicant exposure. *Journal of Experimental Marine Biology and Ecology*, 388, 39–50.

Svensson, J. R., Lindegarth, M. and Pavia, H. (2009). Equal rates of disturbance cause different patterns of diversity. *Ecology*, 90, 496–505.

Swartz, R. C., Ferraro, S. P., Lamberson, J. O., *et al*. (1997). Photoactivation and toxicity of mixtures of polycyclic aromatic hydrocarbon compounds in marine sediment. *Environmental Toxicology and Chemistry*, 16, 2151–2157.

Swartz, W., Sala, E., Tracey, S., Watson, R. and Pauly, D. (2010). The spatial expansion and ecological footprint of fisheries (1950 to present). *PloS ONE*, 5, e15143.

Tamburello, L., Bulleri, F., Bertocci, I., Maggi, E. and Benedetti-Cecchi, L. (2013). Reddened seascapes: experimentally induced shifts in 1/f spectra of spatial variability in rocky intertidal assemblages. *Ecology*, 94, 1102–1111.

Teichberg, M., Fox, S. E., Olsen, Y. S. *et al.* (2009). Eutrophication and macroalgal blooms in temperate and tropical coastal waters: nutrient enrichment experiments with *Ulva* spp. *Global Change Biology*, 16, 2624–2637.

Thompson, R. C. Olsen, Y., Mitchell, R. P. *et al.* (2004). Lost at sea: where is all the plastic? *Science*, 304, 838–838.

Todgham, A. E., Schulte, P. M. and Iwama, G. K. (2005). Cross-tolerance in the tidepool sculpin: the role of heat-shock proteins. *Physiological and Biochemical Zoology*, 78, 133–144.

Underwood, A. J. (1998). Grazing and disturbance: an experimental analysis of patchiness in recovery from a severe storm by the intertidal alga *Hormosira banksii* on rocky shores in New South Wales. *Journal of Experimental Marine Biology and Ecology*, 231, 291–306.

Van Haren, R. J. F. and Kooijman, S. A. L. M. (1993). Application of a dynamic energy budget model to *Mytilus edulis* (l.). *Netherlands Journal of Sea Research*, 31, 119–133.

Vinebrooke, R. D., Cottingham, K. L., Norberg, J. *et al.* (2004). Impacts of multiple stressors on biodiversity and ecosystem functioning: the role of species co-tolerance. *Oikos*, 104, 451–457.

Walker, B. H. (1992). Biodiversity and ecological redundancy. *Conservation Biology*, 6, 18–23.

Wang, J., Zhou, Q., Zhang, Q. and Zhang, Y. (2008). Single and joint effects of petroleum hydrocarbons and cadmium on the polychaete Perinereis aibuhitensis Grube. *Journal of Environmental Sciences*, 20, 68–74.

Yachi, S. and Loreau, M. (1999). Biodiversity and ecosystem productivity in a fluctuating environment: the insurance hypothesis. *Proceedings of the National Academy of Sciences*, 96, 1463–1468.

Young, J. T. F., Gauley, J. and Heikkila, J. J. (2009). Simultaneous exposure of *Xenopus* a6 kidney epithelial cells to concurrent mild sodium arsenite and heat stress results in enhanced hsp30 and hsp70 gene expression and the acquisition of thermotolerance. *Comparative Biochemistry and Physiology Part A: Molecular and Integrative Physiology*, 153, 417–424

5 · Impacts of changing biodiversity on marine ecosystem functioning

TASMAN CROWE

5.1 Introduction

Understanding how changes in biodiversity can lead to changes in the functioning of ecosystems is a critical step for tracing the consequences of human activities through to impacts on ecosystem services. As defined and described in Chapter 1, it is widely recognised that the biodiversity of ecosystems can influence their functioning – the processing of energy and materials – and properties such as stability and total biomass. The relationship between biodiversity and ecosystem functioning has, however, been the subject of extensive and, at times, controversial research over the past two decades, the so-called BEF debate.

The BEF debate rose to prominence in ecology at a conference in Bayreuth in 1992 and has passed through several phases since then, in which the field has expanded in volume, scope, rigour and complexity (Naeem *et al.*, 2009). The first major advances were made in terrestrial ecosystems, particularly grasslands (e.g. Naeem *et al.*, 1994; Tilman *et al.*, 1996). There was, initially, a lag in developing similar levels of understanding for marine ecosystems (Heip *et al.*, 1998), but there is now a substantial body of work covering at least coastal ecosystems (Stachowicz *et al.*, 2007; Naeem *et al.*, 2009). Several authors have described how the structure and functioning of marine systems is different from that of terrestrial systems (e.g. Steele, 1985, 1991; Ormond, 1996; Stachowicz *et al.*, 2007; Naeem, 2012), suggesting that understanding derived from terrestrial systems may not be applicable in a marine context. Indeed, a number of major syntheses have treated the systems separately (e.g. Balvanera *et al.*, 2006; Cardinale *et al.*, 2006). On the other hand, Schmid *et al.* (2009) argue that because their meta-analysis revealed no differences in the proportion of positive BEF relationships in marine versus terrestrial systems, there is no basis to argue that different mechanisms apply (and see Webb, 2012). An analysis by Crowe *et al.* (2012) reveals that in general BEF research in marine and terrestrial ecosystems has differed in spatial and temporal scale and approach, making it difficult to draw direct comparisons. As such, the extent to which findings from terrestrial systems can be applied to marine systems and vice versa remains unclear. What is clear is that relationships between biodiversity and ecosystem functioning are complex and that we are not yet able to predict effectively the long-term consequences of loss of biodiversity for many individual ecosystems.

Given the extent of research and debate on BEF, this chapter will not seek to cover the whole of the field. It will instead consider those aspects of it that are most relevant to the core theme of the book – predicting how particular human-induced changes in biodiversity will affect ecosystem functions that underpin ecosystem services. In the same vein, it will also highlight deficiencies in the existing body of research and recommend ways to improve the knowledge base for predicting changes in ecosystem functioning on the basis of changes in biodiversity.

5.2 Generalised responses of ecosystems to biodiversity change

At the simplest level, ecosystem functioning can be affected by changes to the number of species present, referred to as *richness* or *diversity* effects, or by changes to which species are present, referred to as *identity* effects. However, over 50 different trajectories relating changes in biodiversity to changes in functioning have been proposed (Naeem *et al.*, 2009). Among the early models were *redundancy* and *rivet*, which relate the functioning of ecosystems to changes in numbers of species (i.e. diversity effects) and the *idiosyncratic* model, which is based on identity effects. Redundancy – the capacity of the functioning of a system to remain unchanged in the face of loss of some species because others are present to fulfil their roles (Walker, 1992) – can also be thought of in terms of *biological insurance* – a positive attribute enhancing the long-term stability of the system (Yachi and Loreau, 1999). If functioning is strongly affected by some species and not others (identify effects), predicting the effects of loss of species requires a more detailed understanding of the ecology of the system than if there is a consistent relationship between functioning and the numbers of species present. The most widely cited current formulation divides relationships into *saturating* (derived from redundancy/rivet

Impacts of changing biodiversity on marine ecosystem functioning · 113

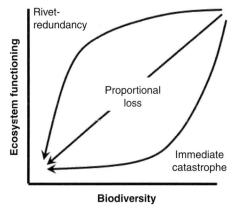

Figure 5.1 Possible relationships between biodiversity and ecosystem functioning. Most predictions about the ecological consequences of loss of biodiversity can be grouped under three general hypotheses: (1) Based on Walker (1992) and Ehrlich and Ehrlich (1981), the rivet–redundancy hypothesis predicts that initial losses of species will have little effect on the functioning of ecosystems because some fraction of species are effectively redundant in the processes they contribute to because other species perform similar roles. However, at some point, loss of species leads to rapid declines in ecological function, much like the loss of one too many rivets can lead to failure of an airplane wing. (2) Others have proposed that the functioning of ecosystems declines in proportion to species loss, and others have argued that (3) even minimal species loss leads to an 'immediate catastrophe', i.e. large declines in the functioning of ecosystems. Based on Cardinale et al. (2011).

models), *linear* and *accelerating*, also referred to as exponential or 'immediate catastrophe' (Kremen, 2005; Cardinale et al., 2011; see Figure 5.1 for illustration and description).

Recent meta-analyses suggest that in general, the relationship between biodiversity and ecosystem functioning is positive and saturating – equivalent to the rivet–redundancy models (Figure 5.1), but also reveal considerable variation (Cardinale et al., 2011), such that the response of a given system to the loss of a given number of species could not be easily predicted. This variation is of key interest to decision makers with responsibility for specific ecosystems. Cardinale et al. (2006) argue that, on average, species loss does influence ecosystem functioning but that the magnitude of effects of species loss depend ultimately on the identities of the species being lost. Reviews of empirical studies in marine systems (e.g. Crowe, 2005; Stachowicz et al., 2007) have tended to highlight a prevalence of identity effects, perhaps because strongly interacting species (e.g. keystone predators or ecosystem engineers) are commonly influential in the systems studied (e.g. Schiel, 2006). However, it should be noted that sometimes diversity effects only become apparent after extended periods (Stachowicz et al., 2008) and that most research into BEF in marine ecosystems has been comparatively short term (Crowe et al., 2012).

Although species richness is often positively correlated with ecosystem functioning, it is important to recognise that some systems function very well with few species. In estuaries, for example, there tend to be few species, but they are present in great abundance and high nutrient inputs make for high productivity in many places. Estuarine species are generally very tolerant to varying environmental conditions, but there is limited functional redundancy (or biological insurance) – if key species are lost, their functional role is unlikely to be replaced by another species, so potential impact on functioning is great. Large areas of the Baltic Sea are naturally low in species as the ecosystem is geologically young and physically harsh due to low salinity and limited circulation (Bonsdorff and Pearson, 1999). Again it could be expected that the loss of one or a few species could have important consequences for functioning and service provision.

Given that we will never be able to study all ecosystems sufficiently to predict consequences of loss of biodiversity from them individually, it would be useful, particularly for environmental decision makers, if some general rules could be established as a basis for initial predictions of the impact of biodiversity loss. Some general trends are being established through meta-analysis. For example, biodiversity generally has positive effects on a wide range of ecosystem functions and services (e.g. Balvanera et al., 2006; Worm et al., 2006; Cardinale et al., 2011; Hooper et al., 2012) and on stability of ecosystem processes (Hooper et al., 2005); in general, the effect of decreasing species richness within a trophic group is to decrease the abundance or biomass of that group, leading to less complete depletion of resources used by that group, although in many cases, multiple species do not outperform the best performing individual species (Cardinale et al., 2006); the loss of 50% of species results on average in a reduction in biomass production of 13% (Hooper et al., 2012); on average, effects of species loss on decomposition are smaller than effects on production (Hooper et al., 2012); different species tend to be influential for different functions (Hector and Bagchi, 2007) and/or in different contexts (Isbell et al., 2011). As described above, these kinds of general patterns often mask a high degree of variability. The next important step will be to establish a framework for predicting effects of biodiversity loss (numbers or types of species lost) in a given context, particularly for specific ecosystems or localities that have not yet been studied in their own right. In other words, the challenge is to identify any trends indicative of where and when different relationships arise, for example in terms of functions, different marine habitats, numbers and types of species in the initial system, etc. It appears that we are still some way short of this capability (and see Balvanera et al., 2006), although a few suggestions can be gleaned from reviews and meta-analyses. For example, Schläpfer et al. (1999) suggested that low diversity systems might be more likely to exhibit idiosyncratic effects, and high diversity systems are more likely to contain redundant species (and see Balvanera et al., 2006); properties such as stability and invasion resistance are more reliably enhanced by biodiversity than standing stocks and flux measures of functions (Srivastava and Vellend, 2005); standing stocks (e.g. biomass) are more likely to be affected by loss of biodiversity than rates of change in such stocks over time (Schmid et al., 2009). In a meta-analysis of 76 studies in aquatic systems, Godbold (2012) found no significant differences in the strength of the BEF relationship in bentho-pelagic, saltmarsh, rocky shore and soft benthos systems.

5.3 Mechanisms underlying relationships between biodiversity and ecosystem functioning

Although predictions can be made on the basis of observed relationships (Peters, 1991), their reliability and generality are greatly improved by understanding of the mechanisms which underlie them. Combinations of species can enhance the functioning of ecosystems through a number of mechanisms. Broadly speaking, these mechanisms can be classified into 'complementarity' and 'selection effects'. Mechanisms of complementarity include the *facilitation* of the functioning of one species by others (e.g. shading by canopy algae creates favourable conditions for photosynthesis by understorey species) and *niche partitioning*, in which different resources are processed by different species leading to greater overall resource use (e.g. different deposit feeders feed in different ways at different levels in the sediment, so diverse assemblages capture more of the deposited material more efficiently). Complementarity can occur in space and/or time. For example, in terrestrial systems pollinators that are active at different times of day correspond to flowering times of different plants so offer a fuller pollination service (Kremen, 2005).

Selection effects arise because the probability of a competitively dominant species with a strong effect on functioning being present increases when greater numbers of species are present (Huston, 1997). This concept is closely tied to the idiosyncratic model described above, essentially encompassing identity effects into a more general predictive BEF framework. Selection effects have been further generalised to encompass any situation in which a species performs differently in a mixture compared to when present alone, in monoculture. Such *sampling effects* can be positive or negative (Hector *et al.*, 2002 cited by Stachowicz *et al.*, 2007).

The presence of multiple species in a system can also enhance its stability (see Chapter 3 for definitions) because different species respond to environmental change in different ways or at different scales and so fluctuate asynchronously such that the aggregate properties of the system are maintained (Hooper *et al.*, 2005; Winfree and Kremen, 2009; Bulleri *et al.*, 2012). From the point of view of ecosystem services, stability is a desirable attribute as more stable ecosystems are more reliable providers of services and are better able to accommodate increased human activity sustainably. The loss of even the weakly interacting species in a system (i.e. those that do not play apparently important roles in community and ecosystem processes) can gradually undermine such stability (O'Gorman *et al.*, 2011).

The mechanisms underpinning the functioning of a particular ecosystem involving a particular combination of species can comprise complex mixtures of positive effects of complementarity and positive or negative selection effects, which can make consequences of biodiversity changes difficult to predict (Stachowicz *et al.*, 2007). Different relationships and mechanisms have different implications for management, however. For example, if selection effects underpin the BEF relationship in a given ecosystem, due to the importance of certain species, then those species need to be identified and their sensitivities should be characterised and taken into account in prioritising interventions.

5.4 Which aspects of biodiversity are most useful for predicting effects on functioning?

Biodiversity is a multifaceted concept. It can be broadly thought of in terms of genetic, organismal and habitat/ecosystem-level diversity (United Nations, 1992; Chapter 1). The BEF debate has focused primarily on organismal diversity and this chapter will reflect that focus (but see Section 5.9.3 for a consideration of the implications of our lack of knowledge of impacts of changing genetic and habitat level diversity). Even with a focus on organismal diversity, there are many different ways of characterising the biodiversity of an ecosystem and a wide range of measures of biodiversity that could be used to predict ecosystem functioning.

Most research to date has focused on species (i.e. taxonomically based classifications of organisms). The generalised relationships between biodiversity and ecosystem functioning described are all based on species-level diversity. Given the degree of variation around the modelled predictions, it could be argued that species richness is a comparatively poor predictor of ecosystem functioning. If we want to make more precise and generalisable predictions of changes in ecosystem processes in response to changes in biodiversity, we may be better served by considering other aspects of biodiversity. In this section, species-level diversity and some of the alternatives will be briefly discussed (see Crowe and Russell, 2009, for a more detailed review).

Species-level diversity itself can be characterised in a number of ways. Diversity indices, such as species richness and Shannon diversity (Magurran, 2004), summarise information about the number of species present and their relative abundances, but information about which species are present and at which level of abundance relative to each other is lost. Multivariate analyses based on matrices of species and their abundances or biomasses can retain the maximum information of this kind and give the most detailed basis for considering how human activities are changing biodiversity and community structure (Clarke, 1993). Given the likelihood of identity effects, particularly in marine ecosystems, in which some species often have disproportionate effects, it is important to be able to consider what kinds of species are present and whether any of them are particularly influential, for example, as keystone consumers or ecosystem engineers (Allison *et al.*, 1996). For example, limpets are particularly influential grazers on many temperate rocky shores (Coleman *et al.*, 2006) and their influence on macroalgal cover cannot be fulfilled by other species, even at increased densities (O'Connor and Crowe, 2005). Doherty *et al.* (2011) showed that richness effects in a restoration project disappeared after 11 years; two species became dominant and made the maximal contribution to functioning.

It can be difficult to identify the most influential key species in an ecosystem. Indeed, it has been argued that it is better to think in terms of a continuum of interaction strength, with 'keystone' species simply being the strongest interactors (Hurlbert, 1997). The effects of losing species, even strong interactors, can be variable in space (e.g. Piraino et al., 2002; Boyer et al., 2009; Crowe et al., 2013) and time (O'Connor and Crowe, 2005), however, with potential for compensation by currently sub-dominant or rare species in the longer term in some cases, particularly under different environmental scenarios (Piraino et al., 2002; Davies et al., 2011; Doherty et al., 2011). Davies et al. (2011) suggested that we seek to characterise the minimum and maximum effects of losing species over the immediate and longer term. They also showed that percentage contributions to biomass can potentially be used to predict percentage impact on functioning if lost. In food webs, stronger and weaker interactors can be identified on the basis of their connectedness to other species (the number of different species they feed on or are fed upon by). Empirical and theoretical work has shown that a prevalence of strong interactors can destabilise food webs and the loss of weak interactors is to be avoided as it can gradually undermine stability (O'Gorman et al., 2011).

When considered in more detail, the attributes of species, their 'traits', have the potential to form a much more effective basis for predicting ecosystem responses to stressors than the numbers or identities of species (Crowe and Russell, 2009; Petchey et al., 2009). Traits include aspects of demography (body size, longevity, etc.), feeding strategies, morphology (e.g. Steneck and Dethier, 1994), responses to environmental or resource changes such as disturbance (Lavorel and Grigulis, 1997; Díaz and Cabido, 2001; Lavorel and Garnier, 2001), adaptive strategies or effects on ecosystem function through biogeochemical cycles (Lavorel and Garnier, 2001). In principle, it should be possible to match traits to functional properties of ecosystems more easily than if working with species' identities. A trait-based approach would also be potentially generalisable across systems because it ignores regional taxonomic differences, focusing on what species do rather than what they are called. Species can either be grouped according to one or a few traits, for example, into morphological functional groups (e.g. Steneck and Dethier, 1994) or trophically based guilds, or different types of ecosystem engineer (Jones et al., 1994), but this approach has limitations (Crowe and Russell, 2009). Methods are now being explored that capture information on a large number of traits for each species such that an overall multivariate matrix of traits can be derived for a given assemblage of organisms. Such matrices can be summarised using indices analogous to diversity indices, such as functional diversity, functional attribute diversity and others (reviewed by Petchey *et al.*, 2009). Alternatively, frameworks like biological traits analysis can be used which draw on the full body of information held in the matrix (Bremner *et al.*, 2003, 2006). This approach is very flexible and offers potential to select traits that underpin particular ecosystem functions and so directly analyse the specific consequences of community changes to inform conservation and management towards stated objectives (Bremner, 2008; Cooper *et al.*, 2008; Frid *et al.*, 2008). If traits are divided into 'response' traits (which determine species' responses to environmental change) and 'effect traits' (Naeem and Wright, 2003), it should in principle be possible to predict how a particular activity causing particular environmental changes will influence both community structure and ecosystem functioning and indeed ecosystem services (Suding *et al.*, 2008; Díaz *et al.*, 2011; Lavorel *et al.*, 2013). If response-and-effect traits are correlated, it is likely that increasing anthropogenic pressure will cause rapid loss of functioning with little scope for functional compensation (Solan *et al.*, 2004; Suding *et al.*, 2008; Chapter 4).

Severe limitations on the potential application of these approaches are imposed by the shortage of trait information for many species, however (see Section 5.9.5). Analyses of plant traits have shown that most can be predicted satisfactorily on the basis of a few readily measurable traits, such as seed size, plant height and specific leaf area (Westoby, 1998; and see Lavorel *et al.*, 2013). Similarly, animal biologists have shown that body size can be an extremely influential trait (Woodward *et al.*, 2005). Thus, some changes in ecosystem processes could potentially be predicted on the basis of changing sizes of organisms alone (Woodward *et al.*, 2005; Emmerson, 2012). Indeed, the effects of some stressors are more easily summarised in terms of effects on traits rather than species; for example, fisheries tend to affect larger species or individuals, which may in itself provide some basis for predicting their ecosystem-level effects (Emmerson, 2012).

Another alternative approach is to consider phylogenetic diversity or taxonomic distinctness, which capture the evolutionary histories or relatedness of different species in assemblages, and a number of metrics are emerging for doing so (Warwick and Clarke, 1995; Clarke and Warwick, 2001; Cadotte *et al.*, 2009; Flynn *et al.*, 2011). Stressors generally reduce taxonomic distinctness – assemblages in degraded systems tend to be dominated by species from the same taxonomic group (e.g. oligochaete worms) rather than representatives of many phyla, classes or families (Warwick and Clarke, 1995). Flynn *et al.* (2011) showed that phylogenetic diversity could be used to predict functioning with a similar degree of success as functional diversity. As phylogenetic information is more readily available than trait information for many species, phylogenetic diversity may offer a more pragmatic way forward and is worthy of further investigation.

5.5 Realistic scenarios of biodiversity change

Much of the experimental research into BEF relationships has been based on random extinctions of species. In reality, some species are more likely to go extinct than others, for example, due to their inherent sensitivity to stressors, small population sizes or traits such as body size and trophic position (Solan et al., 2004). Targeted research has shown that if species are lost in sequences based on these traits, under 'realistic extinction scenarios', the relationship between numbers of species present and rates of ecosystem processes can change quite markedly (Solan et al., 2004; Bracken et al., 2008; Davies et al., 2011). This arises when there is co-variance between traits influencing the likelihood that a species will go extinct and traits related to its functional importance (see Section 5.4). In Solan et al.'s (2004) analysis, a key species had disproportionate influence on the ecosystem function in question, and the timing of its extinction was pivotal to the shape of the relationship between diversity and functioning. Bracken and Low (2012) found that the loss of rare species, which are inherently more likely to go extinct, had a far greater effect than the loss of an equivalent biomass of dominant species.

As described above, realistic extinction scenarios can be predicted in general terms based on broad associations between extinction risk and organismal traits, such as rarity and body size. Larger species tend to be at greater risk of extinction, as do rare species, top predators (Dobson et al., 2006) and specialists (Clavel et al., 2011). Although such knowledge is of value in making predictions, particularly where detailed knowledge is limited, it would be of even greater value to environmental decision makers if extinction scenarios could be derived from the specific sensitivities of key species to particular activities (see Davies et al., 2011, for direct experimental tests of extinction sequences). If the functional roles of those species were also known, the consequences for ecosystem functioning and services of permitting particular activities known to affect certain species in particular areas could be simulated (see Section 5.4). This premise constitutes a key theme of the book and will be further developed in Chapters 6–10, which will review current knowledge of how particular stressors affect biodiversity and ecosystem functioning.

Although the above discussion focuses on extinction, it must be recognised that the number of species in an ecosystem may also change because new species arrive, for example due to climate-related range expansion or invasion. The impacts of invasive species are the focus of extensive research which, by definition, is based on 'realistic' scenarios as it examines impacts of particular invasive species in their invasive range (Chapter 10).

5.6 Changes in density and evenness

In the marine environment, evidence for true, global, extinctions is rare. The loss of species from ecosystems occurs on a local scale – extirpation. In fact, it is far more common for species to change in abundance than for them to be completely extinguished from an area. When one or a few species change in abundance, it changes the overall community structure, the relative proportions of different species, which in itself can have consequences for ecosystem functioning.

When the overall density of individuals or the relative density of individuals belonging to different species changes, the nature, frequency and magnitude of interactions between individuals also changes (Benedetti-Cecchi, 2004). The main experimental designs used to test hypotheses about BEF, additive and substitutive designs, routinely confound changes in richness with changes in either the total or relative density of individuals, respectively (Benedetti-Cecchi, 2004). Experiments that have involved explicit manipulations of density have often shown that density has at least as great an influence on ecosystem processes as species richness and that the relationship can be nonlinear and unpredictable (Maggi et al., 2009, 2011).

The relative abundance of different species can be captured by diversity indices of evenness (the degree of equality in population sizes of different species present) or dominance (the degree of inequality in population sizes of different species present) (Magurran, 2004). Comparatively few studies have explicitly tested the influence of evenness on marine ecosystem functioning and those that have done so have yielded mixed results (Hillebrand et al., 2008, 2012; Maestre et al., 2012). Far more research is needed into the influence of these aspects of diversity on ecosystem functioning and the relationship between functioning and the number of species present.

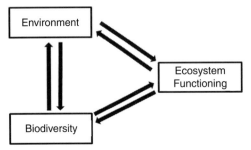

Figure 5.2 Schematic representation of possible interactions between biodiversity, ecosystem functioning and environmental conditions. A change in one of these elements can affect each of the others either directly or indirectly, via changes to the third. Predicting consequences of human-induced environmental change for ecosystem functioning requires knowledge of all of these interactions. The diagram shows equal strength of all interactions. In reality, interaction strength and direction between the three elements varies considerably under different circumstances.

5.7 The importance of environmental conditions

The focus of the BEF debate has been on how biodiversity affects ecosystem functioning, a simple bivariate plot, but this focus ignores a wider and more complex reality (Naeem *et al.*, 2009; Hooper *et al.*, 2012). Ecosystem functioning is also substantially influenced by environmental conditions, such as temperature, salinity, water movements, etc. Under some circumstances, environmental effects (direct and indirect) are most important for influencing functioning; under others, biodiversity is more important for influencing both the local environmental conditions and the functioning of the ecosystem. It is important also to remember that biodiversity responds to ecosystem functioning and environmental conditions as well as influencing them (Hooper *et al.*, 2005), such that there can be complex interrelationships between biodiversity, environmental conditions and ecosystem functioning (Figure 5.2). For example, invasive zebra mussels radically alter both the pelagic and benthic environment in freshwater habitats, directly alter productivity, nutrient cycling and trophic dynamics, and also modify biodiversity both directly through competition and habitat provision and indirectly through effects on pelagic and benthic ecosystem processes (Atalah and Crowe, 2010, and references therein). Understanding feedbacks such as these is critical, but has received comparatively little attention in the context of BEF relationships (but see Cardinale *et al.*, 2009).

In a global synthesis, Hooper et al. (2012) showed that biodiversity loss affected ecosystems to a similar degree as other major drivers of change, such as acidification, elevated CO_2 and nutrient pollution. In some systems, abiotic drivers tend to have a greater influence on some ecosystem processes than others. Hiddink et al. (2009) found that wave stress correlated more strongly with some measures of functioning than biodiversity at subtidal sites in the Irish Sea. Godbold's (2012) meta-analysis showed that abiotic drivers tended to be more important in rocky shore, bentho-pelagic and seagrass ecosystems than in other marine and freshwater ecosystems.

Similarly, the environmental context from which biodiversity is lost has been shown to influence the consequences of that loss for ecosystem functioning (Boyer et al., 2009; McKie et al., 2009; Crowe et al., 2011, 2013) and stability (Bulleri et al., 2012). We do not yet have sufficient data on environmental influences on consequences of biodiversity loss to be able to predict which conditions will lead to more severe consequences of a given loss of species. In assessing how human activities may affect ecosystem functioning and service provision, it is important to recognise that they can affect both biodiversity and the environmental context in which ecosystems function (McKie et al., 2009; Hooper et al., 2012), both of which may affect delivery of ecosystem services.

5.8 Mapping multiple species onto multiple functions and services

The vast majority of work on BEF has focused on subsets of species and ecosystem functions, often just measuring one or a few response variables. This has been leading to some potential underestimates of the role of biodiversity. Recent syntheses and reanalyses have shown that if multiple functions are considered, more species are required for efficient functioning of the system as a whole (Hector and Bagchi, 2007; Gamfeldt et al., 2008; Isbell et al., 2011). Similarly, as described above (Section 5.7), larger numbers of species help to underpin the stability of ecosystem functioning when environmental conditions change (Yachi and Loreau, 2002; Hector and Bagchi, 2007; Isbell et al., 2011). Mouillot et al. (2011) showed that functional divergence among species was more important in predicting multiple ecosystem functions than richness of species per se and argue that consideration needs to be given to the prevalence of specialist taxa if we are to conserve systems capable of delivering multiple services. This is of particular concern given an apparent global decline in specialised species and their replacement by generalists (Clavel et al., 2011). De Bello et al. (2010) propose the concept of trait-service clusters – the identification of sets of traits that contribute to provision of particular services – but contend that most proven trait-service associations relate to plant communities and, to a lesser extent, to soil and freshwater organisms and ecosystems.

5.9 Limitations of research to date as a basis for predicting consequences of human-induced biodiversity change for marine ecosystems

5.9.1 Predominance of research with mesocosms and synthetic assemblages

Most marine BEF research has been undertaken in mesocosms (Crowe et al., 2012). While valuable in indicating potential effects of biodiversity and clarifying mechanisms underpinning those effects, the direct applicability of such results to the real world needs to be verified. As such, there needs to be a greater emphasis on field-based studies. Díaz et al. (2003) made the case that field experiments in which species are removed from natural assemblages are more representative of likely outcomes in nature than 'synthetic assemblage experiments', in which experimenters combine selected species to generate experimental treatments. As discussed above (Section 5.5), experiments based on realistic extinctions are also needed because random extinctions of species can yield misleading results (e.g. Bracken et al., 2008).

5.9.2 Spatial and temporal scales

Most field research has been done at small spatial scales – in small rock pools or 50 × 50 cm plots (Crowe et al., 2012). Most impacts and management interventions take place at much larger scales – whole shores or embayments. Research at larger spatial scales is less likely to be well replicated or controlled, so is less likely to provide a convincing body of evidence on which to base interventions (Raffaelli and Friedlander, 2012). More often, in the absence of alternative data or understanding, findings derived from small-scale research are applied to larger scales. This is potentially misleading and great caution should be applied in making such extrapolations, but a firm foundation for extrapolating between scales remains an important and elusive challenge (but see Chesson, 2012). Naeem (2006) and others have advocated modelling (in silico research) as a tool to do so, but recognise the limitations of large-scale models based on small-scale empirical studies (Naeem, 2012).

Similarly, most marine BEF research has been short term (3 months on average, Crowe *et al.*, 2012). There is good evidence that diversity effects are more likely to be manifested over longer time periods (Grime, 1998; O'Connor and Crowe, 2005; Stachowicz *et al.*, 2008; Cardinale *et al.*, 2011; Isbell *et al.*, 2011) suggesting that diversity effects may currently be underestimated (and see Section 5.8).

5.9.3 Limited knowledge of importance of genetic and habitat diversity

The most widely used definition of biodiversity recognises three main levels at which it should be considered: genetic, organismal and habitat (United Nations, 1992). The current chapter has focused almost exclusively on organismal diversity, mainly because that has been the focus of most BEF research. This could be considered an important deficiency in the field. Population-level genetic diversity can enhance productivity (Hughes *et al.*, 2008), although enhanced genetic diversity in multi-species communities does not necessarily do so (Fridley and Grime, 2010). Similarly, a few authors have commented on or demonstrated the potential importance of habitat diversity and configuration (Hawkins, 2004; Dyson *et al.*, 2007; Griffin *et al.*, 2009). It is therefore important to consider these levels of diversity in assessing the potential for proposed human activities to affect diversity and ecosystem functioning. Some important impacts of built structures, for example, could well be at the level of genetic diversity – changes in the connectivity of populations may change their genetic structure and long-term viability.

5.9.4 Biases in terms of taxa, habitats, localities and functions

Due to the need for complex experiments to test conceptual hypotheses, there has been a strong tendency to focus on tractable taxa, habitat and localities. A lot of work has focused on macrobiota in benthic intertidal habitats. Much less work has been done on plankton, meiofauna, megafauna and pelagic or subtidal systems in general. There is also limited understanding of how impacts of changing biodiversity on functioning are mediated by microbes (but see Green *et al.*, 2012; Naeem, 2012). New techniques are making research on functional ecology of microbes more tractable and may ultimately lead to cost-effective tools for rapid assessment of functional properties of microbial communities (de Menezes et al., 2012; Jansson and Prosser, 2013).

In many ecosystems, the vast majority of species are rare, having very low abundances on local or wider scales (Kunin and Gaston, 1993). Such species generally have a high risk of extinction (see above). In the studies described above, the loss of rare species had comparatively little impact on ecosystem functioning (Solan et al., 2004; Davies et al., 2011), but this can change if biomass is taken into consideration (Bracken et al., 2012). What's more, species that are currently rare can become more abundant and/or more influential under changing environmental circumstances (e.g. Smith and Knapp, 2003). Due to their rarity and the challenges inherent in studying them, we generally know the least about their biology, so have little basis for predicting the consequences of their loss.

BEF relationships at some localities (e.g. Ythan estuary) have been studied extensively. There is, however, limited knowledge of how BEF relationships vary in space and time and whether the knowledge gained at 'research hotspots' is more widely applicable. Structured research is needed in which the extent and nature of spatial variation in responses to biodiversity loss is explicitly tested (e.g. Boyer et al., 2009; Crowe et al., 2012) such that more general predictive frameworks can be developed.

Given the complexity of the experiments required to test BEF hypotheses, the selection of study organisms, systems and functions for BEF research has been driven primarily by scientific pragmatism. Research is needed in which direct societal relevance has a greater influence on the agenda (Raffaelli, 2006; Duffy, 2009), with work focused on species losses that are likely to arise through particular activities (e.g. fishing, see Naeem, 2012) and functions that underpin ecosystem services that are highly valued by society.

5.9.5 Limited knowledge of functional traits

The application of functional trait-based approaches is severely limited by the lack of trait information for most of the species in most ecosystems. There have been some valuable initiatives, collating available trait information (e.g. Marlin's BIOTIC database www.marlin.ac.uk/biotic), but nevertheless, the necessary information is simply not available for collation for most species (Tyler et al., 2012). We have a particularly poor understanding of the roles of rare species and indeed their natural history and basic biology. For many species that do feature in trait databases,

Impacts of changing biodiversity on marine ecosystem functioning · 127

the information available relates mainly to basic demographic and life-history characteristics. Although useful information, for example, on body size, growth form, feeding and movement are listed, we often lack the trait information that is likely to be most useful in directly linking human activities to impacts on ecosystem functioning via response and functional traits, such as tolerances to different pollutants or details of nutrient use by algae and functional roles.

5.10 Key points and recommendations

- In general, having more species in a given ecosystem tends to increase rates and stability of ecosystem processes until a saturation point is reached, particularly if a long-term view is taken and multiple functions are considered. Nevertheless, there is considerable variation in the outcomes of individual studies.
- Mechanisms underpinning BEF relationships include complementarity (in which species either contribute to ecosystem processes in different ways or facilitate each other) and selection or sampling effects (in which ecosystems containing more species are more likely to contain particularly influential species).
- Changes in density or evenness of species are more likely to arise than extinctions of species and can also influence ecosystem processes, but have been less extensively studied.
- Environmental context can modify the influence of biodiversity loss on ecosystem functioning; variation in environmental conditions can sometimes have a stronger influence on ecosystem processes than variation in biodiversity; ecosystem processes can also influence biodiversity and environmental conditions. Thus, there is a three-way interaction between biodiversity, environmental conditions and ecosystem functioning.
- Given that we cannot possibly generate empirical data on all ecosystem functions for all combinations of species in all habitats and localities, it would be of great value to identify the kinds of circumstances under which the loss of particular numbers or types of species would tend to have particular consequences for ecosystem functioning, for example, in terms of the initial numbers of species in the system, the profile of interaction strengths or traits, the physical nature of the system, etc. Some relevant generalisations have been made (see Section 5.2), but we are still some way short of an empirical basis for such a framework.
- A range of approaches is available for predicting functioning based on different aspects of diversity captured in different ways, for example,

in terms of functional traits and phylogenetic diversity. Some generalisations are possible, but complex ecosystems can always respond unpredictably to change.
- Individual species often play key roles for particular functions under particular environmental circumstances. If the species that support functions considered to be particularly important can be identified, they can provide a focus for conservation and management. If policy requires that multiple functions are maintained, then it is appropriate to seek to maximise the number of species in the system, the degree of functional divergence among species and the abundance of specialised species.
- Future research should include (1) long-term field-based studies, ideally at large spatial scales, (2) work on habitat and genetic level diversity, (3) work to characterise relevant functional traits and the roles of rare species, microbes and meiofauna, and (4) stronger alignment with the needs of policy and management.

Subsequent chapters in this book will explore how different anthropogenic stressors modify environmental conditions and particular aspects of biodiversity and may therefore influence ecosystem functioning and ecosystem services.

References

Allison, G. W., Menge, B. A., Lubchenco, J. et al. (1996). Predictability and uncertainty in community regulation: consequences of reduced consumer diversity in coastal rocky ecosystems. In *Functional roles of biodiversity: a global perspective*, ed. H. A. Mooney, J. H. Cushman, E. Medina, O. E. Sala and E.-D. Schulze. New York: John Wiley and Sons Ltd., pp. 371–392.

Atalah, J. and Crowe, T. P. (2010). Combined effects of nutrient enrichment, sedimentation and grazer loss on rock pool assemblages. *Journal of Experimental Marine Biology and Ecology*, 388, 51–57.

Balvanera, P., Pfisterer, A. B., Buchmann, N. et al. (2006). Quantifying the evidence for biodiversity effects on ecosystem functioning and services. *Ecology Letters*, 9, 1146–1156.

Benedetti-Cecchi, L. (2004). Increasing accuracy of causal inference in experimental analyses of biodiversity. *Functional Ecology*, 18, 761–768.

Bonsdorff, E. and Pearson, T. H. (1999). Variation in the sublittoral macrozoobenthos of the Baltic Sea along environmental gradients: a functional-group approach. *Australian Journal of Ecology*, 24, 312–326.

Boyer, K. E., Kertesz, J. S. and Bruno, J. F. (2009). Biodiversity effects on productivity and stability of marine macroalgal communities: the role of environmental context. *Oikos*, 118, 1062–1072.

Bracken, M. E. S., Friberg, S. E., Gonzalez-Dorantes, C. A. et al. (2008). Functional consequences of realistic biodiversity changes in a marine ecosystem. *Proceedings of the National Academy of Sciences of the United States of America*, 105, 924–928.

Bracken, M. E. S. and Low, N. H. N. (2012). Realistic losses of rare species disproportionately impact higher trophic levels. *Ecology Letters*, 15, 461–467.

Bremner, J., Rogers, S. I. and Frid, C. L. J. (2003). Assessing functional diversity in marine benthic ecosystems: a comparison of approaches. *Marine Ecology Progress Series*, 254, 11–25.

Bremner, J., Rogers, S. I. and Frid, C. L. J. (2006). Matching biological traits to environmental conditions in marine benthic ecosystems. *Journal of Marine Systems*, 60, 302–316.

Bremner, J. (2008). Species' traits and ecological functioning in marine conservation and management. *Journal of Experimental Marine Biology and Ecology*, 366, 37–47.

Bulleri, F., Benedetti-Cecchi, L., Cusson, M. et al. (2012). Temporal stability of European rocky shore assemblages: variation across a latitudinal gradient and the role of habitat-formers. *Oikos*, 121, 1801–1809.

Cadotte, M. W., Cavender-Bares, J., Tilman, D. et al. (2009). Using phylogenetic, functional and trait diversity to understand patterns of plant community productivity. *PLoS ONE*, 4, e5695.

Cardinale, B. J., Srivastava, D. S., Duffy, J. E. et al. (2006). Effects of biodiversity on the functioning of trophic groups and ecosystems. *Nature*, 443, 989–992.

Cardinale, B. J., Bennett, D. M., Nelson, C. E. et al. (2009). Does productivity drive diversity or vice versa? A test of the multivariate productivity-diversity hypothesis in streams. *Ecology*, 90, 1227–1241.

Cardinale, B. J., Matulich, K. L., Hooper, D. U. et al. (2011). The functional role of producer diversity in ecosystems. *American Journal of Botany*, 98, 572–592.

Chesson, P. (2012). Scale transition theory: Its aims, motivations and predictions. *Ecological Complexity*, 10, 52–68.

Clarke, K. R. (1993). Non-parametric multivariate analyses of changes in community structure. *Australian Journal of Ecology*, 18, 117–143.

Clarke, K. R. and Warwick, R. M. (2001). A further biodiversity index applicable to species list: variation in taxonomic distinctness. *Marine Ecology Progress Series*, 216, 265–278.

Clavel, J., Julliard, R. and Devictor, V. (2011). Worldwide decline of specialist species: toward a global functional homogenization? *Frontiers in Ecology and the Environment*, 9, 222–228.

Coleman, R. A., Underwood, A. J., Benedetti-Cecchi, L. et al. (2006). A continental scale evaluation of the role of limpet grazing on rocky shores. *Oecologia*, 147, 556.

Cooper, K. M., Frojan, C., Defew, E. et al. (2008). Assessment of ecosystem function following marine aggregate dredging. *Journal of Experimental Marine Biology and Ecology*, 366, 82–91.

Crowe, T. P. (2005). What do species do in intertidal ecosystems? In *The Intertidal Ecosystem: The Value of Ireland's Shores*, ed. J. G. Wilson. Dublin, Ireland: Royal Irish Academy, pp. 115–133.

Crowe, T. P. and Russell, R. (2009). Functional and taxonomic perspectives of marine biodiversity: relevance to ecosystem processes. In *Marine Hard Bottom Communities: Patterns, Dynamics, Diversity, Change,* ed. M. Wahl, Amsterdam: Elsevier, pp. 375–390.

Crowe, T. P., Bracken, M. E. and O'Connor, N. E. (2012). Reality check: issues of scale and abstraction in biodiversity research, and potential solutions. In *Marine Biodiversity Futures and Ecosystem Functioning: Frameworks, Methodologies and Integration,* ed. M. Solan, R. J. A. Aspden and D. M. Paterson. Oxford: Oxford University Press, pp. 185–199.

Crowe, T. P., Cusson, M., Bulleri, F. *et al.* (2013). Large-scale variation in combined impacts of canopy loss and disturbance on community structure and ecosystem functioning. *PLoS ONE,* 8, e66238.

Davies, T. W., Jenkins, S. R., Kingham, R. *et al.* (2011). Dominance, biomass and extinction resistance determine the consequences of biodiversity loss for multiple coastal ecosystem processes. *PLoS ONE,* 6(12), e28362.

de Bello, F., Lavorel, S., Díaz, S. *et al.* (2010). Towards an assessment of multiple ecosystem processes and services via functional traits. *Biodiversity and Conservation,* 19, 2873–2893.

de Menezes, A., Clipson, N. and Doyle, E. (2012). Comparative metatranscriptomics reveals widespread community responses during phenanthrene degradation in soil. *Environmental Microbiology,* 14, 2577–2588.

Díaz, S. and Cabido, M. (2001). Vive la différence: plant functional diversity matters to ecosystem processes. *Trends in Ecology and Evolution,* 16, 646–655.

Díaz, S., Symstad, A. J., Chapin III, F. S. *et al.* (2003). Functional diversity revealed by removal experiments. *Trends in Ecology and Evolution,* 18, 140–146.

Díaz, S., Quetier, F., Caceres, D. M. *et al.* (2011). Linking functional diversity and social actor strategies in a framework for interdisciplinary analysis of nature's benefits to society. *Proceedings of the National Academy of Sciences of the United States of America,* 108, 895–902.

Dobson, A., Lodge, D., Alder, J. *et al.* (2006). Habitat loss, trophic collapse, and the decline of ecosystem services. *Ecology,* 87, 1915–1924.

Doherty, J. M., Callaway, J. C. and Zedler, J. B. (2011). Diversity-function relationships changed in a long-term restoration experiment. *Ecological Applications,* 21, 2143–2155.

Duffy, J. E. (2009). Why biodiversity is important to the functioning of real-world ecosystems. *Frontiers in Ecology and the Environment,* 7, 437–444.

Dyson, K. E., Bulling, M. T., Solan, M. *et al.* (2007). Influence of macrofaunal assemblages and environmental heterogeneity on microphytobenthic production in experimental systems. *Proceedings of the Royal Society B,* 274, 2547–2554.

Ehrlich, P. R. and Ehrlich, A. H. (1981). *Extinction. The Causes and Consequences of the Disappearance of Species.* New York: Random House.

Emmerson, M. (2012). The importance of body size, abundance and food-web structure for ecosystem functioning. In *Marine Biodiversity and Ecosystem Functioning: Frameworks, Methodologies and Integration,* ed. M. Solan, R. Aspden and D. M. Paterson. Oxford: Oxford University Press, pp. 85–100.

Flynn, D. F. B., Mirotchnick, N., Jain, M. et al. (2011). Functional and phylogenetic diversity as predictors of biodiversity–ecosystem–function relationships. *Ecology*, 92, 1573–1581.

Frid, C. L. J., Paramor, O. A. L., Brockington, S. et al. (2008). Incorporating ecological functioning into the designation and management of marine protected areas. *Hydrobiologia*, 606, 69–79.

Fridley, J. D. and Grime, J. P. (2010). Community and ecosystem effects of intraspecific genetic diversity in grassland microcosms of varying species diversity. *Ecology*, 91, 2272–2283.

Gamfeldt, L., Hillebrand, H. and Jonsson, P. R. (2008). Multiple functions increase the importance of biodiversity for overall ecosystem functioning. *Ecology*, 89, 1223–1231.

Godbold, J. A. (2012). Effects of biodiversity-environment conditions on the interpretation of biodiversity-function relations. In *Marine Biodiversity and Ecosystem Functioning: Frameworks, Methodologies and Integration*, ed. M. Solan, R. Aspden and D. M. Paterson. Oxford: Oxford University Press, pp. 101–113.

Green, D. S., Boots, B. and Crowe, T. P. (2012). Effects of non-indigenous oysters on microbial diversity and ecosystem functioning. *PLoS ONE*, 7(10), 1.

Griffin, J. N., Jenkins, S. R., Gamfeldt, L. et al. (2009). Spatial heterogeneity increases the importance of species richness for an ecosystem process. *Oikos*, 118, 1335–1342.

Grime, J. P. (1998). Benefits of plant diversity to ecosystems: immediate, filter and founder effects. *Journal of Ecology*, 86, 902–910.

Hawkins, S. J. (2004). Scaling up: the role of species and habitat patches in functioning of coastal ecosystems. *Aquatic Conservation: Marine and Freshwater Ecosystems*, 14, 217–219.

Hector, A. and Bagchi, R. (2007). Biodiversity and ecosystem multifunctionality. *Nature*, 448, 188–191.

Heip, C., Warwick, R. and d'Ouzville, L. (1998). *A European Science Plan on Marine Biodiversity*. Strasbourg, France: European Science Foundation.

Hiddink, J. G., Davies, T. W., Perkins, M. et al. (2009). Context dependency of relationships between biodiversity and ecosystem functioning is different for multiple ecosystem functions. *Oikos*, 118, 1892–1900.

Hillebrand, H., Bennett, D. M. and Cadotte, M. W. (2008). Consequences of dominance: a review of evenness effects on local and regional ecosystem processes. *Ecology*, 89, 1510–1520.

Hillebrand, H., Burgmer, T. and Biermann, E. (2012). Running to stand still: temperature effects on species richness, species turnover, and functional community dynamics. *Marine Biology*, 159, 2415–2422.

Hooper, D. U., Chapin, F. S., Ewel, J. J. et al. (2005). Effects of biodiversity on ecosystem functioning: a consensus of current knowledge. *Ecological Monographs*, 75, 3–35.

Hooper, D. U., Adair, E. C., Cardinale, B. J. et al. (2012). A global synthesis reveals biodiversity loss as a major driver of ecosystem change. *Nature*, 486, 105–108.

Hughes, A. R., Inouye, B. D., Johnson, M. T. J. et al. (2008). Ecological consequences of genetic diversity. *Ecology Letters*, 11, 609–623.

Hurlbert, S. H. (1997). Functional importance vs. keystoneness: reformulating some questions in theoretical biocenology. *Australian Journal of Ecology*, 22, 369–382.

Huston, M. A. (1997). Hidden treatments in ecological experiments: re-evaluating the ecosystem function of biodiversity. *Oecologia*, 110, 449–460.

Isbell, F., Calcagno, V., Hector, A. et al. (2011). High plant diversity is needed to maintain ecosystem services. *Nature*, 477, 199-U196.

Jansson, J. and Prosser, J. (2013). Microbiology: the life beneath our feet. *Nature*, 494, 40–41.

Jones, C. G., Lawton, J. H. and Shachak, M. (1994). Organisms as ecosystem engineers. *Oikos*, 69, 373–386.

Kremen, C. (2005). Managing ecosystem services: what do we need to know about their ecology? *Ecology Letters*, 8, 468–479.

Kunin, W. E. and Gaston, K. J. (1993). The biology of rarity: patterns, causes and consequences. *Trends in Ecology and Evolution*, 8, 298–301.

Lavorel, S. and Grigulis, K. (1997). How fundamental plant functional trait relationships scale-up to trade-offs and synergies in ecosystem services. *Journal of Ecology*, 100, 128–140.

Lavorel, S. and Garnier, E. (2001). Aardvark to *Zyzyxia:* functional groups across kingdoms. *New Phytologist*, 149, 360–364.

Lavorel, S., Storkey, J., Bardgett, R. D. et al. (2013). A novel framework for linking functional diversity of plants with other trophic levels for the quantification of ecosystem services. *Journal of Vegetation Science*, 24, 942–948.

Maestre, F. T., Castillo-Monroy, A. P., Bowker, M. A. et al. (2012). Species richness effects on ecosystem multifunctionality depend on evenness, composition and spatial pattern. *Journal of Ecology*, 100, 317–330.

Maggi, E., Bertocci, I., Vaselli, S. et al. (2009). Effects of changes in number, identity and abundance of habitat-forming species on assemblages of rocky seashores. *Marine Ecology Progress Series*, 381, 39–49.

Maggi, E., Bertocci, I., Vaselli, S. et al. (2011). Connell and Slatyer's models of succession in the biodiversity era. *Ecology*, 92, 1399–1406.

Magurran, A. E. (2004). *Measuring Biological Diversity*. Oxford: Blackwell.

McKie, B. G., Schindler, M., Gessner, M. O., et al. (2009). Placing biodiversity and ecosystem functioning in context: environmental perturbations and the effects of species richness in a stream field experiment. *Oecologia*, 160, 757–770.

Mouillot, D., Villeger, S., Scherer-Lorenzen, M. et al. (2011). Functional structure of biological communities predicts ecosystem multifunctionality. *PLoS ONE*, 6, e17476.

Naeem, S. (2006). Expanding scales in biodiversity-based research: challenges and solutions for marine systems. *Marine Ecology Progress Series*, 311, 273–283.

Naeem, S. (2012). Ecological consequences of declining biodiversity: a biodiversity-ecosystem function (BEF) framework for marine systems. In *Marine Biodiversity and Ecosystem Functioning: Frameworks, Methodologies and Integration*, ed. M. Solan, R. Aspden and D. M. Paterson. Oxford: Oxford University Press, pp. 34–51.

Naeem, S. and Wright, J. P. (2003). Disentangling biodiversity effects on ecosystem functioning: deriving solutions to a seemingly insurmountable problem. *Ecology Letters*, 6, 567–579.

Naeem, S., Thompson, L. J., Lawler, S. P. et al. (1994). Declining biodiversity can alter the performance of ecosystems. *Nature*, 368, 734–737.

Naeem, S., Bunker, D. E., Hector, A. et al. (2009). *Biodiversity, Ecosystem Functioning, and Human Wellbeing: An Ecological Perspective.* Oxford: Oxford University Press.

O'Connor, N. E. and Crowe, T. P. (2005). Biodiversity and ecosystem functioning: distinguishing between effects of the number of species and their identities. *Ecology*, 86, 1783–1796.

O'Gorman, E. J., Yearsley, J. M., Crowe, T. P. et al. (2011). Loss of functionally unique species may gradually undermine ecosystems. *Proceedings of the Royal Society B: Biological Sciences*, 278, 1886–1893.

Ormond, R. F. G. (1996). Marine biodiversity: causes and consequences. *Journal of the Marine Biological Association of the United Kingdom*, 76, 151–152.

Petchey, O. L., O'Gorman, E. J. and Flynn, D. F. B. (2009). A functional guide to functional diversity measures. In *Biodiversity, Ecosystem Functioning, and Human Wellbeing: an ecological perspective*, ed. S. Naeem, D. E. Bunker, A. Hector, M. Loreau and C. E. Perrings. Oxford: Oxford University Press, pp. 49–59.

Peters, R. H. (1991). *A Critique for Ecology.* Cambridge: Cambridge University Press.

Piraino, S., Fanelli, G. and Boero, F. (2002). Variability of species' roles in marine communities: change of paradigms for conservation priorities. *Marine Biology*, 140, 1067–1074.

Raffaelli, D. G. (2006). Biodiversity and ecosystem functioning: issues of scale and trophic complexity. *Marine Ecology Progress Series*, 311, 285–294.

Raffaelli, D. and Friedlander, A. M. (2012). Biodiversity and ecosystem functioning: an ecosystem-level approach. In *Marine Biodiversity and Ecosystem Functioning: Frameworks, Methodologies and Integration*, ed. M. Solan, R. Aspden and D. M. Paterson. Oxford: Oxford University Press, pp. 149–163.

Schiel, D. R. (2006). Rivets or bolts? When single species count in the function of temperate rocky reef communities. *Journal of Experimental Marine Biology and Ecology*, 338, 233–252.

Schläpfer, F., Schmid, B. and Seidl, I. (1999). Expert estimates about effects of biodiversity on ecosystem processes and services. *Oikos*, 84, 346–352.

Schmid, B., Balvanera, P., Cardinale, B. J. et al. (2009). Consequences of species loss for ecosystem functioning: meta-analyses of data from biodiversity experiments. In *Biodiversity, Ecosystem Functioning, and Human Wellbeing: an ecological perspective*, ed. S. Naeem, D. E. Bunker, A. Hector, M. Loreau, and C. E. Perrings. Oxford: Oxford University Press, pp. 14–29.

Smith, M. D. and Knapp, A. K. (2003). Dominant species maintain ecosystem function with non-random species loss. *Ecology Letters*, 6, 509–517.

Solan, M., Cardinale, B. J., Downing, A. L. et al. (2004). Extinction and ecosystem function in the marine benthos. *Science*, 306, 1177–1180.

Srivastava, D. S. and Vellend, M. (2005). Biodiversity–ecosystem function research: is it relevant to conservation? *Annual Review of Ecology Evolution and Systematics*, 36, 267–294.

Stachowicz, J. J., Bruno, J. F. and Duffy, J. E. (2007). Understanding the effects of marine biodiversity on communities and ecosystems. *Annual Review of Ecology and Systematics*, 38, 739–766.

Stachowicz, J. J., Graham, M., Bracken, M. E. S. et al. (2008). Diversity enhances cover and stability of seaweed assemblages: the role of heterogeneity and time. *Ecology*, 89, 3008–3019.

Steele, J. H. (1985). A comparison of terrestrial and marine ecological systems. *Nature*, 313, 355–358.

Steele, J. H. (1991). Can ecological theory cross the land-sea boundary? *Journal of Theoretical Biology*, 153, 425–436.

Steneck, R. S. and Dethier, M. N. (1994). A functional-group approach to the structure of algal-dominated communities. *Oikos*, 69, 476–498.

Suding, K. N., Lavorel, S., Chapin, F. S. et al. (2008). Scaling environmental change through the community level: a trait-based response-and-effect framework for plants. *Global Change Biology*, 14, 1125–1140.

Tilman, D., Wedin, D. and Knops, J. (1996). Productivity and sustainability influenced by biodiversity in grassland ecosystems. *Nature*, 379, 718–720.

Tyler, E. H. M., Somerfield, P. J., Vanden Berghe, E. et al. (2012). Extensive gaps and biases in our knowledge of a well-known fauna: implications for integrating biological traits into macroecology. *Global Ecology and Biogeography*, 21, 922–934.

United Nations (1992). *Convention on Biological Diversity*. Rome: United Nations.

Walker, B. H. (1992). Biodiversity and ecological redundancy. *Conservation Biology*, 6, 18–23.

Warwick, R. M. and Clarke, K. R. (1995). New 'biodiversity' measures reveal a decrease in taxonomic distinctness with increasing stress. *Marine Ecology Progress Series*, 129, 301–305.

Webb, T. J. (2012). Marine and terrestrial ecology: unifying concepts, revealing differences. *Trends in Ecology and Evolution*, 27, 535–541.

Westoby, M. (1998). A leaf-height-seed plant-ecology strategy scheme. *Plant and Soil*, 199, 213–227.

Winfree, R. and Kremen, C. (2009). Are ecosystem services stabilized by differences among species? A test using crop pollination. *Proceedings of the Royal Society B-Biological Sciences*, 276, 229–237.

Woodward, G., Ebenman, B., Emmerson, M., et al. (2005). Body size in ecological networks. *Trends in Ecology and Evolution*, 20, 402–409.

Worm, B., Barbier, E. B., Beaumont, N., et al. (2006). Impacts of biodiversity loss on ocean ecosystem services. *Science*, 314, 787–790.

Yachi, S. and Loreau, M. (1999). Biodiversity and ecosystem productivity in a fluctuating environment: the insurance hypothesis. *Proceedings of the National Academy of Sciences of the United States of America*, 96, 1463–1468.

Part II
Impacts of human activities and pressures

6 · *Marine fisheries and aquaculture*

ODETTE PARAMOR AND
CHRISTOPHER FRID

6.1 Introduction

Fisheries products are an important source of food for a large proportion of the world's population and account for around 6% of the total protein supply (FAO, 2012; Figures 6.1 and 6.2). Fish (in this chapter the term fish is used inclusively to represent fish, shellfish and other aquatic animals used as foodstuffs) are significantly more important in the diet of people in Low Income Food Deficit Countries (FAO, 2012; Beveridge *et al.*, 2013) but are also important in coastal areas where the majority of the world's population currently resides (Small and Nicholls, 2003; UN, 2010). As over half of the world's population is located within 200 km of the coastline, demand for food from marine systems is only likely to increase with increased urbanisation and an expected human population size exceeding 9 billion by 2050 (FAO, 2009). The question is how an increased food supply might be delivered, its scale, and in particular how it might be done in a sustainable manner cognisant of the need for the protection of biodiversity and healthy functioning ecosystems (Rice and Garcia, 2011; Frid and Paramor, 2012)?

Several decades of overfishing have resulted in dangerously low stock sizes for many of the world's wild commercial fish species and this has partly driven the rise in demand for marine aquaculture. Currently, more than 40% of all fish consumed is derived from aquaculture sources and this is likely to increase in the future (Beveridge *et al.*, 2013). The environmental impacts of both capture fisheries and aquaculture on marine systems, particularly in coastal areas, are significant and this has led to demands for better protection of these systems. However, against a background of an increasing human population size, it has been argued that these conservation demands are incompatible with the increasingly severe

Marine Ecosystems: Human Impacts on Biodiversity, Functioning and Services, eds T. P. Crowe and C. L. J. Frid. Published by Cambridge University Press. © Cambridge University Press 2015.

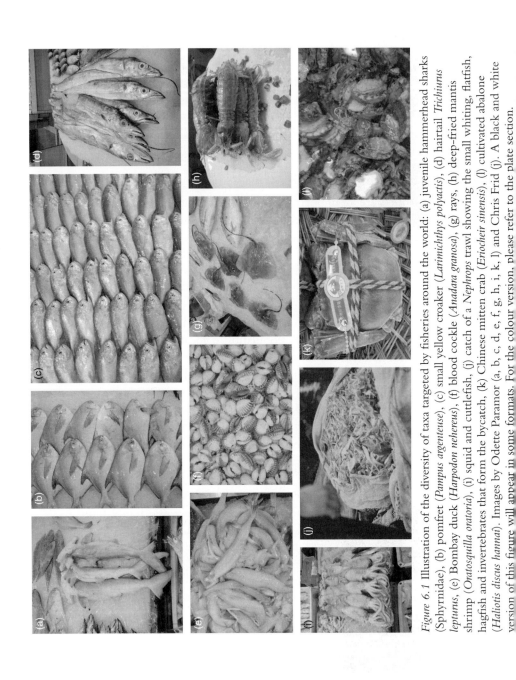

Figure 6.1 Illustration of the diversity of taxa targeted by fisheries around the world: (a) juvenile hammerhead sharks (Sphyrnidae), (b) pomfret (*Pampus argenteuse*), (c) small yellow croaker (*Larimichthys polyactis*), (d) hairtail *Trichiurus lepturus*, (e) Bombay duck (*Harpodon nehereus*), (f) blood cockle (*Anadara granosa*), (g) rays, (h) deep-fried mantis shrimp (*Oratosquilla oratoria*), (i) squid and cuttlefish, (j) catch of a *Nephrops* trawl showing the small whiting, flatfish, hagfish and invertebrates that form the bycatch, (k) Chinese mitten crab (*Eriocheir sinensis*), (l) cultivated abalone (*Haliotis discus hannai*). Images by Odette Paramor (a, b, c, d, e, f, g, h, i, k, l) and Chris Frid (j). A black and white version of this figure will appear in some formats. For the colour version, please refer to the plate section.

Figure 6.2 Some different fisheries and aquaculture approaches. Top row: (a) Scottish trawlers in port at Campbelltown, (b) mussel dredgers on the Irish Sea, (c) bivalve cultivation Ningbo, Zhejiang, PRC. Bottom row: (d) Zhoushan Island, Zhejiang Province, PRC during the closed fishing season, (e) mussels dredged from the Irish Sea, (f) the inner harbour at Mevagissey with lobster/crab pots, inshore fishing boats and pleasure craft. Images by Odette Paramor (c, d), Chris Frid (a, f) and Tasman Crowe (b, e). A black and white version of this figure will appear in some formats. For the colour version, please refer to the plate section.

global food security situation and the strong interest from governments in investigating the potential of marine systems to deliver further food resources (Rice and Garcia, 2011).

Globally, the demand for marine foodstuffs is not equal. The highest increase in the consumption of marine products has been in Oceania and Asia, and especially in China where consumption increased from 11 g to 69 g per person per day over the period 1963–2003 (Stentiford et al., 2012). China currently produces >40 kg y^{-1} of fish protein for each of its citizens and marine food products are extremely important to its national food security (Li, 2013).

6.2 Impacts of capture fisheries on ecosystems

It is generally accepted that there are no major new fishing grounds to be exploited (Godfray et al., 2010) and therefore existing levels and patterns of exploitation are likely to continue. It is impossible to conduct a fishery with no environmental impact; the fishery adds another predator and source of mortality to the fish population. Even moderate rates of exploitation are likely to impact the population dynamics of the exploited species and may even drive evolutionary changes (Laws and Grey, 1989; Law, 2000). However, most fisheries will also impact on other components of the ecosystem either directly (such as by capturing/killing non-target species or altering habitats) or indirectly (by altering food webs, reducing food for, and competing, with higher predators and reducing predation pressure on the prey of the target species).

There have been significant advances in our understanding of the impacts of fishing on marine systems driven by the introduction of policies which have required an 'ecosystem approach' to management, such as the 2002 iteration of the EU Common Fisheries Policy which required that ecosystem-based approaches be used to manage fish stocks (EC, 2002). This has driven much fisheries research to become more interdisciplinary and placed traditional fisheries management in the context of the wider ecosystem.

Many of the effects of fishing on ecosystems are widely known (see Jennings and Kaiser, 1998, for a review) and the key impacts relate to

(1) the biological effects of the removal of a species on the community in which it occurs and
(2) the physical impacts on the habitat. Whilst these impacts and the link between exploitation and production are now well documented, the

wider impacts of fishing and aquaculture on ecological functioning, and hence ecosystem services, are less well understood.

Some effects are measurable, but many are not and knowledge on how particular species contribute to ecological functioning and deliver ecosystem services is poorly understood for many marine taxa, especially infaunal benthos and oceanic plankton and nekton. The direct impacts of fishing on population size and age/size structure are generally the easiest to survey. Changes in habitat extent and 'quality' are much more difficult. We do not understand the ecological implications of, for example, large boulders sitting on the seabed in an area of otherwise fine sediments and the long-term consequences of their removal by fishermen to prevent gear snagging.

6.2.1 Targeted fish stocks

The effects of the removal of large quantities of the adult stages of a specific fish species are well documented. Fishing has led to changes in the size and maturation rate of populations so that in recent years the mean length of populations of many commercial fish are now smaller and the fish mature faster than those at the start of the last century and have a reduced genetic diversity. This is a common observation across the world, for example with cod stocks in the Atlantic (Olsen et al., 2004) and large yellow croaker (*Larimichthys crocea*) in the East China Sea (Lin et al., 2010). The impact of the change in body size on food webs has been modelled and suggests that whilst fishing has increased overall productivity, the species which are benefiting most are the small, short-lived (and often gelatinous) ones that do not contribute to food production services (Pauly et al., 1998; Richardson et al., 2009).

Widespread overexploitation, over several decades, threatens the long-term delivery of food services. Fish stocks that are exploited above their maximum sustainable yield, even if not in danger of collapse, are providing less food than they could to feed a hungry world. The UN Food and Agriculture Organization (1995) has adopted maximum sustainable yield (MSY) as the management objective for all fisheries (FAO, 1995), however it is ecologically impossible for all species to be exploited at MSY due to the ecological linkages between them. For example, exploiting capelin in the North Atlantic at its MSY would lead to the insufficient availability of food for the development of cod stocks to a level where they could support MSY. These species interactions are for the most part poorly understood (cod–capelin being an exception; Bogstad and Gjøsæter, 1994) and will be a challenge in meeting the UN target. Maximising the overall delivery of the food provisioning service from a given area requires that these considerations are taken into account.

6.2.2 Food webs

Many of the most valuable commercial fish species are predators (e.g. cod and tuna) and the targeted removal of large predators from the ecosystem triggers a complex series of indirect responses from the populations of their prey and other competing predators. The effect of fisheries on the size structure of fish communities is well recognised (Rice and Gislason, 1996) as are the consequences for food web dynamics (Pauly et al., 1998).

The physical act of fishing may also change the exposure of various taxa to predation (e.g. discarding of dead or moribund organisms that provide 'easy pickings' for scavengers (Bicknell et al., 2013) and the exposure of benthos on the seafloor to predators (Kaiser et al., 1998b)) or death through their sensitivity to exposure to aerial environments or damage as they pass through the net. The practice of discarding has resulted in increases in the population size and density of species which exploit these predictable anthropogenic food subsidies (PAFS) and the activity has affected a range of ecological processes and trophic levels (Oro et al., 2013). Isotopic evidence has shown that dietary changes in the great skua (*Stercorarius skua*) in the North-east Atlantic were likely to have resulted from an increase in herring and mackerel in their diet, which was attributed to the discarding activities of commercial trawlers (Bearhop et al., 2001). This exploitation of PAFS released populations of the great skua's natural prey from predation for several decades and there are now concerns about how a discard ban being introduced in the latest iteration of the EU Common Fisheries Policy may affect the populations of these smaller bird species (Bicknell et al., 2013).

The impact of fishing on biodiversity and hence ecological functioning is complex and food webs may respond in a variety of ways to this pressure. The removal of large fish predators may cause a competitive release in other predators such as sea birds, and the removal of krill-eating whales from the Antarctic may benefit seals and penguins (Fraser et al., 1992). It may also lead to a predation release in their prey which may lead to increased predation pressure in the next lower trophic level, a so-called 'trophic cascade' (Carpenter et al., 1985).

In addition to the impacts of fishing itself, the fishing industry also contributes large quantities of marine litter to the oceans. This potentially impacts the ecosystem through continued direct mortality from discarded or lost fishing equipment that continues to capture fish (so-called 'ghost fishing') (Matsuoka et al., 2005), while the plastics eventually degrade to microplastic particles that can act as nuclei for the concentration of harmful pollutants (Thompson et al., 2004). There is a high level of uncertainty surrounding the impacts of plastics on marine food webs (Chapter 9). One recent study has shown that species with the highest concentrations of ingested plastic were thought to be primarily mesopelagic and therefore unlikely to come into contact with surface waters where the highest concentrations of plastic debris occur (Choy and Drazen, 2013). Also, the quantities of ingested plastics vary among species, ranging, for example, from <1% of *Gempylus serpens* (snake mackerel) to 58% of *Lampris* sp. (small eye) found with plastics in their gut (Choy and Drazen, 2013).

6.2.3 Benthic systems

The severity of fishing impacts on benthic systems is highly dependent on the gear used, the history of human activity in the area, the natural tolerance of the habitat to physical disturbance and the biological communities it supports.

Different fishing gears affect different parts of the physical marine system. Some gears operate solely in the water column (e.g. pelagic trawls, pole and long lines), with minimal impact on the physical habitat, whilst others operate close to the seafloor with occasional contact (e.g. demersal otter trawls) and some are designed to be dragged across the seafloor with a high level of contact disturbing both the biota and sediments (e.g. scallop dredgers). Some gears have also been designed to deliberately 'dig out' bivalves or encourage flat fish to swim up and into nets (e.g. beam trawls fitted with tickler chains). Those gears with a high level of contact with the seafloor have the highest impact on benthic functioning as they dislodge and/or damage benthic organisms, including key habitat-forming species, and have been observed to alter benthic macroinvertebrate communities leading to increases in the number of scavenger species who feed on the dead, moribund or dazed organisms brought to the surface of the sediment (Kaiser, 1998; Kaiser et al., 1998a; Kaiser and de Groot, 2000; Kaiser et al., 2003; Kaiser et al., 2006). These gears also cause the release of plumes of sediment into the water column and expose subsurface sediments and microbial communities to oxygenated water thereby altering the productivity of the system through increased oxygenation (Hiddink et al., 2006).

Fishing is not a random process as fishers tend to return to the same areas repeatedly (Rijnsdrop et al., 1998). The impact of individual fishing events on benthic habitats in heavily fished areas is likely to be much less than in unfished habitats as it is the first exposure to fishing which causes the most damage to the benthos (Cook et al., 2013). Repeated disturbance events will have a lower individual impact but may result in changes to the benthic communities away from those least tolerant of physical disturbance (e.g. large, physically fragile sessile organisms; Eleftheriou and Robertson, 1992) to those which are more robust (e.g. small organisms such as some worm species; Thrush et al., 1998; Freese et al., 1999). This is one of the reasons why there are so many concerns about the expansion of fishing activities into deep-water areas as the impact of fishing on relatively 'pristine' habitats which have not previously been exposed to this type of physical disturbance is likely to be significant. Similarly, the introduction of closed areas to fishing also needs to be managed sensitively to avoid fishers transferring their effort to relatively undisturbed areas (e.g. the transfer of fishing effort from the centre to the borders of the North Sea plaice box; Beare et al., 2010).

The impact of fishing on benthic communities is likely to be greatest in those habitats with naturally low levels of disturbance, such as those areas with a soft muddy substrate where the movement in the water is so slow as to allow fine particles to drop out of the water column, or in biogenic habitats which are formed by living organisms and which may be irreparably damaged by a single physical impact. The impacts of bottom contact gears are most obvious in habitats with erect, epibenthic forms such as *Lophelia* reefs which show obvious mechanical damage in trawled areas (Fosså et al., 2002) and much attention has been focused on these relatively rare reef habitats (Fosså et al., 2002; Kamenos et al., 2004; Hall-Spencer et al., 2009; Cook et al., 2013). However, it is the more spatially extensive habitats (including mud, muddy sand and sand banks) which are likely to deliver a greater 'quantity' of most ecosystem services simply due to their absolute size (Hussain et al., 2010). The impacts of fishing in these habitats can be much more difficult to detect partly due to a lack of suitable data, but also changes in the communities which occur in highly dynamic sedimentary systems, such as those found in areas of coarse sand, may not be measurable as fishing may create less disturbance than the natural background (Eleftheriou and Robertson, 1992).

The impact of fishing disturbance on the ecological functioning of the seafloor is uncertain although several studies using biological traits analysis (BTA) have shown that there is the capacity for a change in ecological functioning with changes in benthic invertebrate communities (Bremner *et al.*, 2003b; Frid, 2011; De Juan and Demestre, 2012). BTA involves an assessment of a taxa's traits related to their behaviour, body structure, life history and living location over their whole life cycle and making assumptions about how these traits may then influence specific ecological functions such as the carbon cycle. For instance, an active, burrowing, deposit-feeding worm living in the top 10 cm of sediment may impact the carbon cycle by drawing oxygen down into its burrow, defecating in the burrow and bioturbating sediment grains, all of which may increase microbial decomposition of organic matter in subsurface layers and enhance the release of nutrients back into the environment. The traits of benthic invertebrates which are affected by fishing include body size (Bremner *et al.*, 2003a), body design (e.g. round, erect, flat), living location (e.g. surface, subsurface, interface) and movement (e.g. swimming, crawling, burrowing) (Paramor, unpubl.). Thus, if such species are lost or reduced in abundance, the rate of nutrient cycling in the system could be expected to be altered. However, we currently lack empirical evidence directly linking changes in biological traits with changes in functioning.

6.4 Impacts of aquaculture on marine ecosystems

Globally, aquaculture production is predicted to increase over the next few decades, fuelled by the demands of the increasing human population and the inability of wild, capture fisheries to expand. Patterns of demand are likely to be driven by both global food security issues, particularly in the less economically developed regions of the world, and the growth of the middle classes in expanding middle-income countries such as China where there is an increasing demand for high quality aquaculture food and traditional medicine products. In 2008, 80.2% of the fish consumed by the Chinese population was derived from aquaculture, whereas the figure was 23.6% in 1970 (FAO, 2010). Demand is affected by issues such as assurances for the consumer about the sustainability of production (e.g. Aquaculture Stewardship Council, 2013) and the safety of the product for human consumption (e.g. EU Directive 2002/69).

The majority of marine aquaculture occurs in Asia and globally China is the largest single producer of aquaculture products (Figure 6.3). Whilst the diversity of organisms used in marine aquaculture is broad and ranges from the high value commercially cultivated products such as tuna,

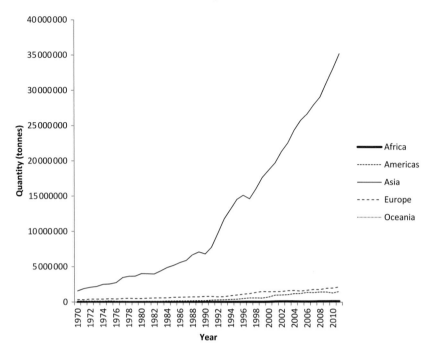

Figure 6.3 Total marine aquaculture production by continent from 1970–2011. Global trends in marine aquaculture production show that Asia has significantly increased its annual production rate since the 1970s, while production increases in the other continents have been more modest and follow in the order of Europe, the Americas, Oceania and Africa. Based on data from FAO (2013a).

salmon, abalone and sea cucumbers to the lower value fish species cultivated by families for their own use, globally, the top five marine groups cultured are (in descending order) marine plants, molluscs, diadromous fish, marine fish and crustaceans (Figure 6.4).

The impact of marine aquaculture on the environment is highly dependent upon the type of organism being cultivated and the location of the activities (e.g. in tanks on land, in open intertidal areas or on/in offshore structures) although there are some common impacts related to waste and husbandry practices required to keep the organisms healthy and disease-free. As aquaculture involves the cultivation of organisms at high densities with the aim of ensuring that the individuals increase in size and weight over relatively short periods of time, this frequently involves the use of artificial nutrients and feeds, and cultivars and breeds that have been developed for their speed of growth. The impacts of

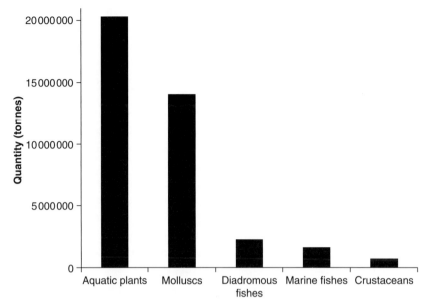

Figure 6.4 The most common globally cultured organisms grown in marine conditions in 2011. Based on data from FAO (2013a).

both of these on the marine environment are well documented and are described below. Again, the natural background level of disturbance and through flow of water will affect the magnitude of the impact (Keeley et al., 2013).

6.4.1 Organic wastes

The cultivation of high densities of organisms results in a high density of waste product, both from the organisms themselves and any unutilised feed (Gowen and Bradbury, 1987; Holby and Hall, 1991; Hall et al., 1992; Cheshuk et al., 2003). These waste products affect both pelagic and benthic habitats, and may cause a decline in water quality with elevated levels of ammonium and reduced dissolved oxygen concentrations and deposition of organic matter over benthic habitats (Kalantzi and Karakassis, 2006; Chapter 8). The impact of these nutrient inputs on the environment and the spatial extent of their impacts are strongly affected by the physical characteristics of the area. It has been observed that the concentrations of chemical pollutants and the impacts of nutrients from fish farms may decline rapidly as a function of distance from the cage edge, with the rate of decline dependent on local water current speeds and directions (Mayor and Solan, 2011; Callier et al., 2013).

It is generally agreed that when aquaculture is conducted in areas with a low water flow, the areas immediately beneath and surrounding the farm are likely to be heavily impacted, but the spatial extent of the impact is limited to a few tens of metres (Lumb, 1989). The observed benthic impacts of aquaculture under these conditions are classically an increased number of opportunistic species and individuals, a decline in species unable to tolerate the higher quantities of particulate matter and the higher nutrient conditions, and often lead to the development of azoic conditions. In terms of functioning, this response would suggest that the natural benthic communities in these low energy sites cannot effectively assimilate the additional organic matter from the aquaculture activities even though their biological traits may be suited to areas with a naturally higher particulate load than communities which occur in higher flow areas.

The impact of aquaculture on areas with a high water flow are generally thought to be less than under low flow conditions as the nutrients and organic materials are dispersed relatively quickly into the wider environment, although any impacts in this case are likely to cover a wider spatial extent as the pollutants are transported over a greater area (Bostock et al., 2010). One of the more extreme examples of a spatially extensive aquaculture impact occurred in the Yellow Sea in China following the deliberate addition of 50 000 tonnes of fermented chicken manure to coastal aquaculture ponds which were adjacent to frames used for the cultivation of the commercially important red alga *Porphyra* spp. (commonly known as nori or laver). This led to the world's largest observed bloom of *Ulva prolifera* in 2008 (Liu et al., 2010). The fermented chicken manure was added to the water in Jiangsu Province from where the nutrients and the initial *Ulva* bloom were transported 180 km up the coast to the site of the Olympic sailing event at Qingdao immediately before the 2008 Olympic games. *Ulva* is a fouling organism and it is estimated that 4900 tonnes were removed in this instance from the *Porphyra* cultivation frames. The cost of the emergency clean-up operation exceeded US$100 million, approximately double the size of the profits from the *Porphyra* industry in Jiangsu Province (Hu et al., 2010). Whilst the phenomenon has been repeated every year since, occurring in the same region, during the same period of the year and with the same cause, the wider environmental impacts of the blooms have not been investigated.

The effects of aquaculture on benthic systems and their functioning in areas with a high water flow are not quite so simple to predict and studies conducted in these environments have reported conflicting results ranging from a relatively low or no measureable impact (Frid and Mercer, 1989; Keeley et al., 2013) to visually observable impacts with organic waste materials becoming trapped in complex structures on the seafloor (Hall-Spencer et al., 2006; Mayor and Solan, 2011). Mayor and Solan (2011) observed an increase in the organic carbon in sediments around the larger fish farms (>900 tonnes annual production) that may have been caused by particles becoming trapped in the coarse seabed substrates which are characteristic under high flow regimes. Taxa supporting traits unsuited to this more sedimentary environment would therefore be impacted by the presence of a farm despite the high flow regime. The organic carbon in sediments surrounding the smaller farms in high flow areas behaved differently and was observed to decrease rapidly away from the sites.

A study from New Zealand that investigated aquaculture sites under a variety of flow regimes found that benthic macrofaunal assemblages under high flow conditions were relatively resilient to enrichment from aquaculture sources (Keeley et al., 2013). The high flow site in this study had extremely high densities of macrofauna (1.5 million individuals per m^2) and supported natural populations of opportunistic species at low to moderate abundances. In this situation, the resilience of the community to enrichment was possibly enhanced because of the natural abundance of opportunistic species that could assimilate the additional organic material.

One of the concerns for conservationists is that areas with high water flow may also support rare and slow-growing organisms, such as maerl, which function as a habitat supporting diverse invertebrate communities and are also important habitat and feeding grounds for commercial species such as scallops (Kamenos et al., 2004; Hall-Spencer et al., 2006; Hall-Spencer and Bamber, 2007). Studies have shown that aquaculture operations located over or near maerl beds have resulted in observable declines in live maerl cover. This resulted from the entrapment of particulate matter and caused significant declines in the diversity of the crustacean communities, particularly ostracods, isopods, tanaids and cumaceans in areas near the fish cages (Hall-Spencer et al., 2006). As maerl is slow-growing, recovery of these sites after the cessation of aquaculture activities, if it occurs at all, would be very slow.

6.4.2 Escapees

The accidental or deliberate release of cultured organisms into the wild can have widespread ecological consequences dependent upon how they behave under natural conditions and how they may alter important fitness-related species traits through interbreeding with potentially small and increasingly rare wild populations (Youngson et al., 2001; Jonsson and Jonsson, 2006). Most of the research in this area has been conducted on the Atlantic salmon (*Salmo salar*) as it is one of the most commonly cultured species in Europe and there have been significant numbers of escapees.

The scale of the loss of cultured individuals from salmon farms can lead to biologically important effects. Samples obtained from the commercial fishery in the Faroe Islands showed that the proportion of formerly cultured fish in the 'wild' catch was relatively low from 1980/1981 to 1986/87, but in 1991/1992 rose to 40% of the total catch before declining back to 20% (Hansen et al., 1999). Of particular concern is the development of transgenic organisms specifically for aquaculture purposes. The US Food and Drug Administration (FDA) is currently considering the introduction of transgenic salmon which grow at twice the speed of their unmanipulated conspecifics (Ledford, 2013). The FDA does not consider that transgenic fish pose a significant environmental threat to the United States provided they are grown and properly secured in landlocked tanks. China is also investing heavily in this technology.

In the laboratory, comparisons of the behaviour of cultured, hybrid and wild Atlantic salmon have shown that the cultured fish had a higher growth rate, matured faster, were more aggressive, bold and likely to take risks compared to the wild individuals (Einum and Fleming, 1997; Fleming and Einum, 1997). They were also less fertile and produced fewer eggs than the wild individuals (Jonsson and Jonsson, 2006). Observations from the wild showed that whilst the survival rate of cultured individuals was less than for wild individuals (Youngson et al., 2001), the two groups did compete for resources such as territory, food and breeding partners (Einum and Fleming, 1997; Jonsson and Jonsson, 2006). Aside from the environmental and genetic impacts, these inflated stock sizes may result in overestimates of the size of the wild stocks of salmon thereby compounding overfishing issues.

The main impact of cultured escapee salmon on marine systems is through the food web (Jonsson and Jonsson, 2006). Stomach content analysis of salmon from the North Atlantic showed dietary differences between the two groups. In coastal areas, amphipods formed the most abundant item in the stomachs of cultured post-smolts, whereas it was krill (euphausids) for the wild post-smolts. Sandeels were twice as abundant in the stomachs of cultured post-smolts than of their wild counterparts. These dietary differences may alter the dynamics of predator–prey interactions particularly if the number of the formerly cultured individuals becomes greater than of the wild salmon. The dietary differences may also expose individuals to different levels of risk from predators during their feeding periods.

6.4.3 Microbial pathogens and parasites

Microbial pathogens cause significant stock losses in aquaculture globally, and their impacts can severely restrict the development of aquaculture in new locations or using new species (Rigos and Katharios, 2010). For instance, viral pathogens are responsible for the loss of up to 40% of tropical shrimp production annually (>US$3 billion) (Stentiford et al., 2012). The impacts of these aquaculture-related pathogens on the wider ecosystem are largely unknown. Although evidence of a transmission route from aquaculture stocks to wild populations is weak, video evidence has shown how wild fish in a fjord system spend significant amounts of time in close proximity to aquaculture farms and could potentially act as vectors for disease both between wild and captive fish stocks and between different farms (Uglem et al., 2009).

Evidence supporting the transmission of parasites between wild and farmed fish stocks is much stronger and marine salmon farming has been correlated with parasitic sea lice infestations and concurrent declines of wild salmonid stocks (Krkošek et al., 2005; Krkošek et al., 2007; Krkošek et al., 2013). Experiments have shown how a single salmon farm altered the transmission dynamics of sea lice to wild juvenile Pacific salmon by increasing the infection pressure by four orders of magnitude above natural levels (Krkošek et al., 2005). This same study showed that the infection pressure in areas immediately around the farm was 73 times greater than ambient levels and exceeded ambient levels for 30 km along the wild salmon migration corridors. The impact was further exacerbated by the production of a second generation of lice which reinfected the juvenile salmon thereby increasing the infection pressure over ambient levels for 75 km of the migration routes.

Potentially, aquaculture stocks can therefore affect wild populations by acting both as vectors of disease if they are introduced to new geographic areas or by changing the host–parasite dynamics of a site by increasing the number of available hosts (Bergh, 2007). These impacts will affect the functioning of natural systems and may be an important limiting factor in the conservation of some of these wild populations.

6.4.4 Antimicrobials and biocides

Husbandry practices to keep cultured organisms healthy include the addition of antimicrobials and biocides to marine systems to prevent and treat the microbial pathogens and parasites which can arise as a consequence of lowered host defences associated with high density aquaculture practices and sub-optimal hygiene. This is a major industry and the global salmonid aquaculture industry alone spends an estimated €300 million annually on biocides to control sea lice infestations (Costello, 2009). Many authors argue that disease will be the limiting factor affecting food supply from the aquaculture sector in the future (Bachère *et al.*, 2004; Robertson *et al.*, 2009; Stentiford *et al.*, 2012).

The release of large quantities of antimicrobials into the environment can potentially seriously affect microbial diversity and functioning. Studies indicate that approximately 80% of the antimicrobials used in aquaculture enter the environment with their activity still intact (Cabello *et al.*, 2013). Once released, the antimicrobials can induce selection for those bacteria whose resistance arises from mutations or for mobile genetic elements containing multiple resistance determinants transmissible to other bacteria (Cabello *et al.*, 2013). A study of a Chilean salmon farm detected significant quantities of antimicrobial resistance at distances up to 1 km away from the aquaculture site (Buschmann *et al.*, 2012). The effect of antimicrobials on the functioning of natural microbial communities is not well studied as most research in this area is focused on the investigating the effectiveness of antimicrobials in aquaculture systems and how they may potentially impact on human health. In addition to this, there is a high level of uncertainty around which antimicrobials are entering marine systems and in what quantities. Regulations, and enforcement of the regulations, surrounding the use of antimicrobials is highly variable globally and is a particular concern for the international trade in seafood products.

The main biocides used to treat sea lice infestations (*Lepeophtheirus* spp. and *Caligus* spp.) in farmed salmonids include organophosphates, pyrethrins, hydrogen peroxide or benzoylphenyl ureas (Robertson et al., 2009). The use of these medicines is highly regulated due to their non-specific toxicity and concerns about the potential environmental impacts. There are a limited number of studies on the effects of these medicines in the environment so knowledge of their impact on functioning is largely unknown. One study on the behaviour of a common sea lice treatment for salmon, emamectin benzoate, suggested that its impact on the macrobenthos was limited with no major changes in community composition resulting from its use (Telfer et al., 2006).

6.4.5 Antifouling

The cages and frames used for aquaculture tend to attract fouling organisms that may then compete with the cultured organisms for resources or compromise the infrastructure and operation of the farm. Significant efforts are made to keep the structures clear of fouling organisms and the direct economic costs of antifouling measures are substantial and estimated to be in the region of 5–10% of the production costs (Lane and Willemsen, 2004). Globally, this equates to costs of US$1.5–3 billion annually (Fitridge et al., 2012).

Fouling organisms tend to be removed mechanically in shellfish and fish aquaculture practices, whilst copper coatings on fish nets are the only consistently effective form of antifouling at an industrial scale (Thomas and Brooks, 2010; Fitridge et al., 2012). There is a huge demand for the development of a non-toxic, low energy antifouling coating for aquaculture purposes (Myan et al., 2013).

Antifouling biocides are commonly used in aquaculture but the active ingredients used cover a wide range of chemicals with different properties and which behave very differently in the environment (Thomas and Brooks, 2010). There have always been concerns about using biocides for aquaculture purposes due to the wider environmental impacts including the bioaccumulation of heavy metals by humans and other predators of the cultured organisms, and the development of antibiotic resistance in bacteria (Guardiola et al., 2012). However, there are a few antifouling biocides which have been used for decades and there is a good understanding of how they and their metabolites behave in the environment (e.g. Irgarol 1051 and Diuron) (Thomas and Brooks, 2010). The behaviour of most aquaculture antifouling biocides in the environment and how they may affect ecosystem functioning are unknown (see Chapter 9 for consideration of impacts of chemical contaminants in general).

6.4.6 Sources of aquaculture feed

Aquaculture is often seen as the solution to overfished wild stocks. However, many of the economically valuable species used in aquaculture require high quality feeds containing high levels of marine lipids and proteins. Manufactured feeds are usually used for the cultivation of piscivorous fish and crustaceans, and the fishmeal and oil used to formulate their diets are derived from several sources, including fisheries targeting species not for human consumption (usually small pelagic fish such as the Peruvian anchoveta (*Engraulis ringens*), landed non-marketable fish (i.e. where discarding is restricted) and waste (trimmings) from fish processing (Frid and Paramor, 2012). Hence, aquaculture and wild capture fisheries are not entirely independent sources of dietary material.

According to the International Fishmeal and Fish Oil Organisation, the global production of fishmeal and fish oil is stable and the use of whole fish in fish feed is decreasing (Natale *et al.*, 2012). In 2009, around 63% of fishmeal and 81% of fish oil was used in aquaculture feed. Fishmeal was used almost equally in the feed for salmonids, marine fish, crustaceans and other species, whereas more than 66% of the fish oil went into salmonid feeds with around a further 20% to marine fish. The same report observed that in 2009 about 25% (around 4.8 million tons) of global fishmeal production was derived from the by-products of marine fish processing.

Some authors have argued that the declining stocks of small pelagic fish which are used for fish feed will hinder the development of certain types of aquaculture, although this may be balanced by the discard ban which has been proposed for the revised EU Common Fishery Policy and may increase the availability of 'bycatch' material for feed purposes (Natale *et al.*, 2012). Regardless of these developments, the demand for alternative feed materials will increase in the near future. Significant efforts have been made in recent years, for example, to investigate alternative sources of high quality n-3 PUFAs for use in aquaculture feed including moves towards the use of terrestrial plant materials (CFFRC, 2012; Liland *et al.*, 2013) and specially cultured rotifers as crab feed (Liu *et al.*, 2013).

The impact of the fishmeal and fish oil industries on the ecosystem and how it functions is significant as they operate at the base of the food chain, reducing the biomass of feed fish which are important prey items for sea birds, piscivorous fish and marine mammals.

6.4.7 Invasive species

The introduction of non-native species for aquaculture purposes may have a significant impact on ecosystem dynamics if they escape or are released and behave as an invasive species. For example, the seaweed *Undaria pinnatifida* (commonly known as wakame) is grown commercially around the world but is highly invasive as it grows rapidly and has the potential to outcompete native species of macroalgae (FAO, 2013b). In Tasmania, its presence may have altered the food resources of herbivores that would normally consume native species. *Undaria* also has the potential to become a problem for marine farms by increasing labour costs due to fouling problems.

Mandatory biosecurity management measures introduced at national (US) and European levels have meant that within these regions there has been a marked decrease in new introductions of marine invasive from aquaculture (half of new marine invasive species in Europe are introduced via shipping) (Katsanevakis *et al.*, 2013). The same is not true for other areas of the world, such as Asia, where monitoring of invasive species is often lacking.

6.5 Key knowledge gaps, their implications and future research needs

Since the introduction of the ecosystem approach to management in European fisheries, significant resources have been devoted to investigating the impacts of fishing on various ecosystem components. The same cannot be said in many other areas of the world, including many that are heavily reliant upon fisheries resources and where there is, therefore, an urgent need to incorporate information collected, and lessons learnt, elsewhere into fisheries management plans.

Demand for food from marine systems is likely to increase in the future and it is important that these needs are met in a sustainable manner. Most commercial wild fish stocks are already fished at or beyond safe limits and the only real space left for expansion is in the aquaculture sector. However, to achieve this in a sustainable manner requires the decoupling of aquaculture from wild systems for the supply of feed and juvenile stock.

Whilst there is a huge body of research on the aquaculture sector, there have been comparatively few studies on how the industry impacts on ecosystems dynamics, particularly in those regions with the highest aquaculture production and often the lowest levels of enforced regulation.

The aquaculture industry is driven by consumer receptiveness to its products and the economics and price of the operation. The perception of aquaculture products is affected by the consumers' belief in the safety, quality and in some cases, the environmental ethics of the products. For instance, since EU Resolution 2002/69 which regulates the levels of dioxins in foodstuffs, Chinese seafood products have been closely scrutinised within Europe and this has led to further scrutiny from its other major markets in the United States, Japan and Korea. The Chinese government's recognition that it needed to improve the public health 'safety' of the food stuff and reduce the environmental impact of aquaculture has led to the introduction of 'Law on the quality of and safety of agricultural products of the P. R. China' (Sun and Che, 2012).

6.5.1 Functioning

The linkages between most marine taxa and the functions they deliver in the ecosystem are still, largely, to be elucidated and there is much uncertainty in this area, even for the most studied species (Cesar and Frid, 2012, Tyler *et al.*, 2012).

There are considerable opportunities for further research investigating the within-sediment impacts of fishing and how it affects biogeochemical cycles. For instance, how has the harvesting of benthic fauna such as cockles affected the chemical and sedimentary processes within estuaries? A limited number of studies have shown how disturbance can trigger alterations in the behaviour of benthic taxa under these conditions and hence the ecological roles they deliver (Cesar and Frid, 2009; Cesar and Frid, 2012).

The impacts of aquaculture on functioning have largely been studied in relation to the organic loading from salmon farms but there have been few studies on the wider impacts of microbial pathogens and biocides on microbial communities and the functions they deliver. Also, while the ecological consequences of organic inputs are well established for some ecosystems, the impacts of the extensive aquaculture systems in Asia remain poorly characterised.

6.5.2 Ecosystem services and economic evaluations

Fundamental understanding of how individual species function within the environment to deliver ecosystem services is largely lacking although there are several projects underway that seek, at least in part, to address this. This means that there is currently a high level of uncertainty associated with the evidence base being used to support the economic evaluation of marine systems (Chapter 2). There is an urgent need to strengthen the evidence base and the quality of the valuations as they are increasingly being presented to policy makers and used to justify policy decisions. Fisheries and aquaculture probably present the most straightforward case in that the economic value of the harvest is directly measureable in the market place. However, this is overly simplistic, certainly for capture fisheries, where the catch, had it been left in the sea, may have contributed to other economically valuable services. For example, the Danish fishery targeting sandeels and other small pelagics for fish meal and oil production is worth around €83 million (2003 value), but the fishery yield was greater prior to restrictions being placed on it to protect breeding populations of kittiwakes. However, those restrictions have improved the breeding success of the sea birds and in turn delivered both direct economic benefits (ecotourism) and potentially economic value indirectly (e.g. existence value). Multi-species fisheries modelling also suggests that a 40% reduction in the take of sandeels in the North Sea would result in larger spawning stock biomasses and greater yields from many fisheries targeting species for human consumption (Gislason and Kirkegaard, 1998).

6.5.3 Integrated spatial management

The temporal and spatial extent of fishing means that this industry has a large footprint that is likely to overlap with many other activities and commercial sectors. As such, there is an urgent need to improve understanding of the combined impacts of fisheries and other activities (Chapter 4) so that informed decisions can be made about which combinations to allow in an area and which to restrict.

Currently, the scale and importance of the aquaculture industry varies considerably as do the regulations controlling its operations and impact on the environment. With the growing global population and increasing proportion of that population living in the coastal zone, the pressures on coastal habitats will increase considerably and bring increased conflict between sectors. This may constrain aquaculture development. Aquaculture requires land or sea space, and water of sufficiently high quality, both of which will be in limited supply and in some cases also required for agriculture. There is a clear scope for aquaculture growth in lower trophic-level species that are not constrained by fisheries dependence (e.g. shellfish). Balancing the demands of aquaculture (e.g. space, feed and processing costs) against other demands made on marine systems is a sensitive issue. The growing interest in marine spatial management, i.e. zoning, may be a key element of future attempts to limit conflict between sectors and ensure protection of ecosystem functioning.

6.5.4 Fundamental understanding of the aquaculture life cycle on global ecosystems

The impacts of aquaculture can be geographically widespread with the extraction of feed fish from one continent feeding the cultured piscivores on another continent before being sold as food to another continent. If aquaculture is to be developed sustainably, then the entire life cycle needs to be assessed to ensure that the industry is operating ethically, fairly and sustainably at each stage of the operation and based on the best available knowledge. Regions that do not have a good history of environmental protection are amongst the regions with the highest intensity of aquaculture. Aquaculture has a long history but has developed rapidly in the last few decades. It can in many ways be seen as an emerging sector that is growing ahead of the ability of regulators/environmental managers to control.

6.5.5 Antimicrobials

Marine food products are one of the most highly traded foodstuffs globally and concerns over the safety of products, particularly from China, still need to be addressed. The use of antimicrobials is not well regulated globally and our understanding of how they affect microbial diversity and the fundamental ecological/biogeochemical processes that underpin system functioning is poorly developed.

6.5.6 Aquaculture feeds

The scope for sustainable growth in the aquaculture of predatory fish is limited. There are three mechanisms for expansion while limiting ecological impacts: (1) better feed conversion ratios and more use of non-fish material in the diet, (2) increased supply of fish material from more effective waste-management strategies, and the removal of the ecologically damaging practice of discarding, and (3) the use of polyculture and the cultivation of low-value species to either feed directly to the stock or for entry into the fishmeal supply chain (e.g. filter-feeding shellfish). Extensive suspended rope aquaculture could, for example, be colocated with existing farms and with offshore wind energy sites, utilising areas of sea otherwise rendered off-limits to fisheries (Folke and Kautsky, 1992).

6.6 Key findings

- Demand for food and medicinal products from marine systems will increase in the future as the global human population continues to expand. As most wild capture fisheries are already exploited at maximum capacity (and in many cases beyond this), efficiencies and alternatives to these fisheries need to be investigated. The challenges facing global aquaculture are diverse but this is a rapidly expanding industry and there is an urgent need to improve both the efficiency and cleanliness of its operation in many geographical locations.
- The direct impacts of fishing and aquaculture on marine systems have been well studied and can be recognised globally, but their indirect impacts, and their impact on ecosystem functioning, are significantly less well understood due to limitations in our understanding of the underpinning biology. This situation is exacerbated in regions which do not have well-established environmental monitoring programmes and where even the direct impacts may not have been fully assessed.
- The impacts of fishing can broadly be divided into two categories; those related to its biological impact such as the removal of biomass and its impact on food webs and genetic diversity, and physical impacts related to how the act of fishing may disturb the seafloor affecting how it functions as a habitat.
- The impacts of aquaculture on marine systems are diverse and include husbandry issues such as the use of antimicrobials and biocides in open-water systems, feed issues, such as the use of wild caught fish in fish meal for aquaculture purposes, as well as issues related to the introduction of non-native species and their impact on local biological diversity.

- Understanding and mitigating the impacts of wild capture fisheries and aquaculture on marine ecosystems are essential for both the long-term sustainability of the industries and the protection of other services delivered by marine systems. This is particularly important as many national governments are increasingly looking at marine ecosystems as an under-exploited source of income and are actively investigating how to extract further revenue ('the marine economy').
- Well-studied systems such as the North Sea have considerable data resources to support their management, but the majority of the world's fisheries and aquaculture operations are working in a much less data-rich environment and the impacts of these industries on the local and regional marine ecosystems are largely unknown. Research collaboration between Europe/United States/Oceania and these areas is essential for the future sustainability of global fish products. The urgency of the challenge of global food security means that management will need to proceed using a much lower level of evidence/certainty than has been the case in North America and Europe. New management frameworks will need to be developed that are ecologically holistic, protective of functioning, socially, economically and ecologically sustainable and include precaution, adaptation and risk management as key tenets.

References

Aquaculture Stewardship Council (2013). http://www.asc-aqua.org/.

Bachère, E., Gueguen, Y., Gonzalez, M. et al. (2004). Insights into the antimicrobial defense of marine invertebrates: the penaeid shrimps and the oyster Crassostrea gigas. *Immunological Reviews*, 198, 149–168.

Beare, D., Rijnsdorp, A., Van Kooten, T. et al. (2010). *Study for the revision of the plaice box: final report*. Wageningen IMARES Report number C002/10. European Commission (DG Maritime Affairs and Fisheries), Brussels.

Bearhop, S., Thompson, D. R., Phillips, R. A. et al. (2001). Annual variation in great skua diets: The importance of commercial fisheries and predation on seabirds revealed by combining dietary analyses. *Condor*, 104, 802–809.

Bergh, Ø. (2007). The dual myths of the healthy wild fish and the unhealthy farmed fish. *Diseases of Aquatic Organisms*, 75, 159–164.

Beveridge, M. C. M., Thilsted, S. H., Phillips, M. et al. (2013). Meeting the food and nutrition needs of the poor: the role of fish and the opportunities and challenges emerging from the rise of aquaculture. *Journal of Fish Biology*, 83, 1067–1084.

Bicknell, A. W. J., Oro, D., Camphuysen, C. J. and Votier, S. C. (2013). Potential consequences of discard reform for seabird communities. *Journal of Applied Ecology*, 50, 649–658.

Bogstad, B. and Gjøsæter, H. (1994). A method for estimating the consumption of capelin by cod in the Barents Sea. *ICES Journal of Marine Science*, 51, 273–280.

Bostock, J., McAndrew, B., Richards, R. *et al.* (2010). Aquaculture: global status and trends. *Philosophical Transactions of the Royal Society B: Biological Sciences*, 365, 2897–2912.

Bremner, J., Frid, C. L. J. and Rogers, S. I. (2003a). Assessing marine ecosystem Health: the long-term effects of fishing on functional biodiversity in North Sea benthos. *Aquatic Ecosystem Health and Management*, 6, 131–137.

Bremner, J., Rogers, S. I. and Frid, C. L. J. (2003b). Assessing functional diversity in marine benthic ecosystems: a comparison of approaches. *Marine Ecology Progress Series*, 254, 11–25.

Buschmann, A. H., Tomova, A., López, A. *et al.* (2012). Salmon aquaculture and antimicrobial resistance in the marine environment. *PLoS ONE*, 7(8): e42724.

Cabello, F. C., Godfrey, H. P., Tomova, A. *et al.* (2013). Antimicrobial use in aquaculture re-examined: Its relevance to antimicrobial resistance and to animal and human health. *Environmental Microbiology*, 15, 1917–1942.

Callier, M. D., Lefebvre, S., Dunagan, M. K. *et al.* (2013). Shift in benthic assemblages and organisms' diet at salmon farms: community structure and stable isotope analyses. *Marine Ecology Progress Series*, 483, 153–167.

Carpenter, S. R., Kitchell, J. F. and Hodgson, J. R. (1985). Cascading trophic interactions and lake productivity. *Bioscience*, 35, 634–639.

Cesar, C. P. and Frid, C. L. J. (2009). Effects of experimental small-scale cockle (*Cerastoderma edule* L.) fishing on ecosystem function. *Marine Ecology*, 30, 123–137.

Cesar, C. P. and Frid, C. L. J. (2012). Benthic disturbance affects intertidal food web dynamics: implications for investigations of ecosystem functioning. *Marine Ecology Progress Series* 466, 35–41.

CFFRC (Crops for the Future Research Centre) (2012).: FishPlus. Available at: http://www.nottingham.edu.my/CFFRC/documents/CFFRCPLUS-FishPlus.pdf.

Cheshuk, B. W., Purser, G. J. and Quintana, R. (2003). Integrated open-water mussel (*Mytilus planulatus*) and Atlantic salmon (*Salmo salar*) culture in Tasmania, Australia. *Aquaculture*, 218, 357–378.

Choy, C. A. and Drazen, J. C. (2013). Plastic for dinner? Observations of frequent debris ingestion by pelagic predatory fishes from the central North Pacific. *Marine Ecology Progress Series*, 485, 155–163.

Cook, R., Fariñas-Franco, J. M., Gell, F. R. *et al.* (2013). The substantial first impact of bottom fishing on rare biodiversity hotspots: a dilemma for evidence-based conservation. *PLoS ONE*, 8, e69904.

Costello, M. J. (2009). The global economic cost of sea lice to the salmonid farming industry. *Journal of Fish Diseases*, 32, 115–118.

De Juan, S. and Demestre, M. (2012). A trawl disturbance indicator to quantify large-scale fishing impact on benthic ecosystems. *Ecological Indicators*, 18, 183–190.

EC (2002). Council Regulation (EC) No 2371/2002 of 20 December 2002 on the conservation and sustainable exploitation of fisheries resources under the Common Fisheries Policy. European Commission, Brussels.

Einum, S. and Fleming, I. A. (1997). Genetic divergence and interactions in the wild among native, farmed and hybrid Atlantic salmon. *Journal of Fish Biology*, 50, 634–651.

Eleftheriou, A. and Robertson, M. R. (1992). The effects of experimental scallop dredging on the fauna and physical environment of a shallow sandy community. *Netherlands Journal of Sea Research*, 30, 289–299.

FAO (1995). *Code of Conduct for Responsible Fisheries.* Rome: FAO.

FAO (2009). *The State of World Fisheries and Aquaculture 2008.* Rome: FAO.

FAO (2010). *World Review of Fisheries and Aquaculture.* Rome: FAO.

FAO (2012). *The State of World Fisheries and Aquaculture 2012.* Rome: FAO.

FAO (2013a). *Fishery Statistical Collections: Global Aquaculture Production.* Rome: FAO.

FAO (2013b). *Species Fact Sheets: Unidaria pinnatifida.* Rome: FAO.

Fitridge, I., Dempster, T., Guenther, J. and de Nys, R. (2012). The impact and control of biofouling in marine aquaculture: A review. *Biofouling*, 28, 649–669.

Fleming, I. A. and Einum, S. (1997). Experimental tests of genetic divergence of farmed from wild Atlantic salmon due to domestication. *ICES Journal of Marine Science*, 54, 1051–1063.

Folke, C. and Kautsky, N. (1992). Aquaculture with its environment: prospects for sustainability. *Ocean and Coastal Management*, 17, 5–24.

Fosså, J. H., Mortensen, P. B. and Furevik, D. M. (2002). The deep-water coral Lophelia pertusa in Norwegian waters: distribution and fishery impacts. *Hydrobiologia*, 471, 1–12.

Fraser, W. R., Trivelpiece, W. Z., Ainley, D. G. and Trivelpiece, S. G. (1992). Increases in Antarctic penguin populations: reduced competition with whales or a loss of sea ice due to environmental warming? *Polar Biology*, 11, 525–531.

Freese, L., Auster, P. J., Heifetz, J. and Wing, B. L. (1999). Effects of trawling on seafloor habitat and associated invertebrate taxa in the Gulf of Alaska. *Marine Ecology Progress Series*, 182, 119–126.

Frid, C. L. J. (2011). Temporal variability in the benthos: does the sea floor function differently over time? *Journal of Experimental Marine Biology and Ecology*, 400, 99–107.

Frid, C. L. J. and Mercer, T. S. (1989). Environmental monitoring of caged fish farming in macrotidal environments. *Marine Pollution Bulletin*, 20, 379–383.

Frid, C. L. J. and Paramor, O. A. L. (2012). Feeding the world: What role for fisheries? *ICES Journal of Marine Science*, 69, 145–150.

Gislason, H. and Kirkegaard, E. (1998). Is the industrial fishery in the North Sea sustainable? In *Northern Waters: Management Issues and Practice*, ed. D. Symes. London: Fishing News Books, pp. 195–207.

Gowen, R. J. and Bradbury, N. B. (1987). The ecological impact of salmonid farming in coastal waters: a review. *Oceanography and Marine Biology: Annual Review*, 25, 563–575.

Guardiola, F. A., Cuesta, A., Meseguer, J. and Esteban, M. A. (2012). Risks of using antifouling biocides in aquaculture. *International Journal of Molecular Sciences*, 13, 1541–1560.

Hall-Spencer, J. and Bamber, R. (2007). Effects of salmon farming on benthic Crustacea. *Ciencias Marinas* 33, 353–366.

Hall-Spencer, J., White, N., Gillespie, E., Gillham, K. and Foggo, A. (2006). Impact of fish farms on maerl beds in strongly tidal areas. *Marine Ecology Progress Series*, 326, 1–9.

Hall-Spencer, J. M., Tasker, M., Soffker, M. *et al.* (2009). Design of marine protected areas on high seas and territorial waters of Rockall bank. *Marine Ecology Progress Series*, 397, 305–308.

Hall, P. O. J., Holby, O., Kollberg, S. and Samuelsson, M. O. (1992). Chemical fluxes and mass balances in a marine fish cage farm. IV. Nitrogen. *Marine Ecology Progress Series*, 89, 81–91.

Hansen, L. P., Jacobsen, J. A. and Lund, R. A. (1999). The incidence of escaped farmed Atlantic salmon, *Salmo salar* L., in the Faroese fishery and estimates of catches of wild salmon. *ICES Journal of Marine Science*, 56, 200–206.

Hiddink, J. G., Jennings, S., Kaiser, M. J. *et al.* (2006). Cumulative impacts of seabed trawl disturbance on benthic biomass, production, and species richness in different habitats. *Canadian Journal of Fisheries and Aquatic Sciences*, 63, 721–736.

Holby, O. and Hall, P. O. J. (1991). Chemical fluxes and mass balances in a marine fish cage farm. IV. Nitrogen. *Marine Ecology Progress Series*, 70, 263–272.

Hu, C. D., Li C., Chen, J. *et al.* (2010). On the recurrent *Ulva prolifera* blooms in the Yellow Sea and East China Sea. *Journal Geophysical Research,* 115, C05017.

Hussain, S. S., Winrow-Giffin, A., Moran, D. *et al.* (2010). An ex ante ecological economic assessment of the benefits arising from marine protected areas designation in the UK. *Ecological Economics*, 69, 828–838.

Jennings, S. and Kaiser, M. J. (1998). The effects of fishing on marine ecosystems. *Advances in Marine Biology*, 34, 201–352.

Jonsson, B. and Jonsson, N. (2006). Cultured Atlantic salmon in nature: a review of their ecology and interaction with wild fish. *ICES Journal of Marine Science*, 63, 1162–1181.

Kaiser, M. J. (1998). Significance of bottom-fishing disturbance. *Conservation Biology*, 12, 1230–1235.

Kaiser, M. J., Armstrong, P. J., Dare, P. J. and Flatt, R. P. (1998a). Benthic communities associated with a heavily fished scallop ground in the English Channel. *Journal of the Marine Biological Association of the United Kingdom*, 78, 1045–1059.

Kaiser, M. J., Clarke, K. R., Hinz, H. *et al.* (2006). Global analysis of response and recovery of benthic biota to fishing. *Marine Ecology Progress Series*, 311, 1–14.

Kaiser, M. J., Collie, J. S., Hall, S. J., Jennings, S. and Poiner, I. R. (2003). Impacts of fishing gear on marine benthic habitats. In *Responsible Fisheries in the Marine Ecosystem*, eds. M. Sinclair and G. Valdimarsson. Rome: FAO, pp. 197–217.

Kaiser, M. J. and de Groot, S. J. (eds) (2000). *Effects of Fishing on Non-target Species and Habitats*. Oxford: Blackwell Scientific.

Kaiser, M. J., Edwards, D. B., Armstrong, P. J. *et al.* (1998b). Changes in megafaunal benthic communities in different habitats after trawling disturbance. *ICES Journal of Marine Science*, 55, 353–361.

Kalantzi, I. and Karakassis, I. (2006). Benthic impacts of fish farming: meta-analysis of community and geochemical data. *Marine Pollution Bulletin*, 52, 484–493.

Kamenos, N. A., Moore, P. G. and Hall-Spencer, J. M. (2004). Maerl grounds provide both refuge and high growth potential for juvenile queen scallops

(*Aequipecten opercularis* L.). *Journal of Experimental Marine Biology and Ecology*, 313, 241–254.

Katsanevakis, S., Zenetos, A., Belchior, C. and Cardoso, A. C. (2013). Invading European seas: assessing pathways of introduction of marine aliens. *Ocean and Coastal Management*, 76, 64–74.

Keeley, N. B., Forrest, B. M. and Macleod, C. K. (2013). Novel observations of benthic enrichment in contrasting flow regimes with implications for marine farm monitoring and management. *Marine Pollution Bulletin*, 66, 105–116.

Krkošek, M., Ford, J. S., Morton, A. *et al*. (2007). Declining wild salmon populations in relation to parasites from farm salmon. *Science*, 318, 1772–1775.

Krkošek, M., Lewis, M. A. and Volpe, J. P. (2005). Transmission dynamics of parasitic sea lice from farm to wild salmon. *Proceedings of the Royal Society B: Biological Sciences*, 272, 689–696.

Krkošek, M., Revie, C. W., Gargan, P. G. *et al*. (2013). Impact of parasites on salmon recruitment in the Northeast Atlantic Ocean. *Proceedings of the Royal Society B: Biological Sciences*, 280, 20122359.

Lane, A. and Willemsen, P. R. (2004). Collaborative effort looks into biofouling. *Fish Farming International*, September 2004, 34–35.

Law, R. (2000). Fishing, evolution and phenotypic evolution. *ICES Journal of Marine Science*, 57, 659–668.

Laws, R. and Grey, D. R. (1989). Life-history evolution and sustainable yields from populations with age specific cropping. *Evolutionary Ecology*, 3, 343–359.

Ledford, H. (2013). Transgenic salmon nears approval. *Nature*, 497, 17–18.

Li, R. (2013). National and regional socio-economic dependence on the fishery sector in mainland China. *Fisheries Management and Ecology*, doi: 10.1111/fme.12055.

Liland, N. S., Rosenlund, G., Berntssen, M. H. G. *et al*. (2013). Net production of Atlantic salmon (FIFO, Fish in Fish out < 1) with dietary plant proteins and vegetable oils. *Aquaculture Nutrition*, 19, 289–300.

Lin, L. S., Ling, J. Z., Cheng, J. H. and Yu, L. F. (2010). *Current Condition of Yellow Croaker and Recommendation*. Beijing: China Academic Journal Electronic Publishing House.

Liu, D., Keesing, J. K., Dong, Z. *et al*. (2010). Recurrence of the world's largest green-tide in 2009 in Yellow Sea, China: *Porphyra yezoensis* aquaculture rafts confirmed as nursery for macroalgal blooms. *Marine Pollution Bulletin*, 60, 1423–1432.

Liu, F., Pang, S., Chopin, T. *et al*. (2013). Understanding the recurrent large-scale green tide in the Yellow Sea: temporal and spatial correlations between multiple geographical, aquacultural and biological factors. *Marine Environmental Research*, 83, 38–47.

Lumb, C. M. (1989). Self-pollution by Scottish salmon farms. *Marine Pollution Bulletin*, 20, 375–379.

Matsuoka, T., Nakashima, T. and Nagasawa, N. (2005). A review of ghost fishing: scientific approaches to evaluation and solutions. *Fisheries Science*, 71, 691–702.

Mayor, D. J. and Solan, M. (2011). Complex interactions mediate the effects of fish farming on benthic chemistry within a region of Scotland. *Environmental Research*, 111, 635–642.

Myan, F. W. Y., Walker, J. and Paramor, O. A. L. (2013). The interaction of marine fouling organisms with topography of varied scale and geometry: a review. *Biointerphases*, 8, 30.

Natale, F., Hofherr, J., Fiore, G. and Virtanen, J. (2012). Interactions between aquaculture and fisheries. *Marine Policy*, 38, 205–213.

Olsen, E. M., Heino, M., Lilly, G. R. *et al.* (2004). Maturation trends indicative of rapid evolution preceded the collapse of northern cod. *Nature*, 428, 932–935.

Oro, D., Genovart, M., Tavecchia, G., Fowler, M. S. and Martinez-Abra, A. (2013). Ecological and evolutionary implications of food subsidies from humans. *Ecology Letters*, 16, 1501–1514.

Pauly, D., Christensen, V., Dalsgaard, J., Forese, R. and Torres, F. (1998). Fishing down marine food webs. *Science*, 279, 860–863.

Rice, J. and Gislason, H. (1996). Patterns of change in the size spectra of numbers and diversity of the North Sea fish assemblage, as reflected in surveys and models. *ICES Journal of Marine Science*, 53, 1214–1225.

Rice, J. C. and Garcia, S. M. (2011). Fisheries, food security, climate change, and biodiversity: characteristics of the sector and perspectives on emerging issues. *ICES Journal of Marine Science*, 68, 1343–1353.

Richardson, A. J., Bakun, A., Hays, G. C. and Gibbons, M. J. (2009). The jellyfish joyride: causes, consequences and management responses to a more gelatinous future. *Trends in Ecology and Evolution*, 24, 312–322.

Rigos, G. and Katharios, P. (2010). Pathological obstacles of newly-introduced fish species in Mediterranean mariculture: a review. *Reviews in Fish Biology and Fisheries*, 20, 47–70.

Rijnsdorp, A. D., Buys, A. M., Storbeck, F. and Visser, E. G. (1998). Micro-scale distribution of beam trawl effort in the southern North Sea between 1993 and 1996 in relation to the trawling frequency of the sea bed and the impact on benthic organisms. *Ices Journal of Marine Science*, 55, 403–419.

Robertson, P. K. J., Black, K. D., Adams, M. *et al.* (2009). A new generation of biocides for control of Crustacea in fish farms. *Journal of Photochemistry and Photobiology B: Biology*, 95, 58–63.

Small, C. and Nicholls, R. J. (2003). A global analysis of human settlement in coastal zones. *Journal of Coastal Research* 19, 584–599.

Stentiford, G. D., Neil, D. M., Peeler, E. J. *et al.* (2012). Disease will limit future food supply from the global crustacean fishery and aquaculture sectors. *Journal of Invertebrate Pathology*, 110, 141–157.

Sun, C. and Che, B. (2012). The influence of marine aquaculture on the seafood supply chain in China. *Aquaculture Economics and Management* 16, 117–135.

Telfer, T. C., Baird, D. J., McHenery, J. G. *et al.* (2006). Environmental effects of the anti-sea lice (Copepoda: Caligidae) therapeutant emamectin benzoate under commercial use conditions in the marine environment. *Aquaculture*, 260, 163–180.

Thomas, K. V. and Brooks, S. (2010). The environmental fate and effects of antifouling paint biocides. *Biofouling*, 26, 73–88.

Thompson, R. C., Olsen, Y., Mitchell, R. P. *et al.* (2004). Lost at sea: where is all the plastic? *Science*, 304, 838–838.

Thrush, S. F., Hewitt, J. E., Cummings, V. J. et al. (1998). Disturbance of the marine benthic habitat by commercial fishing: impacts at the scale of the fishery. *Ecological Applications*, 8, 866–879.

Tyler, E. H. M., Somerfield, P. J., Berghe, E. V. et al. (2012). Extensive gaps and biases in our knowledge of a well-known fauna: Implications for integrating biological traits into macroecology. *Global Ecology and Biogeography*, 21, 922–934.

Uglem, I., Dempster, T., Bjørn, P. A., Sanchez-Jerez, P. and Økland, F. (2009). High connectivity of salmon farms revealed by aggregation, residence and repeated movements of wild fish among farms. *Marine Ecology Progress Series*, 384, 251–260.

UN (2010). Atlas of the Oceans. Available at: http://www.oceansatlas.org/servlet/CDSServlet?status=ND0xODc3JjY9ZW4mMzM9KiYzNz1rb3M~.

Youngson, A. F., Dosdat, A., Saroglia, M. and Jordan, W. C. (2001). Genetic interactions between marine finfish species in European aquaculture and wild conspecies. *Journal of Applied Ichthyology*, 17, 153–162.

7 · Artificial physical structures

FABIO BULLERI AND M. GEE CHAPMAN

7.1 Infrastructure in marine habitats in the third millennium

Coastal armouring occurs because of the construction of artificial structures along natural shorelines. These include revetments, docks, groynes and seawalls, in addition to numerous minor structures (Table 7.1). This coastal infrastructure not only provides important onshore functions, but also brings shipping onto the shoreline, facilitating loading and unloading. Structures running alongshore, such as revetments and seawalls, are often built to protect shorelines against erosion, or to provide easy access to ships. They are usually steep and constructed of material designed to withstand erosion and wear, and are placed above the high tide level, inter- or subtidally, or offshore. Other structures, built perpendicular to the shore, for example, groynes and jetties, are often built to prevent movement of sand, or to gain access to boats in deeper water. Structures such as drilling platforms and wind turbines have become common features in offshore waters. Dugan et al. (2011) provide a detailed summary of the types and extent of infrastructure common on many shores today.

Shoreline alteration of this sort is not a new phenomenon. It has occurred since people started aggregating into coastal settlements which became centres for trade, many thousands of years ago. Over 2000 years ago, harbours along the Mediterranean coast were protected by built breakwaters and seawalls, and areas vulnerable to flooding, such as the Netherlands, have been protected by walls for hundreds of years (Rippon, 2000; Charlier et al., 2005).

With increased human populations and the worldwide expansion of oceanic trade, shoreline development and armouring has increased in both industrialised and more rural countries. In some harbours, such as Singapore, there are no or few natural rocky habitats, although much

Table 7.1 Purposes and characteristics of common urban infrastructures deployed in nearshore waters (after Bulleri and Chapman, 2010).

Type of structure	Action and purposes	Materials used respect to the shore	Positioning/orientation the sea surface	Position respect to	Wave exposure
Breakwaters	Reduce the intensity of wave-forces in inshore waters; used for protecting ports, harbours and marinas and as coastal defences	Sandstone; geotextile; granite; sand-bags; concrete; wood	Not connected to shore parallel or fish-tail	Emergent; low crested; submerged	Exposed
Groynes	Reduce alongshore transport of sediments; used in coastal defence schemes, often in association with breakwaters	Sandstone; geotextile; granite; concrete; wood; sand-bags	Connected to shore perpendicular	Emergent; low crested; submerged	Exposed
Jetties	Reduce wave- and tide-generated currents; used for developing, ports, harbours, marinas and as constituents of coastal defence schemes	Sandstone; geotextile; granite; concrete; wood; sand-bags	Connected to shore perpendicular	Emergent; low crested; submerged	Exposed
Seawalls Bulkheads	Reduce the impact of waves on shore; used as a tool against coastal erosion and as a constituent of ports, docks and marinas	Sandstone; geotextile; granite; concrete; steel vinyl; sand-bags; wood	Onshore parallel on open coasts, but variable in enclosed waters	Emergent	Exposed to sheltered
Pilings	Sustain infrastructure, such as bridges, piers, docks and for the mooring of vessels	Concrete, wood; fibreglass; metal	Onshore to offshore	Emergent	Exposed to sheltered
Floating docks	Create boating facilities	Concrete; wood; plastic; fibreglass; metal	Connected to shore varying orientation	Emergent	Sheltered
Ropes-Poles Cages-Nets	Constituents of aquaculture facilities	Fabric; plastic; wood; fibreglass; metal	Not connected to shore varying orientation	Emergent; submerged	Moderately exposed to sheltered

of the coastline is composed of extensive concrete walls (P. Todd, pers. comm.). In 1990, 21% of 759 km of coastal Florida was armoured, with up to 50% in developed areas. More than 1500 km of Mediterranean coastlines are now predominantly concrete. Along the north-west Adriatic coast, about 300 km of the sandy shoreline are protected by an almost uninterrupted stretch of breakwaters (Cencini et al., 1998). Even in countries with relatively small populations, such as Australia, seawalls can dominate shorelines locally (Chapman, 2003).

More than 75% of people are expected to live within 100 km of a coast by 2025 – a worldwide phenomenon (EEA, 2006). This trend will further impact coastal landscapes, which are already extremely altered. Anthropogenically created climate changes will exacerbate current impacts. For instance, 140 000 km^2 of land in Europe is 1 m or less above sea level, so a sea-level rise of 50 cm will result in 1.1–1.3 million people being exposed to flooding each year (EC, 2009). In Europe, between 2000 and 17 000 km^2 of coastal land may be lost by 2080, with an estimated cost of €1.8 billion (EC, 2009). In addition, storm intensity is predicted to rise as a consequence of warming (Knutson et al., 2010), with enormous social and financial implications. In Europe, the cost of a 100-year storm event has been predicted to double to €40 billion, with average storm losses increasing by 16–68% by 2080 (EC, 2009). The need to protect infrastructure and properties will undoubtedly accelerate current rates of armouring of coastlines, although alternative policies, such as the 'no further armouring' in California, are increasingly being considered for the management of coastal areas under future climate scenarios (Hanak and Moreno, 2012).

In addition to infrastructure on the coastline, more and more structures are being deployed offshore. Offshore groynes, such as along much of the coast of Italy, have been used for some time to prevent loss of beaches by shoreline movement of sand (Cencini et al., 1998). Many harbours have shipping facilities, marinas and ship moorings offshore. Since the end of the nineteenth century, oil platforms have proliferated in sheltered embayments and, more recently, miles offshore. The number of active oil rigs worldwide, currently estimated at 7500, is set to increase to fulfil the increasing demand of oil (Macreadie et al., 2011). With a growing interest in renewable energy, offshore infrastructure will soon include increasing numbers of wind farms and other structures, either floating or fixed to the bottom, generating energy from surface water heat, waves, tides or currents (Pelc and Fujita, 2002). Finally, further infrastructure will be introduced in offshore waters to sustain the demand of marine fish farming, a rapidly growing industry worldwide (Wu, 1995).

7.2 Environmental effects of built structures on the coast

Along with important, yet often under-appreciated ecological relevance, sandy beaches are valuable sites for recreation and tourism. Coastal habitats, such as mudflats, reefs and mangrove forests, fuel coastal economies worldwide (Klein et al., 2004). Armouring shorelines to protect urban infrastructure or halt erosion has become common practice in the management of many coastal areas. Such practice is, however, unfortunately associated with dramatic alterations to environmental conditions and, thus, to biodiversity and ecosystem functioning of these transformed shorelines.

7.2.1 Changes to currents and wave action

Coastal armouring can have detrimental effects on intertidal and subtidal habitats, especially soft sediments which can be lost to erosion (Hall and Pilkey, 1991). Beaches seaward of alongshore revetments and seawalls are particularly vulnerable to loss of sediment, due to the placement of the structure itself and erosion as the waves impact on a hard vertical structure rather than the natural shoreline (Kraus and McDougal, 1996; Griggs, 2005). The walls reflect rather than dissipate wave energy and the seawall itself prevents the shoreline migrating inland as it shrinks in area (so-called coastal squeeze).

Many coastal structures are built specifically to alter water movement and currents. Thus, the primary purpose of groynes is often to reduce alongshore movement of sediments, often for recreation, such as retaining the amenity of beaches (Airoldi et al., 2005a). Under some coastal conditions, the tidal currents created can permanently remove sediment from adjacent downstream habitats (Dugan et al., 2011). These structures are also used to reduce erosion in areas where beaches create a buffer between the land and sea. Offshore groynes, jetties and other structures can create novel rip currents, which change the morphology of subtidal benthic habitats, often by creating deep holes (Pattiaratchi et al., 2009).

Other artificial structures cause accumulation of sediments. When groynes running perpendicular to the shore are associated with breakwaters running parallel to shoreline (Figure 7.1), there can be large accumulation of fine organic sediments on the landward side of the

Figure 7.1 Top panel: Enclosed watershed along a sandy coastline at Marina di Pisa (Tuscany, Italy), defended through the deployment of breakwaters and groynes that run parallel and perpendicular to the shore, respectively. Bottom panel: Accumulation of fine sediments on the sea-bottom on the landward side of breakwaters. Benthic assemblages are dominated by filamentous macroalgae and by the invasive seaweed, *Caulerpa racemosa*, with scattered plants of *Codium* sp. Image by F. Bulleri. For the colour version, please refer to the plate section.

breakwaters (Martin *et al.*, 2005; Lee *et al.*, 2012). The height and porosity of artificial structures such as breakwaters, determines the magnitude of alterations to hydrodynamic conditions and the depositional regime. For example, low-crested structures (LCS) running parallel to the shoreline can be submerged or regularly overtopped by waves. This type of coastal defence causes less dramatic changes to hydrodynamics, sediments and benthic assemblages than do emergent breakwaters (Lamberti and Zannutigh, 2005).

Offshore installations can alter currents and wave action in a number of ways (Pelc and Fujita, 2002). For instance, wave-power plants, acting as wave breakers, dampen hydrodynamism on exposed shores while tidal plants, often located at river mouths, can alter inflow and outflow currents and, hence, impact local hydrology.

7.2.2 Artificial lighting

Artificial structures, such as marinas, wharves, pontoons, jetties, oil platforms and wind farms, are usually brightly lit by artificial lighting. This phenomenon, known as 'light pollution' at a larger scale (Longcore and Rich, 2004), is widespread in urbanised environments. It has thus been predicted that in cities such as Milan and Rome, brightness of the sky at night will be nine times greater than natural conditions within a few decades (Cinzano *et al.*, 2000) and that natural night lighting will disappear over much of Europe. This is a developing ecological challenge as light regimes can affect biological rhythms of many animals and plants, changing reproductive patterns and the efficiency of photosynthesis.

There are few data on the extent of artificial lighting on many built coastal structures, but lighting on offshore oil rigs can increase by 10–1000 times the intensity of light on the surface of the surrounding water (Keenen *et al.*, 2007). When lights on shoreline structures are combined with that emanating from the surrounding streets and buildings, the surface of coastal waters at night in and around large cities would be brightly lit compared to natural waters.

On the other hand, the introduction of artificial structures can reduce solar irradiance levels through shading, both in intertidal (Blockley and Chapman, 2006) and subtidal environments (Glasby, 1999, Lindegarth, 2001). This can alter distribution of fishes among nearshore habitats (Able *et al.*, 1998, 2013). Likewise, steep or vertical artificial surfaces (e.g. seawalls) are lit differently from the horizontal or gently sloping natural rocky surfaces they often replace (Chapman and Bulleri, 2003).

7.2.3 Provision of novel habitats

Novel habitat can be provided by artificial coastal structures in two ways. First, the artificial structures are novel habitats in themselves. This is typified in harbours where seawalls dominate the shoreline, creating artificial 'rocky' intertidal habitat. Often, these hard substrata occur in areas normally dominated by sediments, creating not only a novel habitat, but adding a new type of habitat (hard substrata) into areas that are naturally soft sediments.

Second, they can create novel environmental conditions in the surrounding areas. Thus, large extents of groynes and breakwaters along a shore cause accumulation of fine organic sediments on their landward sides, but also reduce exchange between the landward and seaward waters. This often leads to stagnant water and anoxic conditions on the landward side of the breakwaters (Martin *et al.*, 2005; Munari *et al.*, 2011).

7.3 Effects of built coastal structures on marine biodiversity

The deployment of coastal defence structures has multiple implications for ecological assemblages (Bulleri, 2005a). These can be divided into two broad – but interconnected – categories. First, the changed environmental conditions can affect extant assemblages through modifications in water flow, sedimentation rates, nutrient loading and illumination. Second, species may be attracted to the area by the provision of new hard substrata. These species may then establish populations on nearby hard natural habitat or use surrounding soft-bottoms as feeding grounds (Davis *et al.*, 1982; Barros *et al.*, 2001; Martin *et al.*, 2005).

7.3.1 Ecological effects of changes to currents and wave action

The loss of soft sediments due to erosion caused by coastal modification obviously can have major impacts on the flora and fauna using these sediments. Sandy beaches and mudflats support diverse invertebrate assemblages, provide nesting sites for turtles and birds and deliver key ecological services to humans, such as seawater filtration and nutrient recycling (Schlacher *et al.*, 2007). Several features of artificial structures contribute to their overall effects on assemblages in adjacent sediments, including their positioning in respect to the shoreline, extent, porosity, the amount of novel habitat provided and the distances among the structures and to natural rocky reefs (Airoldi *et al.*, 2005a).

For instance, Walker *et al.* (2008) found that alterations in accretion–erosion dynamics caused by a single 95 m groyne were limited to 15 m each side of the structure. The ecological footprint of a single structure can be, however, massive. The built island Palm Jumeirah, a megaconstruction in Dubai on the Persian Gulf, extends 6 km off the original coastline. It has an effect on sediments and benthic assemblages up to 800 m from the structure (Sale *et al.*, 2011). Sediments in the impacted area are mainly silt and fine particles (<63 μm), rather than the larger particles (≥ 125 μm) that typify natural surface sediments. As a consequence, the infaunal communities are dominated by stress-tolerant species, mostly polychaetes, while crustaceans (Amphipoda, Isopoda) dominate further from the structure (Sale *et al.*, 2011). Similarly, the extensive areas dominated by groynes and breakwaters in the north Adriatic, have caused the infauna of landward, wave-sheltered areas to be dominated by species that are typical of lagoons (Martin *et al.*, 2005; Munari *et al.*, 2011).

Seawalls or bulkheads at the landward edge of soft-sedimentary habitats, such as beaches, mudflats, saltmarsh or mangrove forests, will prevent inland migration of these habitats following sea-level rises. This may result in the loss of valuable nursery and foraging grounds for fish and birds, or breeding sites for turtles (Hulme, 2005; Gilman *et al.*, 2007; Dugan *et al.*, 2008; Schleupner, 2008).

In addition, sheltered conditions can alter competitive hierarchies among species colonising artificial hard surfaces. For instance, along the coast of Tuscany (north-west Mediterranean), mussels can exclude macroalgae on the seaward side of breakwaters, while the opposite happens on wave-sheltered surfaces on the landward side (Vaselli *et al.*, 2008). In this case, let alone the ecological implications of attracting hard-bottom species, the provision of sheltered habitat facilitates the outbreak of certain types of macroalgae (mostly filamentous) that create a nuisance for beach activities.

At broader spatial scales, coastal infrastructure disrupting alongshore water circulation (Cavalcante *et al.*, 2011), together with changes to currents caused by groynes (Burcharth *et al.*, 2007), could cause the retention of larvae and propagules in artificially created sheltered areas on an exposed coast. This may have large effects on sink populations elsewhere, which may, in turn, alter connectivity within metapopulations by changing the relative proportions of source and sink populations because connectivity among marine populations is generally maintained by movement of larvae and propagules (Kinlan and Gaines, 2003). Although our understanding of the degree of connection of marine populations is currently limited, populations of many species may not be as open as generally thought, with populations spread over large areas being maintained by relatively few breeding adults (Swearer *et al.*, 1999). Thus, any structures which change patterns of water flow and, hence, movement of propagules may cause alterations to biodiversity at a larger scale than the area where the structures are actually deployed.

7.3.2 Ecological effects of artificial lighting

Negative effects of urban lighting on nesting and foraging activities of turtles and birds have been widely demonstrated (Verheijen, 1985; Longcore and Rich, 2004), whilst less is known about potential impacts of artificial light on behaviour or physiology of fish and invertebrates (Nightingale *et al.*, 2005; Able *et al.*, 2013). This gap of knowledge calls for more research to assess how artificial lighting may contribute to the impacts of these structures on aquatic ecosystems.

It has been known for decades that marine organisms are attracted by light. Light traps have been used as a means of sampling fish for many years (Doherty, 1987; Gyekis *et al.*, 2006), controlling populations of pest species (Stamplecoskie *et al.*, 2012) and increasing catches in commercial fishing (Yamashita *et al.*, 2012). In fact, a brief glance at the internet shows pages of advertisements for lights to attract fish for both recreational and commercial fisheries. Invertebrates, too, respond to light as many are positively phototactic. Light has thus been used to divert pest invertebrates away from populated areas of lakes (Goretti *et al.*, 2011) and lights on oil rigs attract and concentrate many invertebrates that constitute the prey of fish, fish larvae and other plankton (Keenen *et al.*, 2007). Likewise, recent work has shown that artificial lights associated with infrastructure altered the behaviour of small shoaling fishes (aggregation and prolonged residence in lit areas), enhancing their susceptibility to predation (Becker *et al.*, 2013).

Whether artificial lighting can change reproductive patterns of algae and marine animals, invertebrate or vertebrate, as has been reported for many terrestrial plants, invertebrates and vertebrates (Longcore and Rich, 2004) has yet to be investigated. Many marine invertebrates may be particularly vulnerable to changes in patterns of polarised light caused by urbanisation because they use polarised light for navigation and foraging (Horváth *et al.*, 2009).

We know, however, that reduction of natural light levels caused by the shading by artificial structures, such as floating pontoons or docks can cause major alterations to soft-bottom and fish assemblages (Able et al., 1998, 2013; Lindegarth, 2001). Likewise, shading by wharves has been shown to affect recruitment on intertidal seawalls, promoting the dominance by sessile invertebrates (e.g. bryozoans, serpulid polychaetes, solitary ascidians and sponges) over macroalgae (Blockley and Chapman, 2006).

7.3.3 Provision of novel hard substrata

Artificial structures are directly used as habitat by a suite of epibenthic organisms, including macroalgae, invertebrates and fish (Connell and Glasby, 1999; Davis et al., 2002; Bulleri and Chapman, 2010). The ecological impacts of epibiota colonising coastal infrastructure in shallow coastal waters vary according to the nature of the surrounding habitats (Bulleri, 2005a). The deployment of artificial structures on hard-bottoms has often not been considered as a major alteration, in particular when they are built with the same material as the natural rocky habitats. This view is based on the assumption that artificial structures support epibiotic assemblages analogous to those occurring on adjacent natural habitats (Southward and Orton, 1954; Hawkins et al., 1983; Thompson et al., 2002). There is, however, increasing consensus that there are major differences between the structure of epibiotic and fish assemblages supported by natural hard substrata and those found on a variety of coastal infrastructures, such as seawalls, breakwaters, pier pilings and floating docks (Lincoln-Smith et al., 1994; Glasby and Connell, 1999; Rilov and Benayahu, 2000; Perkol-Finkel and Benayahu, 2004; Bulleri et al., 2005; Moschella et al., 2005; Clynick et al., 2008; Lam et al., 2009).

Several intrinsic features of artificial structures can influence the development of associated assemblages. For instance, the ability of some species to get established on artificial structures can be reduced in comparison to natural rocky habitats, as a consequence of their steep slope (Whorff et al., 1995; Benedetti-Cecchi et al., 2000). In addition, species on vertical artificial surfaces are squeezed within a narrower intertidal band compared to more gently sloping natural rocky habitats; this can create unnaturally large densities, ultimately increasing the strength of both intra- and interspecific interactions (Bulleri et al., 2000). Some artificial structures are built with unnatural material (e.g. concrete, plastic, metal, wood); others are mobile (e.g. floating docks, buoys), influencing patterns of colonisation and the intensity of competitive interactions (Connell, 2000; Edwards and Smith, 2005; Iveša et al., 2008; Perkol-Finkel et al., 2008).

Artificial structures may also lack appropriate microhabitats (e.g. rock pools, over-hangs) that function as refuges against predators or stressful environmental conditions (e.g. wave action or desiccation in intertidal habitats) compared to natural hard surfaces, decreasing post-settlement survival of plant or animal propagules (Bulleri, 2005b; Moschella et al., 2005).

Thus, altered key ecological processes, such as recruitment (Bulleri, 2005b, c; Perkol-Finkel and Benayahu, 2007), competition (Moreira et al., 2006; Iveša et al., 2008; Klein et al., 2011), predation (Kirk et al., 2007) and behaviour (Bulleri et al., 2004) would ultimately lead to the establishment of epibiotic and fish assemblages on artificial structures that are distinct from those occurring on adjacent natural hard substrata. Under these circumstances, artificial structures cannot be viewed as surrogates of natural rocky shores.

The provision of hard substrata in areas dominated by soft sediments attracts a new suite of species (Davis et al., 2002; Bacchiocchi and Airoldi, 2003). These colonisers can have important ecological and economic impacts at varying spatial scales. For example, a variety of coastal installations (e.g. coastal defences, aquaculture installations, wind farms) provide suitable habitat for the sessile stages of scyphozoans and hydromedusae which may contribute to the ongoing expansion of jellyfish (Purcell, 2012). Likewise, the proliferation of gas and oil platforms in the Gulf of Mexico ($\sim$3600 offshore platforms) has promoted the expansion of scleractinian corals across the region (Sammarco et al., 2004). Thus, at a regional scale, enhanced availability of hard substrata due to the deployment of coastal infrastructure can result in altered patterns of species dispersal.

Since the primary objective of built structures, either for coastal defence and urban/industrial infrastructure, is not that of supporting or enhancing marine life (in contrast to artificial reefs), the attraction of hard-bottom species is generally considered a side-effect. There are no a-priori expectations about which species should or should not get established (Bulleri, 2005a). In areas lacking natural hard substrata (i.e. a reference condition against which to measure the impact of the introduced structures), it is not possible to assess to what extent built structures act as surrogates of natural rocky habitats. The implications of the development of epibiota on the artificial surfaces is often decided by local social or economic demands, rather than scientific criteria (Bulleri, 2005a), but an increase in biodiversity (measured as numbers of species) is not necessarily a desirable effect when it was not planned. The attraction of a novel suite of species is an alteration to natural patterns of biodiversity. When many of the species using the habitat are from the regional species pool, the artificial structures may represent an extension of natural rocky habitats, albeit into an area where they do not naturally exist, but when they allow non-native species to flourish (Bulleri and Airoldi, 2005; Glasby et al., 2007) or promote unnatural densities of native species (e.g. macroalgal blooms; Bacchiocchi and Airoldi, 2003; Airoldi and Bulleri, 2011), they can obviously have major impacts on biodiversity (Chapter 10).

7.3.4 Ecological effects of different regimes of disturbance

Levels of disturbance experienced by epibiota are likely to be greater on artificial structures than on natural rocky shores. This applies to natural as well as anthropogenic disturbances. In order to fulfil the primary objective for which they are built, artificial shorelines are often built in areas with extreme hydrodynamic forces, such as waves. Thus, any assemblages living on them are likely to be subjected to strong wave action.

Further disturbance on epibenthic organisms is caused by repair and maintenance works that are necessary to ensure efficient functioning and safety of artificial coastlines. For example, maintenance of breakwaters in the north Adriatic Sea requires repositioning of extant blocks and the addition of new units, which is done yearly (Airoldi and Bulleri, 2011). Over-turning of blocks and the provision of new surfaces favours the dominance of ephemeral and exotic (*Codium fragile* ssp. *tomentosoides*) macroalgae at the expense of organisms susceptible to mechanical disturbance, such as mussels and oysters (Figure 7.2). These negative effects are particularly severe on the landward side of breakwaters, due to less recruitment of mussels (Figure 7.2).

At a smaller scale, similar effects are generated by recreational harvesting of shellfish (Airoldi et al., 2005b). In the northern Adriatic Sea, about 2.7 tons of mussels were found to be collected over each summer from a series of breakwaters protecting 800 m of shore, but this disrupts the low-shore assemblages on these habitats over an area of $\sim$144 m^2 (Airoldi et al., 2005b). This is a conservative estimate since it does not include mussels harvested in the subtidal areas. When scaled up to the over 300 km of breakwaters that have been deployed along this coast, the amount of bare space potentially made available for exotic macroalgae through harvesting mussels is impressive. Similarly, filling crevices on

Artificial physical structures · 179

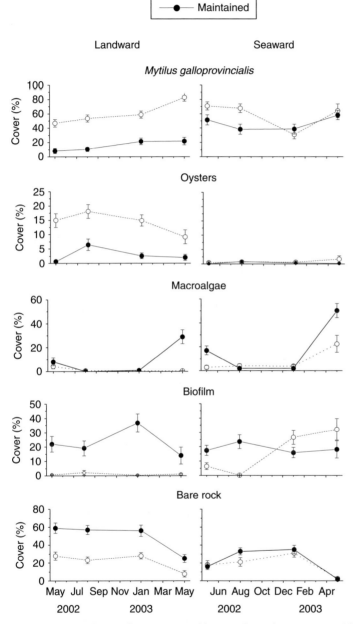

Figure 7.2 Abundance of main taxa and bare rock (rock not occupied by visible macroscopic forms) at the landward and seaward sides of control (non-maintained) and maintained (in April 2002) breakwaters at Cesenatico, in the Adriatic Sea. After Airoldi and Bulleri (2011).

seawalls to meet engineering or aesthetic criteria can remove habitat for some of the intertidal species that can live on seawalls (Moreira et al., 2007).

The net outcome of these repeated disturbances is to halt or slow down ecological succession, so many assemblages on artificial habitats are kept at an early stage of succession. Because they are often dominated by large volumes of ephemeral algae, storms and other disturbances often remove large amounts of biomass from the substratum, which causes accumulation of algal wreck on beaches. This is a major nuisance to recreational and tourist activities and local authorities may have to spend large amounts of money to remove this material and preserve amenities.

7.3.5 Evolutionary effects: effects on genetic flow

Potential genetic consequences of alterations to natural patterns of species distribution caused by the proliferation of artificial structures have received attention only recently, despite the fact that evidence for genetic effects of coastal armouring on intertidal populations was first highlighted about two decades ago (Johannesson and Warmoes, 1990). Artificial structures can decrease or enhance genetic connectivity, either by isolating or connecting populations. On the Belgian coast, populations of the periwinkle *Littorina saxatilis* established on walls which had less genetic diversity than did populations on natural substrata (Johannesson and Warmoes, 1990). This species lacks a planktonic life stage. Similarly, in the Adriatic Sea, the genetic diversity of populations of the limpet *Patella caerulea* and the serpulid polychaete *Pomatoceros triqueter* were smaller on artificial structures (breakwaters and gas platforms, respectively) than on natural substrata (Mauro et al., 2001; Fauvelot et al., 2009, 2012). Reduced genetic diversity is expected in small, isolated, recently established populations. Fauvelot et al. (2009), in contrast, found no evidence of recent bottlenecks in *P. caerulea* populations on artificial structures along the Adriatic coast, suggesting that variations in genetic diversity between natural and artificial surfaces reflects variations in selection pressure from place to place.

Artificial structures, particularly when they create patches of hard substrata in areas where these substrata are not found, can increase connectivity and gene flow among discrete populations, by providing stepping stones on which populations can establish. It has been proposed that this will reduce local adaptation of species, ultimately influencing evolutionary processes and the ability of species to withstand further disturbances (Airoldi et al., 2005a). To date, there is, however, no empirical evidence of such evolutionary effects of artificial structures, but in times of such rapid environmental change, these suggested long-term impacts must raise concern.

7.3.6 Facilitating the establishment and spread of alien species

There is mounting evidence that artificial structures, both coastal and offshore, are susceptible to biological invasion (Bulleri and Airoldi, 2005; Glasby et al., 2007; Page et al., 2007). The large numbers of exotic species that are often found on such structures might be due to the fact that such structures are often deployed in areas with poor environmental conditions (i.e. organic and inorganic pollution), frequent disturbances or those which host commercial and recreational shipping, boating and fishing. Nevertheless, recent work suggests that artificial structures intrinsically provide habitat particularly suitable for the establishment of exotic species. Artificial structures seem, therefore, to be intrinsically vulnerable to invasion.

Glasby et al. (2007) found that pontoons and pilings hosted greater numbers of exotic species than did adjacent rocky reefs and their frequency of occurrence on these structures was greater than would be expected given the local species pool. Similarly, exotic algae can dominate assemblages on artificial structures, such as breakwaters and groynes (Bulleri and Airoldi, 2005; Vaselli et al., 2008) and the mussel *Mytilus galloprovincialis* is frequently widespread on seawalls in Sydney Harbour, although quite rare on natural shores (M. G. Chapman, pers. observ.).

Features of habitat, such as chemistry and topography of the substratum, slope, shading or wave exposure, differ from natural rocky shores and many exotic species might be able to take advantage of these novel features. For example, the sheltered environments landward of breakwaters on wave-exposed coasts are key for the establishment and persistence of some invasive macroalgae (Bulleri and Airoldi, 2005; Vaselli et al., 2008). Filter-feeding invertebrates transported as fouling on ships' hulls are well adapted to withstand shear forces and can outcompete natives on moving substrata on the coast, such as floating docks (Dafforn et al., 2009). In fact, Tyrrell and Byers (2007) have proposed that novel surfaces, differing in some way from their natural counterparts, can cause the loss of the competitive advantage that native species should have over newly introduced exotics. Exotic species are often characterised by opportunistic traits, which may make them particularly capable of dominating assemblages on novel surfaces. The exotic tunicates, *Botrylloides* and *Botryllus* spp. are competitively superior to native species, but can only achieve dominance in assemblages developing on artificial surfaces, not in natural habitats (Tyrrell and Byers, 2007).

Key processes regulating the success of exotic species in a new region, such as recruitment, reproduction and feeding, can differ between artificial and natural marine habitats. For example, assemblages on coastal defence structures on sandy shores are often characterised by small numbers of native species (Bacchiocchi and Airoldi, 2003; Vaselli *et al.*, 2008). According to some theories of biological invasions, exotic species might be more successful in establishing on such artificial habitats, because they would be subjected to weak competition from resident species (the biotic resistance hypothesis; Elton, 1958), or would suffer less mortality from native predators (the enemy release hypothesis; Keane and Crawley, 2002). These potential explanations remain, however, conjectural as more experimental work is needed to assess the mechanisms underpinning the susceptibility of artificial marine habitats to invasion.

Although research on the role of artificial structures in promoting establishment and spread of exotic species is limited, interesting patterns have started to emerge. Proliferation of artificial structures can function as corridors for the expansion of exotic species, thus removing constraints to dispersal set by the availability of suitable habitat. Using artificial hard substrata as stepping stones, propagules of exotic species might thus spread across areas lacking suitable habitat (Glasby and Connell, 1999). For instance, in the northern Adriatic Sea, an almost uninterrupted stretch of breakwaters has promoted the expansion of the seaweed *Codium fragile* ssp. *tomentosoides* across the region (Bulleri and Airoldi, 2005).

In other cases, artificial structures, functioning as prime sites for the establishment of exotic species, can promote secondary invasion of natural environments. In the Derwent Estuary (Australia), the predatory seastar *Asterias amurensis* living under wharves has gonads almost double the size of those of seastars living elsewhere. This is thought to be due to the large numbers of mussels (their prey) growing on the wharves (Ling *et al.*, 2012). Thus, artificial habitats, can function as invasion hotspots, a gateway for the further invasion of natural environments.

Artificial structures can have both direct and indirect effects on the same species. In Sydney Harbour, the kelp *Ecklonia radiata* growing on pier pilings supports greater covers of the exotic bryozoan *Membranipora membranacea* than on kelps living on nearby natural reefs (Marzinelli *et al.*, 2011). Pilings influence bryozoans both directly, through shading,

Figure 3.1 Some examples of human activities and pressures potentially impacting on marine ecosystems: (a) shipping, (b) aquaculture, (c) fisheries, (d) marine litter, (e) power generation. Images by C. Frid (a, c) and T. Crowe (b, d, e). A black and white version of this figure will appear in some formats.

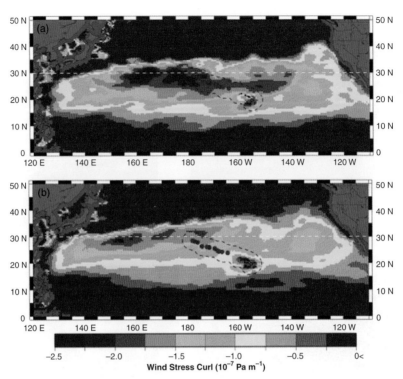

Figure 4.2 The 'Great Pacific Garbage Patch' is the name given to the accumulation of litter and debris in the area of convergent surface flows (blue and violet). The differential atmosphere–ocean conditions in El Niño years (a) result in a more intense and more extensive garbage patch and a greater spatial overlap between the convergent zone and the range of the Hawaiian monk seal (delimited by the black dashed line) than in non-El Niño years (b). The white dashed line at 30°N is for ease of comparison and the main islands of Hawaii are represented by red dots not to scale. Reprinted from Donohue and Foley (2007). A black and white version of this figure will appear in some formats.

Figure 6.1 Illustration of the diversity of taxa targeted by fisheries around the world: (a) juvenile hammerhead sharks (Sphyrnidae), (b) pomfret (*Pampus argenteus*), (c) small yellow croaker (*Larimichthys polyactis*), (d) hairtail *Trichiurus lepturus*, (e) Bombay duck (*Harpodon nehereus*), (f) blood cockle (*Anadara granosa*), (g) rays, (h) deep-fried mantis shrimp (*Oratosquilla oratoria*), (i) squid and cuttlefish, (j) catch of a *Nephrops* trawl showing the small whiting, flatfish, hagfish and invertebrates that form the bycatch, (k) Chinese mitten crab (*Eriocheir sinensis*), (l) cultivated abalone (*Haliotis discus hannai*). Images by Odette Paramor (a, b, c, d, e, f, g, h, i, k, l) and Chris Frid (j). A black and white version of this figure will appear in some formats.

Figure 6.2 Some different fisheries and aquaculture approaches. Top row: (a) Scottish trawlers in port at Campbelltown, (b) mussel dredgers on the Irish Sea, (c) bivalve cultivation Ningbo, Zhejiang, PRC. Bottom row: (d) Zhoushan Island, Zhejiang Province, PRC during the closed fishing season, (e) mussels dredged from the Irish Sea, (f) the inner harbour at Mevagissey with lobster/crab pots, inshore fishing boats and pleasure craft. Images by Odette Paramor (c, d), Chris Frid (a, f) and Tasman Crowe (b, e). A black and white version of this figure will appear in some formats.

Figure 7.1 Accumulation of sediments on the sea-bottom on the landward side of breakwaters. Benthic assemblages are dominated by filamentous macroalgae and by the invasive seaweed, *Caulerpa racemosa*, with scattered plants of *Codium* sp. Image by F. Bulleri.

Figure 7.3 Saltmarsh planted in the middle section of an intertidal seawall in Sydney Harbour. Image by M. Gee Chapman.

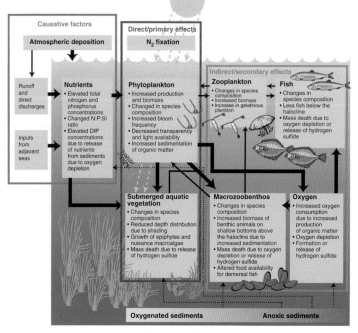

Figure 8.1 Conceptual model of eutrophication. The arrows indicate the interactions between different ecological compartments. Nutrient enrichment causes changes in the structure and function of marine ecosystems, as indicated with bold lines. Dashed lines indicate the release of hydrogen sulfide (H_2S) and phosphorus, which both occur under conditions of oxygen depletion. Adapted with permission from HELCOM (2010). A black and white version of this figure will appear in some formats.

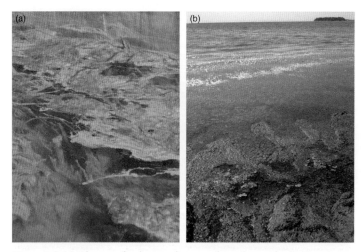

Figure 8.2 (a) Bloom of ephemeral green algae. Image by S. Korpinen. (b) Decaying filamentous ephemeral algae being washed ashore, destroying the recreational and ecological values of sandy beaches. Image by E. Bonsdorff. A black and white version of this figure will appear in some formats.

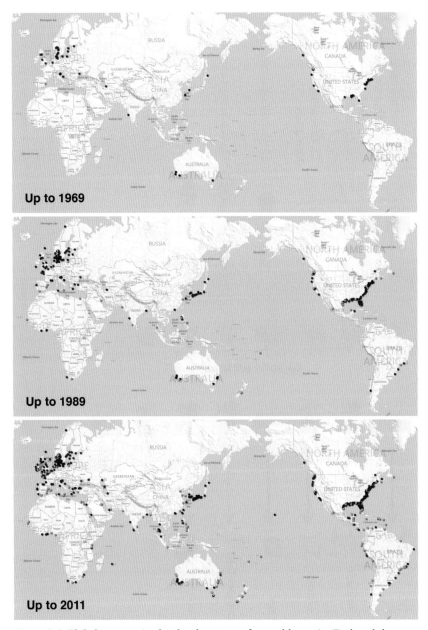

Figure 8.3 Global patterns in the development of coastal hypoxia. Each red dot represents a documented case of hypoxia related to human activities. Blue dots indicate previously hypoxic sites that have improved and yellow dots indicate eutrophic sites. Numbers of sites are cumulative through time. Based on Díaz et al. (2010), adapted with permission from http://www.wri.org/project/eutrophication. A black and white version of this figure will appear in some formats.

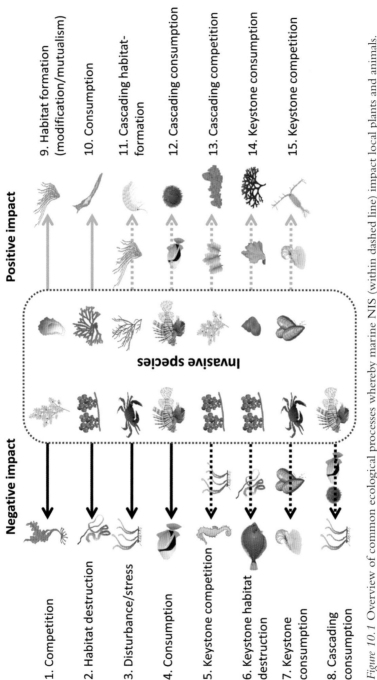

Figure 10.1 Overview of common ecological processes whereby marine NIS (within dashed line) impact local plants and animals. These processes ultimately cascade up to impacts on communities, patterns of biodiversity (cf. Figure 10.2a and D3 on Figure 10.3) and ecosystem functions (cf. Figure 10.2b and C4 on Figure 10.3). Black and green arrows = negative and positive invasion effects, respectively. Solid and dashed arrows = direct and indirect invasion effects, respectively. See Section 10.2 for examples and details on each process. Images used in the figure courtesy of the Integration and Application Network, University of Maryland Center for Environmental Science (ian.umces.edu/symbols/). A black and white version of this figure will appear in some formats.

and indirectly, by altering the density of the sea urchin *Holopneustes purpurascens*, a species that grazes on the laminae of the kelp (Marzinelli et al., 2011).

The presence of exotic species is a particularly pressing issue when it comes to the case of decommissioning offshore artificial structures, such as oil rigs and platforms, wind-, tidal- and wave-power installations (Page et al., 2006; Brito et al., 2011). This is particularly so since the concept of the 'rigs-to-reefs programs' suggests that these structures be deployed as a tool for enhancing biological productivity, restoring overexploited fish stocks or mariculture (Page et al., 2006; Macreadie et al., 2011).

7.4 Impacts on ecosystem functioning

How ecosystem functioning compares between artificial structures and natural rocky shores remains largely unexplored, despite a well-defined theoretical framework for such research. According to current biodiversity–ecosystem functioning theories (Hooper et al., 2005), considerable variation in ecosystem functioning can be predicted from widely documented differences in the structure of benthic assemblages supported by natural versus artificial surfaces. For instance, reduced species diversity on seawalls compared to natural rocky substrata can be expected to result in reduced productivity and temporal stability (Tilman, 1996, Lehman and Tilman, 2000). A greater number of species can enhance productivity through overyielding (i.e. a more efficient use of available resources due to either complementarity or facilitation) or sampling effects (i.e. the probability of including key species is greater in species-rich assemblages; Hector et al., 1999; Hooper et al., 2005; Chapter 5). Likewise, species richness can sustain the temporal stability of aggregate community properties via compensatory dynamics (i.e. asynchronous species fluctuations generating negative species covariances), the portfolio effect (if biodiversity tends to reduce mean abundance of individual species, then, due to the mean–variance relationship, those reduced abundances may result in greater than proportional reductions in variance) and, again, overyielding (Lehman and Tilman, 2000). Unfortunately, the long-term data sets to test these models do not yet exist for artificial structures.

Even when species diversity is similar between artificial and natural surfaces, altered patterns of relative abundance and evenness of species, along with variations in the strength of key processes, such as competition and predation (Iveša et al., 2008; Klein et al., 2011), might lead to substantial differences in important functions, which are themselves key in the maintenance of biodiversity. Dominance of exotic species in an assemblage can affect trophic pathways and productivity, as shown by Page et al. (2007; and see Chapter 10). Similarly, large populations of exotic prey on oil platforms enhanced numbers of predatory fish compared to natural structures (see also Keenen et al., 2007), which may have cascading effects on surrounding habitats.

Although preliminary, there is evidence to suggest that the introduction of artificial structures alters the functioning of adjacent natural environments. Artificially sheltered wave-exposed rock platforms can change from consumer- (barnacles, limpets) to producer- (ephemeral macroalgae) dominated assemblages, with important consequences at the ecosystem level (Martins et al., 2009). Goodsell et al. (2007) showed changes in species diversity on rocky shores fragmented and abutted by seawalls, compared to those surrounded by natural rocky shores, which would affect the functioning of these shores. Decreased water circulation, caused by the construction of dykes, can cause enrichment of organic matter in the surface sediments, enhancing benthic–pelagic coupling (Lee et al., 2012). When structures are deployed along sedimentary coasts, the new species attracted to artificial structures may increase local productivity, by supporting additional species, but such gains need to be considered against environmental losses. So, enhancing the biomass and productivity of ephemeral macroalgae is hardly an environmental improvement, especially in areas with reduced water exchange, as it is part of the process of eutrophication (Chapter 8).

In addition, shore armouring can depress resistance and resilience of natural habitats to disturbance, for example, by promoting beach erosion through the disruption of sediment transport dynamics (Dugan et al., 2011). Slower recovery from storms as a consequence of armouring is not limited to abiotic features of shorelines, but extends also to the biotic compartment. Lucrezi et al. (2010) found that storm-induced decline in ghost crab densities was greater in an armoured than in an unarmoured section of a sandy beach. This may have been due to the seawall preventing crab migration towards safer habitats during the storm (Lucrezi et al., 2010).

7.5 Management of artificial shorelines

7.5.1 Softening the edge

In many parts of the world, scientists are investigating how to soften the environmental effects of armoured shorelines (reviewed by Chapman and

Underwood, 2011). This can involve removing the artificial shoreline altogether, allowing the sea to find its own natural shoreline, a form of restoration called managed realignment (French, 2008). This is the ultimate in 'soft' engineering. Removing a seawall to create habitat and prey for juvenile salmon was successful (Toft et al., 2008), but removal of artificial structures has not been universally successful in creating new viable habitat, at least in the short term (Hughes et al., 2009).

Some programmes plan to remove hard armouring, but replace it with dunes or marshes to continue to protect the land from the sea with more natural habitat. The concept of 'The Living Shoreline' (Smith, 2006) recognises that erosion is natural, but suggests that it can be dealt with using natural habitats if wave action is not too strong. This approach needs room for development of a relatively large intertidal area needed to dampen wave action or prevent flooding, if one is not to build a steep sloping structure. The need to actively manage such created habitats, for example, by stabilizing sediments in dunes until the vegetation is established, may, however, have impacts on adjacent natural habitats (Brown and McLachlan, 2002).

Hybrid designs, combining engineered structures, such as seawalls, with more natural habitats have been suggested for more wave-exposed areas. Woody vegetation (Davis et al., 2006) or boulder fields (Green et al., 2012) can be placed offshore to dampen wave action, while providing additional habitat for marine invertebrates and fish – species that may have had their natural habitat reduced by building the wall in the first place. Unconsolidated rocks deployed at the level of low water which allow overflowing during high spring tides, can protect inshore marshes and support some levels of natural diversity (Currin et al., 2008). In Sydney, Australia, saltmarsh has been planted in a 'garden' in a horizontal section of a boulder seawall (Figure 7.3). This supports local populations of many saltmarsh plants, but the continued viability of such small patches of habitat is not known.

7.5.2 Modifying the structure of artificial habitat

Obviously, when hard artificial structures are placed in areas dominated by sediments, there is nothing that one can do to make the new habitat (hard, rocky) more like the former (sedimentary). But when it comes to many artificial structures in areas with nearby natural rocky shores and reefs, there are ways in which the artificial structures can be modified to become more similar to natural rocky reefs, while still retaining their primary function of protection and provision of facilities (reviewed in

Figure 7.3 Saltmarsh planted in the middle section of an intertidal seawall in Sydney Harbour. Image by M. Gee Chapman. For the colour version, please refer to the plate section.

Chapman and Underwood, 2011; Dugan *et al.*, 2011). To date, most of the research in this direction has been done on seawalls.

Leaving crevices between the blocks of a wall (Moreira *et al.*, 2007), or using riprap rather than constructing smooth vertical walls (Davis *et al.*, 2002) can each improve the value of intertidal habitat for some species. Reducing the slope of artificial habitat (Chapman and Underwood, 2011) increases intertidal areas and creates additional habitats. This will, however, increase the size of the 'footprint' of the wall on the substratum on which it is built. If this substratum is extensive, the loss of some of it may not be an important ecological issue compared to the advantage that a sloping or riprap wall might provide. In contrast, if the underlying habitat is itself limited, for example a small width of intertidal beach, the benefit of increasing the area of hard intertidal habitat will have to be weighed against the loss of the other habitat.

Although it seems obvious that sloping walls of unconsolidated boulders would create more area of types of intertidal habitat and more types of habitats, they are not suitable for areas where there is considerable wave action (Jones and Johnson, 1982). Habitat may, however, be added elsewhere in conjunction with the modified area. Iverson and Bannerot (1984) suggested deploying small patches of boulders in marinas to replace some of the habitat lost to the construction. Many species that do not live on seawalls in Sydney Harbour (Chapman, 2003), some of which are rare, rapidly colonise new patches of boulders (Chapman, 2012, 2013). Thus, quarried boulders placed near seawalls (Green et al., 2012) can provide habitat for a larger variety of species than those that can live on the seawalls themselves. Davis et al. (2006) similarly showed that adding concrete blocks, vegetation, oyster shells and woody debris offshore from seawalls in San Diego Bay increased local species richness.

7.5.3 Building artificial structures as ecological habitats

There has been considerable research on building artificial reefs as habitat, mainly for fish (reviewed by Baine, 2001), but as many of these are built de novo there is considerable scope to design them to maximise biodiversity. Coastal defences, either on- or offshore, have a primary function of protecting the shoreline, or providing facilities, so there are limitations on how they can be altered because they need to retain their primary functions. In addition, many are already in place and cannot be altered in any major way. Nevertheless, adding small pits or crevices into the structures (Martins et al., 2010; Chapman and Underwood, 2011) has been shown to be a cheap and easy way of increasing populations of particular species, or the number of species living on seawalls. In areas with many common or exotic species living on the walls, these increases in diversity may not be maintained in the long term if the small pits and crevices rapidly fill up with a few common species, removing the habitat for other rarer species (Chapman and Underwood, 2011).

Some experiments have tested the value of engineered habitats on seawalls for some of the larger and less common species of intertidal animals (Chapman and Underwood, 2011). Thus, Chapman and Blockley (2009) built cavities into sandstone seawalls. These had a lip over the entrance, so that they retained water during low tide. Although they did not resemble rock pools in many ways, they were colonised by numerous taxa that naturally live in pools, increasing the diversity of intertidal species that the wall supported. Because such structures can only be built where walls are constructed of blocks, Browne and Chapman (2011) used custom-built 'flower pots' attached to intertidal walls to evaluate whether they could be used by species that usually live in rock pools on natural shores. They, too, provided habitat for many species that cannot normally live on artificial structures.

7.6 Opportunities offered by the proliferation of artificial shorelines

7.6.1 Restoration of lost or degraded habitat

Because rocky shores, both intertidal and subtidal, are not generally considered severely threatened (Thompson *et al.*, 2002), it is unlikely that artificial shorelines would be considered as a means of restoring lost or degraded habitat, although they may provide valued habitat for individual species (see below). Where combined with marshes, dunes or mangroves, they may assist in the protection of these habitats on a small scale, but generally they are intrusions into the natural world. The natural habitat would always be a better option if it were able to persist.

Nevertheless, to many people living in urbanised environments – an increasing proportion of the world's population – the closest they get to 'nature' is already very compromised. To increase people's awareness of and concern for conservation, it is important to conserve environments in areas where 'people live and work' (Miller and Hobbs, 2002). As artificial habitats proliferate and replace natural habitats, parks, gardens and remnant patches of bushland have become important components of any conservation programme in urban areas. Although impacts on marine habitats are not yet so great as in terrestrial areas, they are increasing as coastal urban populations increase. Thus, artificial structures in the marine environment, like parks and gardens in the terrestrial environment, do support many species and should, in the long term, become incorporated into conservation efforts around cities. This will require understanding of what impacts such structures have on the environment, in addition to what values they may provide as habitat. We are not yet at the point in which the answers to these important issues are well understood.

7.6.2 Artificial structures as marine protected areas (MPAs) for endangered species

Human activities on and around coastal infrastructure which is associated with military, civil (e.g. airport runways, ports, marinas, coastal defence, aquaculture facilities) or industrial installations (e.g. oil refineries, power and nuclear plants, wind farms, oil/gas platforms) are often strictly regulated or totally banned for security or health reasons. When human disturbance is reduced, coastal infrastructure has been shown to provide suitable habitat for declining fish populations (Guidetti *et al.*, 2005), coral reefs (Feary *et al.*, 2011) or invertebrate populations (Wen *et al.*, 2007; Espinosa *et al.*, 2009). For example, in the western Mediterranean, populations of the endangered limpet *Patella ferruginea* have been found thriving on breakwaters under custody (private or military areas), including individuals larger than those occurring in natural areas (Espinosa *et al.*, 2009). These artificial structures may, therefore, function as analogues of marine protected areas (MPAs) for some species, providing a unique opportunity to restore overexploited species, without having to set aside further reserves. As suggested by García-Gómez *et al.* (2011), the key to promoting such a positive side-effect of coastal infrastructure appears to be the restriction of human access. Without enforcement, disturbance, such as harvesting, can be severe and will reduce any conservation value that the artificial habitats may have (Airoldi *et al.*, 2005b).

In addition to protection, another requisite for such structures to function as a conservation tool is that environmental conditions (e.g. water quality, sedimentation rates) must be suitable to support those species in need of protection. Artificial structures located in the proximity of urban or industrial settings generally support benthic assemblages dominated by opportunistic or ephemeral species (Bacchiocchi and Airoldi, 2003; Moschella *et al.*, 2005; Vaselli *et al.*, 2008). Such species do not need protection and, in fact, they themselves may, in some cases, reduce biodiversity. Lessening human pressure on these structures could, however, protect fish (e.g. sea-breams, sea-basses, mullets) that are intensely targeted by recreational (i.e. angling and spearfishing) and commercial fisheries (Guidetti *et al.*, 2005; Clynick *et al.*, 2007). The presence of private marinas or military bases along relatively pristine coasts might, on the other hand, provide refuge to species that are susceptible to the degradation of environmental conditions (García-Gómez *et al.*, 2011).

Marine renewable energy installations are likely to be enclosed within areas where public access and extraction (including trawling) is banned and, hence, also act as MPAs (Inger *et al.*, 2009). Poor environmental conditions might be not an issue in the case of offshore installations, unless there are impacts associated with the structures themselves. As for other coastal infrastructure, their ability to generate indirect benefits to biodiversity depends upon the characteristics of the novel surfaces being introduced and on the value and extent of habitat/s being protected.

Since the siting of many of these installations is selected to maximise energy production (i.e. wind speed; strength of tidal current and wave energy) – with an eye to the minimisation of ecological impacts – natural habitats protected as a side-effect might be of little conservation value because those particular habitats are probably currently little impacted.

Echoing García Gómez et al. (2011), we would like to note that MPAs consisting of artificial structures should not, by any means, be established in lieu of MPAs in natural settings. They are not a means of mitigation prior to habitat destruction. Well-defined objectives should underpin the selection of natural sites to be assigned the status of MPAs (Banks and Skilleter, 2007), whilst ecological criteria are not generally incorporated in the spatial planning and construction of infrastructure whose primary objective is that of absorbing wave force, or other societal function. In addition, modern criteria to select sites to be protected go beyond safeguarding single species, but prioritise the conservation of biodiversity at different levels of organisation (i.e. from species to habitats to ecosystems) and the maintenance of ecosystem functioning (Zacharias and Roff, 2000). Although the amount of artificial habitat that may be able to be protected in urban settings may be too small to foster the recovery of overexploited species (Claudet et al., 2008), networks of artificial marine micro reserves (AMMR; García-Gómez et al., 2011) might succeed, given the ubiquity of coastal artificial structures.

7.6.3 Important sites for research

Ecological research is fundamental to the understanding and protection of natural habitats and their innumerable species. Research in urban and agricultural areas has been fundamental in understanding not only the impacts of humans on natural habitats and environments, but also in measuring and understanding declining populations (Short and Burdick, 1996), biotic homogenisation (McKinney and Lockwood, 1999), interactions between physical, chemical and biological features of the environment (Pickett et al., 2001), to name but a few. Whether we like it or not, the pristine parts of the world are getting fewer and smaller. Conservation in the long term will depend on conserving other species in a human-dominated landscape. Thus, research in highly altered environments – so-called emergent or novel ecosystems (Hobbs et al., 2006) – is considered essential to long-term conservation goals.

In the sea, the problem may seem less serious, but there are no parts of the ocean that have not already been impacted by human activities

(Jackson, 2001; Seaman, 2007). This is bound to increase with larger populations of people becoming more and more crowded onto the coast. Thus, just like in the terrestrial environment, it will be necessary to understand the impacts of multiple human disturbances on marine populations and assemblages (Chapter 4), so that tools to improve their chances of persistence are developed. The urban environment, with its multitude of artificial structures and ongoing disturbances, is just the place to do this research (Bulleri, 2006).

7.7 Conclusions and future research directions

Research carried out in the last two decades has provided compelling evidence for the deployment of coastal infrastructure to be among the major drivers of change in marine biodiversity (Bulleri and Chapman, 2010, Dugan et al., 2011). Whether alterations caused to the structure of assemblages, both on soft- and hard-bottoms, have been widely documented, our understanding of how the introduction of artificial structures can alter the complex set of ecological processes, such, as competition, recruitment, predation and behaviour, underpinning ecosystem functioning, remains scant. There is, therefore, an urgent need to move beyond documenting patterns to investigate processes and, more generally, from the production of confirmatory to novel evidence (Chapman and Underwood, 2011). A straightforward way to advance our knowledge of the alterations in the functioning of coastal ecosystems caused by the proliferation of coastal infrastructure would be that of framing research into broad ecological theory. For instance, a formidable research effort has been and is being devoted to assess the relationship between biodiversity and ecosystem functioning (Chapter 5). Nonetheless, except for the case of biological invasions, most of this theory has been developed and tested exclusively in natural habitats and key ecological questions, such as how species diversity influences the productivity, stability and biogeochemical cycles, are yet to be posed, let alone addressed, in highly altered marine environments. Likewise, the effects of artificial structures have, so far, not been investigated within the frameworks of multiple stressors and global climate changes (Chapter 4). Thus, whether the proliferation of artificial structures is going to magnify or dampen the effects of other stressors in coastal areas remains utterly unexplored. Answering these questions will be crucial for preserving marine biodiversity in coastal areas under forecasted local and global changes and to promote ecological engineering (Schulze, 1996).

This discipline, merging engineering theory and practice with ecological understanding, may allow building artificial structures that, in addition to fulfilling their primary role, minimise changes to natural assemblages and create improved habitats. In an era characterised by the pervasiveness of the human footprint, leading to the coining of the term Anthropocene, assessing the potential of infrastructure to exacerbate or mitigate the loss of ecosystem services from impaired natural ecosystems will be paramount to sustain the economies and social welfare of coastal countries.

7.8 Key findings and recommendations

- Artificial structures are becoming ubiquitous features of coastal and offshore environments, modifying currents, wave action and light levels and providing novel habitat.
- The introduction of artificial structures alters patterns of abundance and distribution of organisms, regimes of disturbance and genetic flow and facilitates the establishment and spread of non-indigenous species.
- Features of artificial structures can be modified to reduce their environmental impact or to fulfil restoration or conservation needs, while still retaining their primary function of protection and provision of facilities.
- The human footprint on natural systems is set to expand. Thus, research in highly modified environments, the so-called emergent or novel systems, will be key to fulfilling long-term conservation goals.
- Framing research into broad ecological theories and taking into account the compounded nature of multiple stressors that characterise coastal areas will foster our understanding of how the proliferation of artificial structures may affect the functioning of marine ecosystems.

References

Able, K. W., Manderson, J. P. and Studholme, A. L. (1998). The distribution of shallow water juvenile fishes in an urban estuary: the effects of manmade structures in the lower Hudson river. *Estuaries*, 21, 731–744.

Able, K. W., Grothues, T. M. and Kemp, I. M. (2013). Fine-scale distribution of pelagic fishes relative to a large urban pier. *Marine Ecology Progress Series*, 476, 185–198.

Airoldi, L., Abbiati, M., Beck, M. W. et al. (2005a). An ecological perspective on the deployment and design of low-crested and other hard coastal defence structures. *Coastal Engineering*, 52, 1073–1087.

Airoldi, L., Bacchiocchi, F., Cagliola, C., Bulleri, F. and Abbiati, M. (2005b). Impact of recreational harvesting on assemblages in artificial rocky habitats. *Marine Ecology Progress Series*, 299, 55–66.

Airoldi, L. and Bulleri, F. (2011). Anthropogenic disturbance can determine the magnitude of opportunistic species responses on marine urban infrastructures. *PLoS ONE* 6(8): e22985.

Bacchiocchi, F. and Airoldi, L. (2003). Distribution and dynamics of epibiota on hard structures for coastal protection. *Estuarine Coastal and Shelf Science*, 56, 1157–1166.

Baine, M. (2001). Artificial reefs: a review of their design, application, management and performance. *Ocean and Coastal Management*, 44, 241–259.

Banks, S. A., and Skilleter, G. A. (2007). The importance of incorporating fine-scale habitat data into the design of an intertidal marine reserve system. *Biological Conservation*, 138, 13–29.

Barros, F., Underwood, A. J. and Lindegarth, M. (2001). The influence of rocky reefs on structure of benthic macrofauna in nearby soft-sediments. *Estuarine Coastal and Shelf Science*, 52, 191–199.

Becker, A., Whitfield, A. K., Cowley, P. D., Järnegren, J. and Næsje, T. F. (2013). Potential effects of artificial light associated with anthropogenic infrastructure on the abundance and foraging behaviour of estuary-associated fishes. *Journal of Applied Ecology*, 50, 43–50.

Benedetti-Cecchi, L., Bulleri, F. and Cinelli, F. (2000). The interplay of physical and biological factors in maintaining mid-shore and low-shore assemblages on rocky coasts in the north-west Mediterranean. *Oecologia*, 123, 406–417.

Blockley, D. J. and Chapman, M. G. (2006). Recruitment determines differences between assemblages on shaded or unshaded seawalls. *Marine Ecology Progress Series*, 327, 27–36.

Brito, A., Clemente, S. and Herrera, R. (2011). On the occurrence of the African hind, *Cephalopholis taeniops*, in the Canary Islands (eastern subtropical Atlantic): Introduction of large-sized demersal littoral fishes in ballast water of oil platforms? *Biological Invasions*, 13, 2185–2189.

Brown, A. C. and McLachlan, A. (2002). Sandy shore ecosystems and the threats facing them: some predictions for the year 2025. *Environmental Conservation*, 29, 62–77.

Browne, M. A. and Chapman, M. G. (2011). Ecologically informed engineering reduces loss of intertidal biodiversity on artificial shorelines. *Environmental Science and Technology*, 45, 8204–8207.

Bulleri, F. (2005a). The introduction of artificial structures in soft- and hard-bottoms: the ecological value of epibiota. *Environmental Conservation*, 32, 101–102.

Bulleri, F. (2005b). Role of recruitment in causing differences between intertidal assemblages on seawalls and rocky shores. *Marine Ecology Progress Series*, 287, 53–64.

Bulleri, F. (2005c). Experimental evaluation of early patterns of colonisation of space on rocky shores and seawalls. *Marine Environmental Research*, 60, 355–374.

Bulleri, F. (2006). Is it time for urban ecology to include the marine realm? *Trends in Ecology and Evolution*, 21, 658–659.

Bulleri, F. and Airoldi, L. (2005). Artificial marine structures facilitate the spread of a non-indigenous green alga, *Codium fragile ssp. tomentosoides*, in the north Adriatic Sea. *Journal of Applied Ecology*, 42, 1063–1072.

Bulleri, F. and Chapman, M. G. (2004). Intertidal assemblages on artificial and natural habitats in marinas on the north-west coast of Italy. *Marine Biology*, 145, 381–391.

Bulleri, F. and Chapman, M. G. (2010). The introduction of coastal infrastructure as a driver of change in marine environments. *Journal of Applied Ecology*, 47, 26–35.

Bulleri, F., Chapman, M. G. and Underwood, A. J. (2004). Patterns of movement of the limpet *Cellana tramoserica* on rocky shores and retaining seawalls. *Marine Ecology Progress Series*, 281, 121–129.

Bulleri, F., Chapman, M. G. and Underwood, A. J. (2005). Intertidal assemblages on seawalls and vertical rocky shores in Sydney Harbour, Australia. *Austral Ecology*, 30, 655–667.

Bulleri, F., Menconi, M., Cinelli, F. and Benedetti-Cecchi, L. (2000). Grazing by two species of limpets on artificial reefs in the northwest Mediterranean. *Journal of Experimental Marine Biology and Ecology*, 255, 1–19.

Burcharth, H. F., Hawkins, S. J., Zanuttigh B. and Lamberti, A. (2007) *Environmental Design Guidelines for Low Crested Coastal Structures*. Amsterdam: Elsevier.

Cavalcante, G. H., Kjerfve, B., Feary, D. A., Bauman, A. G. and Usseglio, P. (2011). Water currents and water budget in a coastal megastructure, Palm Jumeirah Lagoon, Dubai, UAE. *Journal of Coastal Research*, 27, 384–393.

Cencini, C. (1998). Physical processes and human activities in the evolution of the Po delta, Italy. *Journal of Coastal Research*, 14, 774–793.

Chapman, M. G. (2003). Paucity of mobile species on constructed seawalls: effects of urbanization on biodiversity. *Marine Ecology Progress Series*, 264, 21–29.

Chapman, M. G. (2006). Intertidal seawalls as habitats for molluscs. *Journal of Molluscan Studies*, 72, 247–257.

Chapman, M. G. (2012). Restoring intertidal boulder-fields as habitat for 'specialist' and 'generalist' animals. *Restoration Ecology*, 20, 277–285.

Chapman, M. G. (2013). Constructing replacement habitat for specialist and generalist molluscs: the effect of patch size. *Marine Ecology Progress Series*, 473, 201–214.

Chapman, M. G. and Blockley, D. J. (2009). Engineering novel habitats on urban infrastructure to increase intertidal biodiversity. *Oecologia*, 161, 625–635.

Chapman, M. G. and Bulleri, F. (2003). Intertidal seawalls: new features of landscape in intertidal environments. *Landscape and Urban Planning*, 62, 159–172.

Chapman, M. G. and Underwood, A. J. (2011). Evaluation of ecological engineering of 'armoured' shorelines to improve their value as habitat. *Journal of Experimental Marine Biology and Ecology*, 400, 302–313.

Charlier, R. H., Chaineux, M. C. P. and Morcos, S. (2005). Panorama of the history of coastal protection. *Journal of Coastal Research*, 21, 79–111.

Cinzano, P., Falchi, F., Elvidge, C. D. and Baugh, K. E. (2000). The artificial night sky brightness mapped from DMSP satellite Operational Linescan System measurements. *Monthly Notices of the Royal Astronomical Society*, 318, 641–657.

Claudet J., Osenberg, C. W., Benedetti-Cecchi, L. et al. (2008). Marine reserves: size and age matter. *Ecology Letters,* 11, 481–489.

Clynick, B. G., Chapman, M. G. and Underwood, A. J. (2007). Effects of epibiota on assemblages of fish associated with urban structures. *Marine Ecology Progress Series,* 332, 201–210.

Clynick, B. G., Chapman, M. G. and Underwood, A. J. (2008). Fish assemblages associated with urban structures and natural reefs in Sydney, Australia. *Austral Ecology,* 33, 140–150.

Connell, S. D. (2000). Floating pontoons create novel habitats for subtidal epibiota. *Journal of Experimental Marine Biology and Ecology,* 247, 183–194.

Connell, S. D. and Glasby, T. M. (1999). Do urban structures influence local abundance and diversity of subtidal epibiota? A case study from Sydney Harbour, Australia. *Marine Environmental Research,* 47, 373–387.

Currin, C. A., Delano, P. C. and Valdes-Weaver, L. M. (2008). Utilization of a citizen monitoring protocol to assess the structure and function of natural and stabilized fringing saltmarshes in North Carolina. *Wetlands Ecology and Management,* 16, 97–118.

Dafforn, K. A., Johnston, E. L. and Glasby, T. M. (2009). Shallow moving structures promote marine invader dominance. *Biofouling,* 25, 277–287.

Davis, N., Vanblaricom, G. R. and Dayton, P. K. (1982). Man-made structures on marine sediments: effects on adjacent benthic communities. *Marine Biology,* 70, 295–303.

Davis, J. L. D., Levin, L. A. and Walther, S. M. (2002). Artificial armored shorelines: sites for open-coast species in a southern California bay. *Marine Biology,* 140, 1249–1262.

Davis, J. L. D., Takacs, R. L. and Schnabel, R. (2006). Evaluating ecological impacts of living shorelines and shoreline habitat elements: an example from the upper western Chesapeake Bay. In *Management, Policy, Science, and Engineering of Non-structural Erosion Control in the Chesapeake Bay.* Publ. No. 08–164. Chesapeake Bay, VA: CRC, pp. 55–61.

Doherty, P. J. (1987). Light-traps: selective but useful devices for quantifying the distributions and abundances of larval fishes. *Bulletin of Marine Science,* 41, 423–431.

Dugan, J. E., Hubbard, D. M., Rodil, I. F. et al. (2008). Ecological effects of coastal armoring on sandy beaches. *Marine Ecology: An Evolutionary Perspective,* 29, 160–170.

Dugan, J. E., Airoldi, L., Chapman, M. G., Walker, S. and Schlacher, T. A. (2011) Estuarine and coastal structures: environmental effects. A focus on shore and nearshore structures. In *Treatise on Estuarine and Coastal Science,* Vol. 8. Waltham, MA: Academic Press, pp. 17–41.

EC (2009). Regions 2020. The climate change challenge for European regions. European Commission, Directorate General for Regional Policy. Available at http://ec.europa.eu/regional_policy/sources/docoffic/working/regions2020/pdf/regions2020_climat.pdf.

EEA (2006). The Changing Faces of Europe's Coastal Areas. EEA Report 6 / 2006. OPOCE, Luxembourg. Available at: http://www.eea.europa.eu/publications/eca-report-2006-6, accessed 9 December 2009.

Edwards, R. A. and Smith, S. D. A. (2005). Subtidal assemblages associated with a geotextile reef in south-east Queensland, Australia. *Marine and Freshwater Research*, 56, 133–142.

Elton, C. S. (1958). *The Ecology of Invasions by Animals and Plants*. London: Methuen.

Espinosa, F., Rivera-Ingraham, G. A., Fa, D. and Garcia-Gomez. J. C. (2009). Effect of human pressure on population size structures of the endangered Ferruginean limpet: toward future management measures. *Journal of Coastal Research*, 25, 857–863.

Fauvelot, C., Bertozzi, F., Costantini, F., Airoldi, L. and Abbiati, M. (2009). Lower genetic diversity in the limpet *Patella caerulea* on urban coastal structures compared to natural rocky habitats. *Marine Biology*, 156, 2313–2323.

Fauvelot, C., Costantini, F., Virgilio, M. and Abbiati, M. (2012). Do artificial structures alter marine invertebrate genetic makeup? *Marine Biology*, 159, 2797–2807.

Feary, D. A., Burt, J. A. and Bartholomew, A. (2011). Artificial marine habitats in the Arabian Gulf: review of current use, benefits and management implications. *Ocean and Coastal Management*, 54, 742–749.

French, J. R. (2008). Hydrodynamic modelling of estuarine flood defence realignment as an adaptive management response to sea-level rise. *Journal of Coastal Research*, 24, 1–12.

García-Gómez, J. C., López-Fe, C. M., Espinosa, F, Guerra-García, J. M. and Rivera-Ingraham, G. A. (2011). Marine artificial micro-reserves: a possibility for the conservation of endangered species living on artificial substrata. *Marine Ecology: an Evolutionary Perspective*, 32, 6–14.

Gilman, E., Ellison, J. and Coleman, R. (2007). Assessment of mangrove response to projected relative sea-level rise and recent historical reconstruction of shoreline position. *Environmental Monitoring and Assessment*, 124, 105–130.

Glasby, T. M. (1999). Effects of shading on subtidal epibiotic assemblages. *Journal of Experimental Marine Biology and Ecology*, 234, 275–290.

Glasby, T. M. and Connell, S. D. (1999). Urban structures as marine habitats. *Ambio*, 28, 595–598.

Glasby, T. M., Connell, S. D., Holloway, M. G. and Hewitt, C. L. (2007). Non-indigenous biota on artificial structures: could habitat creation facilitate biological invasions? *Marine Biology*, 151, 887–895.

Goodsell, P. J., Chapman, M. G. and Underwood, A. J. (2007). Differences between biota in anthropogenically fragmented habitats and in naturally patchy habitats. *Marine Ecology Progress Series*, 351, 15–23.

Goretti, E., Coletti, A., Di Veroli, A., Di Giulio, A. M. and Gaino, E. (2011). Artificial light device for attracting pestiferous chironomids (Diptera): a case study at Lake Trasimeno (Central Italy). *Italian Journal of Zoology*, 78, 336–342.

Green, D. S., Chapman, M. G. and Blockley, D. J. (2012). Ecological consequences of the type of rock used in the construction of artificial boulder-fields. *Ecological Engineering*, 46, 1–10.

Griggs, G. B. (2005). The impacts of coastal armoring. *Shore and Beach*, 73, 13–22.

Guidetti, P., Bussotti, S. and Boero, F. (2005). Evaluating the effects of protection on fish predators and sea urchins in shallow artificial rocky habitats: a case study in the northern Adriatic Sea. *Marine Environmental Research*, 59, 333–348.

Gyekis, K. F., Cooper, M. J. and Uzarski, D. G. (2006) A high-intensity LED light source for larval fish and aquatic invertebrate floating quatrefoil light traps. *Journal of Freshwater Ecology*, 21, 621–626.

Hall, M. J. and Pilkey, O. H. (1991). Effects of hard stabilization on dry beach width for New Jersey. *Journal of Coastal Research*, 7, 771–785.

Hanak, E. and Moreno, G. (2012). California coastal management with a changing climate. *Climatic Change*, 111, 45–73.

Hawkins, S. J., Southward, A. J. and Barrett, R. L. (1983). Population structure of *Patella vulgata* L. during succession on rocky shores in Southwest England. *Oceanologica Acta*, Special volume, 103–107.

Hector, A., Schmid, B., Beierkuhnlein, C. et al. (1999). Plant diversity and productivity experiments in European grasslands. *Science*, 286, 1123–1127.

Hobbs, R. J., Arico, S., Aronson, J. et al. (2006). Novel ecosystems: theoretical and management aspects of the new ecological world order. *Global Ecology and Biogeography*, 15, 1–7.

Hooper, D. U., Chapin III, F. S., Ewel, J. J. et al. (2005). Effects of biodiversity on ecosystem functioning: a consensus of current knowledge. *Ecological Monographs*, 75, 3–35.

Horváth, G., Kriska, G., Malik, P. and Robertson, B. (2009). Polarized light pollution: a new kind of ecological photopollution. *Frontiers in Ecology and the Environment*, 7, 317–325.

Hughes, R. G., Fletcher, P. W. and Hardy, M. J. (2009). Successional development of saltmarsh in two managed realignment areas in SE England, and prospects for saltmarsh restoration. *Marine Ecology Progress Series*, 384, 13–22.

Hulme, P. E. (2005). Adapting to climate change: is there scope for ecological management in the face of a global threat? *Journal of Applied Ecology*, 42, 784–794.

Inger, R., Attrill, M. J., Bearhop, S. et al. (2009). Marine renewable energy: potential benefits to biodiversity? An urgent call for research. *Journal of Applied Ecology*, 46, 1145–1153.

Iverson, E. S. and Bannerot, S. P. (1984). Artificial reefs under marine docks in southern Florida. *North American Journal of Fisheries Management*, 4, 294–299.

Iveša, L., Chapman, M. G., Underwood, A. J. and Murphy, R. J. (2008). Differential patterns of distribution of limpets on intertidal seawalls: experimental investigation of the roles of recruitment, survival and competition. *Marine Ecology Progress Series*, 407, 55–69.

Jackson, J. B. C. (2001). What was natural in the coastal oceans? *Proceedings of the National Academy of Sciences of the United States of America*, 98, 5411–5418.

Johannesson, K. and Warmoes, T. (1990). Rapid colonization of Belgian breakwaters by the direct developer, *Littorina saxatilis* (Olivi) (Prosobranchia, Mollusca). *Hydrobiologia*, 193, 99–108.

Jones, C. P. and Johnson, L. T. (1982). Coastal construction practices. *Florida Cooperative Extension Marine Advisory Bulletin*, 1–20.

Keane, R. M. and Crawley, M. J. (2002). Exotic plant invasions and the enemy release hypothesis. *Trends in Ecology and Evolution*, 17, 64–74.

Keenan, S. F., Benfield, M. C. and Blackburn, J. K. (2007). Importance of the artificial light field around offshore petroleum platforms for the associated fish community. *Marine Ecology Progress Series*, 331, 219–231.

Kinlan, B. P. and Gaines, S. D. (2003). Propagule dispersal in marine and terrestrial environments: a community perspective. *Ecology*, 84, 2007–2020.

Kirk, M., Esler, D. and Boyd, W. S. (2007). Morphology and density of mussels on natural and aquaculture structure habitats: implications for sea duck predators. *Marine Ecology Progress Series*, 346, 179–187.

Klein, Y. L., Osleeb, J. P. and Viola, M. R. (2004). Tourism-generated earnings in the coastal zone: a regional analysis. *Journal of Coastal Research*, 20, 1080–1088.

Klein, J. C., Underwood, A. J. and Chapman, M. G. (2011). Urban structures provide new insights into interactions among grazers and habitat. *Ecological Applications*, 21, 427–438.

Knutson, T. R., McBride, J. L., Chan, J. et al. (2010). Tropical cyclones and climate change. *Nature Geoscience*, 3, 157–163.

Kraus, N. C. and McDougal, W. G. (1996). The effects of seawalls on the beach .1. An updated literature review. *Journal of Coastal Research*, 12, 691–701.

Lam, N. W. Y., Huang, R. and Chan, B. K. K. (2009). Variations in intertidal assemblages and zonation patterns between vertical artificial seawalls and natural rocky shores: a case study from Victoria Harbour, Hong Kong. *Zoological Studies*, 48, 184–195.

Lamberti, A. and Zanuttigh, B. (2005). An integrated approach to beach management in Lido di Dante, Italy. *Estuarine Coastal and Shelf Science*, 62, 441–451.

Lee, J. S., Kim, K. H., Shim, J. et al. (2012). Massive sedimentation of fine sediment with organic matter and enhanced benthic-pelagic coupling by an artificial dyke in semi-enclosed Chonsu Bay, Korea. *Marine Pollution Bulletin*, 64, 153–163.

Lehman, C. L. and Tilman, D. (2000). Biodiversity, stability and productivity in competitive communities. *American Naturalist*, 156, 534–552.

Lincoln-Smith, M.P., Hair, C. A. and Bell, J. D. (1994). Man-made rock breakwaters as fish habitats: comparisons between breakwaters and natural reefs within an embayment in southeastern Australia. *Bulletin of Marine Science*, 55, 1344.

Lindegarth, M. (2001). Assemblages of animals around urban structures: testing hypotheses of patterns in sediments under boat-mooring pontoons. *Marine Environmental Research*, 51, 289–300.

Ling, S. D., Johnson, C. R., Mundy, C. N., Morris, A. and Ross, D. J. (2012). Hotspots of exotic free-spawning sex: man-made environment facilitates success of an invasive seastar. *Journal of Applied Ecology*, 49, 733–741.

Longcore, T. and Rich, C. (2004). Ecological light pollution. *Frontiers in Ecology and the Environment*, 2, 191–198.

Lucrezi, S., Schlacher, T. A. and Robinson, W. (2010). Can storms and shore armouring exert additive effects on sandy-beach habitats and biota? *Marine and Freshwater Research*, 61, 951–962.

Macreadie, P. I., Fowler, A. M. and Booth, D. J. (2011). Rigs-to-reefs: will the deep sea benefit from artificial habitat? *Frontiers in Ecology and the Environment*, 9, 455–461.

Martin, D., Bertasi, F., Colangelo, M. A. et al. (2005). Ecological impact of coastal defence structures on sediment and mobile fauna: evaluating and forecasting consequences of unavoidable modifications of native habitats. *Coastal Engineering*, 52, 1027–1051.

Martins, G. M., Amaral, A. F., Wallenstein, F. M. and Neto, A. I. (2009). Influence of a breakwater on nearby rocky intertidal community structure. *Marine Environmental Research*, 67, 237–245.

Martins, G. M., Thompson, R. C., Neto, A. I., Hawkins, S. J. and Jenkins, S. R. (2010). Enhancing stocks of the exploited limpet *Patella candei* d'Orbigny via modifications in coastal engineering. *Biological Conservation*, 143, 203–211.

Marzinelli, E. M., Underwood, A. J. and Coleman, R. A. (2011). Modified habitats influence kelp epibiota via direct and indirect effects. *PLoS ONE*, 6(7), e21936.

Mauro, A., Parrinello, N. and Acruleo, M. (2001). Artificial environmental conditions can affect allozyme genetic structure of the marine gastropod *Patella caerulea*. *Journal of Shellfish Research*, 201059–1063.

McKinney, M. L. and Lockwood, J. L. (1999). Biotic homogenization: a few winners replacing many losers in the next mass extinction. *Trends in Ecology and Evolution*, 14, 450–453.

Miller, J. R. and Hobbs, R. J. (2002). Conservation where people live and work. *Conservation Biology*, 16, 330–337.

Moreira, J., Chapman, M. G. and Underwood, A. J. (2006). Seawalls do not sustain viable populations of limpets. *Marine Ecology Progress Series*, 322, 179–188.

Moreira, J., Chapman, M. G. and Underwood, A. J. (2007). Maintenance of chitons on seawalls using crevices on sandstone blocks as habitat in Sydney Harbour, Australia. *Journal of Experimental Marine Biology and Ecology*, 347, 134–143.

Moschella, P. S., Abbiati, M., Aberg, P. et al. (2005). Low-crested coastal defence structures as artificial habitats for marine life: Using ecological criteria in design. *Coastal Engineering*, 52, 1053–1071.

Munari, C., Corbau, C., Simeoni, U. and Mistri, M. (2011). Coastal defence through low crested breakwater structures: jumping out of the frying pan into the fire? *Marine Pollution Bulletin*, 62, 1641–1651.

Nightingale, B., Longcore, T. and Simenstad, C. A. (2005). Artificial night lighting and fishes. In *Ecological Consequences of Artificial Night Lighting*. Washington: Island Press, pp. 257–271.

Page, H. M., Dugan, J. E., Culver, C. S. and Hoesterey, J. C. (2006). Exotic invertebrate species on offshore oil platforms. *Marine Ecology Progress Series*, 325, 101–107.

Page, H. M., Dugan, J. E., Schroeder, D. M. et al. (2007). Trophic links and condition of a temperate reef fish: comparisons among offshore oil platform and natural reef habitats. *Marine Ecology Progress Series*, 344, 245–256.

Pattiaratchi, C. B., Olsson, D., Hetzel, Y. and Lowe, R. (2009). Wave-driven circulation patterns in the lee of groynes. *Continental Shelf Research*, 29, 1961–1974.

Pelc, R. and Fujita, R. M. (2002). Renewable energy from the ocean. *Marine Policy*, 26, 471–479.

Perkol-Finkel, S. and Benayahu, Y. (2004). Community structure of stony and soft corals on vertical unplanned artificial reefs in Eilat (Red Sea): comparison to natural reefs. *Coral Reefs*, 23, 195–205.

Perkol-Finkel, S. and Benayahu, Y. (2007). Differential recruitment of benthic communities on neighboring artificial and natural reefs. *Journal of Experimental Marine Biology and Ecology*, 340, 25–39.

Perkol-Finkel, S., Zilman, G., Sella, I., Miloh, T. and Benayahu, Y. (2008). Floating and fixed artificial habitats: Spatial and temporal patterns of benthic communities in a coral reef environment. *Estuarine Coastal and Shelf Science*, 77, 491–500.

Pickett, S. T. A., Cadenasso, M. L., Grove, J. M. *et al.* (2001). Urban ecological systems: linking terrestrial ecological, physical, and socioeconomic components of metropolitan areas. *Annual Review of Ecology and Systematics*, 32, 127–157.

Purcell, J. E. (2012). Jellyfish and ctenophore blooms coincide with human proliferations and environmental perturbations. *Annual Review of Marine Science*, 4, 209–235.

Rilov, G. and Benayahu, Y. (2000). Fish assemblage on natural versus vertical artificial reefs: the rehabilitation perspective. *Marine Biology*, 136, 931–942.

Rippon, S. (2000). *The Transformation of Coastal Wetlands*. Oxford: British Academy.

Sale, P. F., Feary, D. A., Burt, J. A. *et al.* (2011). The growing need for sustainable ecological management of marine communities of the Persian Gulf. *Ambio*, 40, 4–17.

Sammarco, P. W., Atchison, A. D. and Boland, G. S. (2004). Expansion of coral communities within the northern Gulf of Mexico via offshore oil and gas platforms. *Marine Ecology Progress Series*, 280, 129–143.

Schlacher, T. A., Dugan, J., Schoeman, D. S. *et al.* (2007). Sandy beaches at the brink. *Diversity and Distributions*, 13, 556–560.

Schleupner, C. (2008). Evaluation of coastal squeeze and its consequences for the Caribbean island Martinique. *Ocean and Coastal Management*, 51, 383–390.

Schulze, P. C. (1996). *Engineering Within Ecological Constraints*. Washington DC: National Academy Press.

Seaman, W. (2007). Artificial habitats and the restoration of degraded marine ecosystems and fisheries. *Hydrobiologia*, 580, 143–155.

Short, F. T. and Burdick, D. M. (1996). Quantifying eelgrass habitat loss in relation to housing development and nitrogen loading in Waquoit Bay, Massachusetts. *Estuaries*, 19, 730–739.

Smith, K. M. (2006). Integrating habitat and shoreline dynamics into living shoreline applications. In *Management Policy, Science, and Engineering of Nonstructural Erosion Control in the Chesapeake Bay*, CRC Publ. No. 08–164. Gloucester, VA: CRC Publications, pp. 9–11.

Southward, A. J. and Orton, J. H. (1954). Effects of wave action on the distribution and numbers of the commoner plants and animals living on the Plymouth breakwater. *Journal of the Marine Biological Association of the United Kingdom*, 33, 1–19.

Stamplecoskie, K. M., Binder, T. R., Lower, N. *et al.* (2012). Response of migratory sea lampreys to artificial lighting in portable traps. *North American Journal of Fisheries Management*, 32, 563–572.

Swearer, S. E., Caselle, J. E., Lea, D. W. and Warner, R. R. (1999). Larval retention and recruitment in an island population of a coral-reef fish. *Nature*, 402, 799–802.

Thompson, R. C., Crowe, T. P. and Hawkins, S. J. (2002). Rocky intertidal communities: past environmental changes, present status and predictions for the next 25 years. *Environmental Conservation*, 29, 168–191.

Tilman, D. (1996). Biodiversity: population versus ecosystem stability. *Ecology*, 77, 350–363.

Toft, J., Cordell, J., Heerhartz, S. and Armbrust, E. (2008). Olympic Sculpture Park: Results from Year 1 post-construction monitoring of shoreline habitats. Report SAFS-UW- 0801. Available at: http://ww.ecy.wa.gov/programs/sea/ .../OlympicSculptureParkMonitoring.pdf.

Tyrrell, M. C. and Byers, J. E. (2007). Do artificial substrates favor nonindigenous fouling species over native species? *Journal of Experimental Marine Biology and Ecology*, 342, 54–60.

Vaselli, S., Bulleri, F. and Benedetti-Cecchi, L. (2008). Hard coastal defence structures as habitats for native and exotic rocky-bottom species. *Marine Environmental Research*, 66, 395–403.

Verheijen, F. J. (1985). Photopollution: artificial-light optic spatial control-systems fail to cope with incidents, causations, remedies. *Experimental Biology*, 44, 1–18.

Walker, S. J., Schlacher, T. A. and Thompson, L. M. C. (2008). Habitat modification in a dynamic environment: The influence of a small artificial groyne on macrofaunal assemblages of a sandy beach. *Estuarine Coastal and Shelf Science*, 79, 24–34.

Wen, K. C., Hsu, C. M., Chen, K. S. *et al.* (2007). Unexpected coral diversity on the breakwaters: potential refuges for depleting coral reefs. *Coral Reefs*, 26, 127–127.

Whorff, J. S., Whorff, L. L. and Sweet, M. H. (1995). Spatial variation in an algal turf community with respect to substratum slope and wave height. *Journal of the Marine Biological Association of the United Kingdom*, 75, 429–444.

Wu, R. S. S. (1995). The environmental impact of marine fish culture: towards a sustainable future. *Marine Pollution Bulletin*, 31, 159–166.

Yamashita, Y., Matsushita, Y. and Azuno, T. (2012). Catch performance of coastal squid jigging boats using LED panels in combination with metal halide lamps. *Fisheries Research*, 113, 182–189.

Zacharias, M. A. and Roff, J. C. (2000). A hierarchical ecological approach to conserving marine biodiversity. *Conservation Biology*, 14, 1327–1334.

8 · Eutrophication and hypoxia: impacts of nutrient and organic enrichment

SAMULI KORPINEN AND
ERIK BONSDORFF

8.1 Introduction

Long before industrialisation, human activities had already greatly altered the landscapes and water quality of our planet. However, the shift to artificial fertilisers in agriculture in the twentieth century significantly changed the level of productivity in aquatic systems – ponds, rivers, lakes, fjords and even entire marginal seas – and raised the need to coin a new term 'eutrophication'. Eutrophication is a process of introducing nutrients to ecosystems, fostering increases of plant biomass to excessive levels. It has become extremely widespread and can have a wide range of damaging consequences for marine ecosystems. In this chapter, we outline the eutrophication process and the main sources of nutrient input to the marine environment, before reviewing impacts of eutrophication on marine biodiversity and ecosystem processes in detail for each of the main coastal ecosystems. We include comments on the role of nutrients and organic matter in contributing to the global proliferation of hypoxia and the incidence of harmful algal blooms. After considering how eutrophication is interacting with fisheries, we conclude with some comments about the global occurrence of marine eutrophication, the socioeconomic consequences of eutrophication (which are further elaborated in Chapter 11) and recommendations for environmental decision makers. Although eutrophication can be a natural process without anthropogenic drivers, in this chapter we refer only to anthropogenic eutrophication.

Marine Ecosystems: Human Impacts on Biodiversity, Functioning and Services, eds T. P. Crowe and C. L. J. Frid. Published by Cambridge University Press. © Cambridge University Press 2015.

8.2 The eutrophication process

Two essential elements enhancing plant growth are nitrogen (N) and phosphorus (P). Inputs of N and P correlate strongly with observed concentrations of N and P in both coastal and offshore areas (HELCOM, 2009). P is introduced to the marine environment directly from the coastline or via rivers and more recently from various human activities, such as municipal sewage, production and use of agricultural fertilisers and fish farming. N has similar anthropogenic sources but atmospheric deposition also has a significant role. In addition, N can be directly captured from atmospheric N_2 by N-fixing cyanobacteria and released back to the atmosphere through denitrification by other bacteria. Plants take up P and N in their inorganic forms phosphate (PO_4^-), nitrate (NO_3^-), nitrite (NO_2^-), ammonium (NH_4^+) and dinitrogen (N_2).

The trophic state of aquatic systems can be estimated on the basis of concentrations of total N, total P, oxygen and chlorophyll *a*, and water transparency, with oligotrophic, mesotrophic and eutrophic waters having low, mid and high levels of nutrients, respectively (Smith *et al.*, 1999). The first signs of altered productivity as eutrophication develops are seen as increased biomass of phytoplankton, macroalgae and microphytobenthos, changes in nutrient ratios with subsequent shifts in phytoplankton community composition and seasonal patterns, shift towards opportunistic and fast-growing macroalgae, and increases in the frequency of toxic and harmful algal blooms (Cloern, 2001; Figures 8.1, 8.2). Increased plant biomass allows more food for first level consumers which usually are able to respond quickly to the new level of productivity, and may be able to limit plant growth, remineralising the excessive plant biomass, if environmental conditions remain suitable. It may be, however, more usual that the stimulation of biomass accumulation of phytoplankton, as well as ephemeral macroalgae, increases vertical fluxes of algal-derived organic matter to bottom waters and the sediments and develops hypoxia or anoxia (Figure 8.1). The formation of hypoxia and the responses of ecosystems in general depend on physical and geomorphological characteristics of the coastal area and therefore some areas show clearer symptoms of eutrophication than others (Cloern, 2001; Karlson *et al.*, 2002; Conley *et al.*, 2009a). As eutrophication development in many areas has coincided with the increased exploitation of higher-level consumers, marine ecosystems have experienced dramatic changes all over the world (Worm *et al.*, 2006).

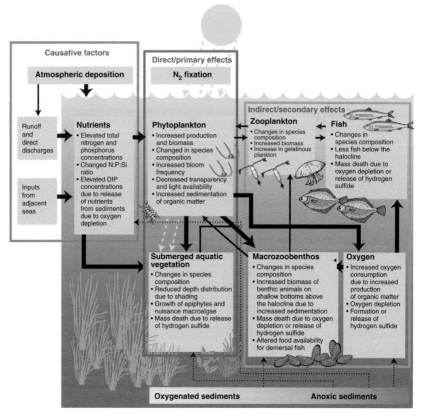

Figure 8.1 Conceptual model of eutrophication. The arrows indicate the interactions between different ecological compartments. Nutrient enrichment causes changes in the structure and function of marine ecosystems, as indicated with bold lines. Dashed lines indicate the release of hydrogen sulfide (H_2S) and phosphorus, which both occur under conditions of oxygen depletion. Adapted with permission from HELCOM (2010). A black and white version of this figure will appear in some formats. For the colour version, please refer to the plate section.

Indirect ecosystem responses to increased productivity often include a series of changes in abiotic and biotic parameters. Nutrient composition, dissolved organic matter, phytoplankton biomass and water transparency affect the distribution and abundance of benthic macrophytes, which are primary habitat-forming biotic features in the littoral zone. In deeper areas, oxygen conditions are affected by benthic metabolism and that influences the nutrient cycling of the sediments, changing them from nutrient sinks to nutrient sources or vice versa. Sudden changes in

Figure 8.2 (a) Bloom of ephemeral green algae. Image by S. Korpinen.
(b) Decaying filamentous ephemeral algae being washed ashore, destroying the recreational and ecological values of sandy beaches. Image by E. Bonsdorff. A black and white version of this figure will appear in some formats. For the colour version, please refer to the plate section.

oxygen conditions affect physical habitats for algae, invertebrates and fishes including catastrophic disturbances that cause mass mortality of animals and subtle changes in the seasonal patterns of key ecosystem functions such as primary production (Cloern, 2001; Karlson et al., 2002; Diáz and Rosenberg, 2008).

Accumulating scientific evidence shows a remarkable consistency in the positive relationships between algal biomass and nutrient enrichment in marine ecosystems, but there are also clear differences in impacts of eutrophication among aquatic systems, also between marine coastal and pelagic areas (Cloern, 2001). The N and P loadings in estuaries or coastal areas do not consistently explain the observed amount of primary production or phytoplankton biomass (Cloern, 2001). Compared to lakes, the net chlorophyll production in estuaries can be ten times lower per unit N input. The observed differences between lakes and estuaries have been found to depend on geomorphological features of the estuaries, which influence the nutrient availability and the formation of benthic hypoxia. Enclosed coastal areas and marginal seas are prone to oxygen deficiency as a result of slow water renewal and strong vertical stratification of the water column (Turner and Rabalais, 1994; Bonsdorff et al., 1997; Conley et al., 2009a). In a global meta-analysis, Elser et al. (2007) concluded that the strongest responses, especially to N or N + P enrichment, are for phytoplankton and to some extent for macro- and microalgae in rocky intertidal, temperate reefs and coral reefs.

8.3 The relationship between eutrophication and nutrient cycling

Marine nutrient cycling is an important ecosystem service supporting the productivity of areas, including terrestrial coastal ecosystems, and the diversity of aquatic food webs (Chapter 2). Eutrophication disrupts nutrient cycling by changing the physical and chemical conditions of the environment. In shallow coastal areas with long water residence times, altered nutrient cycling processes form positive feedback mechanisms, which enhance the eutrophication process. While there is little evidence so far that the acceleration of the global nitrogen cycle by humans has led to detectable changes in the marine N cycle on the global scale (Duce et al., 2008), the perturbation of the coastal N cycle has been well documented.

P enters the marine environment almost solely from land-based sources; a small portion may deposit from anthropogenic P emissions. In P-limited marine systems, such as in subtropical or tropical areas and coastal areas characterised by freshwater inputs, P enrichment causes clear increase in productivity and potential for problematic eutrophication. Phosphorus availability is, however, a eutrophicating factor also in N-limited systems as shown in several coastal areas and marginal seas (Tyrrel, 1999; Cloern 2001). Marine phosphorus ends up mainly in oxygenated sediments, where it is bound to iron (III) to form ferric phosphate. A small part also enriches terrestrial ecosystems as a result of grazing and predation in coastal areas by terrestrial animals and enrichment by the land-cast wreck. Oxygen deficiency decreases the redox potential of the sediment, reducing iron (III) to iron (II) which releases iron-bound P to the water. Significant amounts of P can be released from sediments. For example in the Baltic Sea, this so-called internal loading is an order of magnitude larger than anthropogenic inputs (HELCOM, 2009).

N enters the marine system from land-based inputs, human activities at sea and from the atmosphere. The two former inputs provide nitrogen as NH_4^+, NO_3^-, NO_2^- or in an organic form, whereas atmospheric N originates from anthropogenic emissions (NH_4^+, NO_2^- or NO_3^-) or as

N fixed by nitrogen-fixing cyanobacteria. NH_4^+ is the preferred source of nitrogen for plants because its assimilation does not involve a redox reaction and therefore requires little energy, but NO_3^- can be utilised as well with the help of nitrate reductase enzyme. The N bound to organisms falls to the seafloor where the organic nitrogen is mineralised to NH_4^+ by bacteria. In aerobic conditions NH_4^+ is nitrified to NO_3^-. NH_4^+ and NO_3^- are returned to primary producers in the euphotic water column by vertical mixing and upwelling.

In anaerobic conditions, bacteria reduce nitrates back into gaseous N_2 in a process called denitrification or directly convert nitrite, nitrate and ammonium into gaseous N_2 in anaerobic ammonium oxidation (anammox). Thus, both processes remove N from the marine environment. The denitrification and anammox processes are dependent on NO_2^- and NO_3^- produced in aerobic conditions by nitrification. Denitrification has traditionally been considered the most significant N removal process in oceans and coastal waters, but the role of anammox has been argued to be higher than thought (Kuenen, 2008). Direct reduction from nitrate to ammonium, a process called dissimilatory nitrate reduction to ammonium (DNRA), retains N as NH_4^+. In oceanic shelf waters, N removal by denitrification is greater than the land-based and atmospheric inputs (Hansell and Follows, 2008), keeping oceanic waters oligotrophic and N-limited.

In deeper coastal areas, denitrification increases in proportion to N loads during the early to mid-stages of eutrophication, but declines strongly as sediments become anoxic and highly sulfidic due to high biomass of decaying organic matter. In such conditions, more NH_4^+ is retained in the system due to the reduced denitrification but also because of greater mineralisation, and elevated levels of DNRA (Middelburg and Levin, 2009). As N retention in the sediment increases, the P:N ratio increases and consequently phytoplankton communities shift towards cyanobacterial dominance. Anammox process can, however, remove part of the NH_4^+ in the water column as N_2. Denitrification and anammox form major sinks for N and these losses occur both in the sediments and the water column. However, the recent observations of massive amounts of denitrification occurring in the water column of hypoxic zones in the open ocean have challenged the view of reduced efficiency of denitrification in low oxygen conditions (Deutsch et al., 2007; Vahtera et al., 2007).

In shallow coastal areas, denitrification is not a good nutrient filter because primary producers outcompete bacteria for DIN (dissolved inorganic nitrogen), thus inhibiting the denitrification process

(Dalsgaard, 2003). There does not appear to be a large difference in the effect of different primary producers on denitrification rates. So, a shift in the biological structure of the autotrophic community will not measurably affect denitrification rates. Also macroalgal mats suppress nitrification and denitrification but the effect is balanced by the shifting of the process from sediment surface to the mat itself.

8.4 Key sources of nutrients and organic matter to marine environments

Anthropogenic sources of nitrogen and phosphorus to marine waters are surprisingly similar in different marine areas of the world. The key factors behind the few observed differences in nutrient sources are population density, land use, traffic, use of artificial fertilisers, industrial emissions and other combustions, and progress in sewage management. Human activities have more than doubled the global availability of N and P to biological processes (Vitousek et al., 1997).

The primary cause of eutrophication worldwide has been the rapid intensification of agriculture (Matson et al., 1997). According to the FAO (FAO, 2011), total global fertiliser (N + P_2O_5 + K_2O) consumption, estimated at 170.7 million tonnes in 2010, was forecast to reach 175.7 million tonnes in 2011. With a successive growth of 2.0% per year, it is expected to reach 190.4 million tonnes by the end of 2015. The global production of agricultural fertilisers alone released <10 million metric tonnes of nitrogen in 1950, but reached 109 million metric tonnes in 2013 and may exceed 135 million metric tonnes of N by the year 2030 (Vitousek et al., 1997; FAO, 2011). The demand for phosphate was 42 million metric tonnes in 2011 with the expected annual growth rate of 1.9% (FAO, 2011). The highest demand for fertilisers is in East Europe and Central Asia (annual growth 3.4%), Latin America (3.3%) and West Asia (3.2%), whereas North America and West Europe have the lowest demand for fertilisers (FAO, 2011). East Asia is the largest consumer of fertilisers in the world (37% of all consumption), and Asia as a whole consumes 60% of the world's demand. The consumption in Europe (13%), North America (13%), South America (10%) and Oceania (1.6%) is smaller. Surplus N and P from agricultural fertilisers accumulate in soils and partly moves into surface waters via land erosion, surface runoff and melt waters. The total amount of P exported in runoff from the landscape to surface waters increases linearly with the soil P content. N may also migrate into groundwaters or enter the atmosphere via ammonia volatilisation and nitrous oxide production.

A growing source of agricultural N and P is animal farming, producing manure, waste waters and gaseous nitrogen emissions. Global meat consumption and production have increased 3% per year since the 1960s (FAO, 2006), the highest increase of 6.8% occurring in East Asia. Although the change has been obvious in developing countries, the actual consumption is still greatest in developed countries where annual per capita consumption of meat is 90 kg (<20 kg in most developing countries) (FAO, 2006). The consequent industrial-scale production and separation of crop and animal production has brought new challenges in the placement of manure and treatment of waste waters and ammonia.

Fish farming in the sea is a very high point source of nutrients directly to the system; a net cage farm of rainbow trout introduces as much phosphorus to the system as a moderate-sized city. Bonsdorff et al. (1997) estimated that the fish farms in the Åland islands polluted the sea 15 times more than the municipalities in the region. Most of the fish feed is imported from outside the farm areas.

Human waste waters from municipalities and single houses are significant sources of N and P. Although modern wastewater treatment plants remove over 95% of the phosphorus cost-efficiently by chemical (ferrosulfate binding) or biological (bacterial assimilation) treatment, N removal is more costly and is often not even aimed at in the treatment process. Quite often N is removed only as a side product in the removal of organic matter in the wastewater treatment and rates of removal rarely reach more than 50%. With modern N removal technologies, based on nitrification–denitrification treatment in large sludge pools, the removal of 90% of N can be achieved.

Waterborne nutrient inputs have greatly increased in coastal waters as a result of anthropogenic alterations to the hydrology of watersheds. To increase the area available for agriculture, humans have dried wetlands and deepened water channels resulting in faster runoff of water and associated organic matter. In northern forested areas, the paper and pulp industry straightened and deepened river channels to improve log transport and the large clear-felled areas have increased the runoff of organic matter. Globally, the surface of estuarine wetlands has decreased, as a result of urban development, and hence an important nutrient filter has been weakened.

Atmospheric deposition of nitrogen may contribute between 1% and 40% of the total nitrogen inputs to coastal ecosystems and is most pronounced near to emission sources (Howarth, 2008). For example, atmospheric deposition to the Baltic Sea is approximately 25% of the total, but sub-basin-scale differences are large (HELCOM, 2012). In Chesapeake Bay, atmospheric depositions comprise 12% and 6.5% of

TN (total nitrogen) and TP (total phosphorus) inputs, respectively (Kemp et al., 2005).

Elevated concentrations of dissolved organic and inorganic N (DON, DIN) are found in oceanic coastal and shelf regions which are influenced by freshwater inputs from continents and atmospheric deposition (reviewed in Gruber, 2008; Hansell and Follows, 2008). While most of this N is denitrified within the coastal area (Gruber, 2008), some material is also transported off-shelf, releasing N to the phytoplankton community. Waterborne N is more limited to shelf regions, but atmospheric inputs also enrich the high seas. However, today the most important source of N is biological N_2-fixation, even though the role of N deposition may grow due to increased anthropogenic N emissions (Gruber, 2008).

8.5 Impacts of eutrophication on different marine ecosystems

In shallow coastal areas the most important effect of nutrient enrichment on primary producers is likely the shift from the dominance of perennial macroalgae and seagrasses toward dominance of ephemeral macroalgae and pelagic microalgae. In sheltered or partly enclosed coastal areas such regime shifts may lead to alternative steady states which maintain altered species composition and new food web dynamics, and can be difficult to reverse with management measures (Valiela et al., 1997; McGlathery et al., 2007). There have been, however, some positive experiences of the reversal of eutrophication effects by reducing nutrient inputs leading to recolonisation of eelgrass, improved water transparency, lower DIN and DIP concentrations, reduced phytoplankton blooms and increased oxygen concentrations in coastal bottom waters (Cloern, 2001). In the following sections, we discuss the impacts of eutrophication in more detail in temperate ecosystems and deal separately with hard- and soft-substrate ecosystems, photic and aphotic zones and pelagic ecosystems.

8.5.1 Impacts of eutrophication on photic hard substratum ecosystems

Nutrient availability has been considered as the driving force in the community control of attached microalgae (periphyton) and macroalgae, increasing productivity and biomass, and altering species composition (Lapointe, 1997; Valiela et al., 1997; Hillebrand et al., 2000). On temperate rocky shores, eutrophication typically leads to reduced cover of perennial macroalgae, for example, a 40% reduction of *Ascophyllum nodosum* and *Fucus vesiculosus* in the Bay of Fundy, Canada (Worm, 2000). The winning competitors for the space are annual algae (e.g. *Ulva lactuca* or *Pylaiella littoralis*), which have highly dynamic life strategies, for example, propagule banks and fast growth rates (Korpinen *et al.*, 2007a, b) or filter feeders (*Mytilus edulis*, *Balanus* spp.), which benefit from the increased availability of food (Westerbom, 2006). The effects are greater if herbivore abundance is reduced as a result of commercial exploitation or habitat loss (e.g. Lotze and Milewski, 2004). Increased algal growth has also reduced the health and size of marine coral populations (McCook *et al.*, 2001).

Fast-growing filamentous species are capable of quickly taking advantage of nutrient pulses or otherwise high nutrient concentrations. The species are characterised by high surface to volume ratios enabling fast nutrient uptake, as in sea lettuce (*U. lactuca*) which has a thin and foliose thallus or *Ectocarpus* spp. and *Cladophora* spp. which have long filamentous thalli. The reproductive propagules of opportunistic species are more or less continuously available in the crevices of the hard substrate as a propagule bank (Worm *et al.*, 2006). These species compete efficiently for every available space, thereby threatening the colonisation of slow-growing perennial species, which usually have a very limited recruitment window. Thus, eutrophication affects macroalgal communities by altering the competitive environment.

Opportunistic algae species can also settle and grow as epiphytes on perennial macroalgae under eutrophied conditions (Korpinen *et al.*, 2007b). Increased biomass of the attached epiphytes increases drag forces on wave-exposed shores, which detach the ephemeral algae and cause loose-lying blooms of *Ulva intestinalis* and other opportunistic species (Valiela *et al.*, 1997; Sundbäck *et al.*, 2003). These drifting algal mats continue primary production and can accumulate in other habitats where they become entangled with macrophytes, provide a temporal habitat for grazers and consume oxygen from the underlying seabed (Norkko *et al.*, 2000).

Decreased light penetration into the water column is an obvious reason for the increased limitation of algae to shallow depths closer to the surface and altered zonation of macroalgal species. Scattering and absorption of light is dependent on the amount of particulate matter and dissolved substances in the water. The material in the water is typically living or dead organic particles (e.g. phytoplankton), small inorganic particles, or a combination of both, and dissolved coloured substances (e.g. humic substances). Also epiphytic growth by ephemeral algae and drifting algal mats shadow the growth of underlying species. For example, the lower depth limit of *F. vesiculosus* in the central Baltic Sea has decreased from 11.5 m in the 1940s to 8.5 m in the 1980s, in the eastern Kattegat from 12 m in the 1960s to 5 m in the 1990s and in the Estonian Archipelago from 8 m in the 1960s to 2–5 m in the late 1990s (Schubert and Schories, 2008; HELCOM, 2009). Long-term monitoring of macroalgae in the Baltic Sea has shown decreased growing depth of several red algae (e.g. *Polysiphonia elongata*, *Furcellaria lumbricalis* and *Cystoclonium purpureum*) and brown algae (*F. vesiculosus*, *F. serratus* and *Saccharina latissima*) (reviewed in Schubert and Schories, 2008). Significant correlations have been established between the depth limitation of bladderwrack (*F. vesiculosus*) and eelgrass (*Zostera marina*) and the increased N and P concentrations in several sub-basins of the Baltic Sea (summarised in HELCOM, 2009).

The decline of perennial algae at photic depths can, however, be caused by more complex aspects of the eutrophication process than water transparency alone. A well-studied example is the recruitment of bladderwrack (*F. vesiculosus*). The species has a limited recruitment time in mid-summer and its zygotes are only able to colonise space within a couple of metres of the parent plants. Thus, the species is vulnerable to the pre-emption of bare space by deposited matter and ephemeral algae. Increased abundance of mesograzers, direct nutrient effects on rhizome growth and allelopathic metabolites of the ephemeral macroalga *P. littoralis* also inhibit recruitment and growth of the species (Korpinen *et al.*, 2007b).

Nutrients have been found to change macroalgal species composition and to affect species richness of the macroalgal communities (Worm *et al.*, 2002; Korpinen *et al.*, 2007a). A large meta-analysis has recently shown that the reason for decreased species richness following nitrogen enrichment in terrestrial plant communities is both abundance-based, referring to the random loss hypothesis, and trait-based, referring to the hypothesis of competitive exclusion (Suding *et al.*, 2005). The former means that fertilisation increases plant biomass, therefore causing community-level thinning and the death of small individuals of all species. This random mortality is more likely to lead to the extinction of rare than of common species, thus reducing species richness. The trait-based mechanism works through interspecific competition for resources and is thus based on species' traits to adapt to changing nutrient conditions. In marine benthic environments, the mechanisms may be similar, although given the lack of below-ground competition among macroalgae, competition for recruitment space is more relevant. A meta-analysis by Worm *et al.*

(2002) found that the relationship between the plant species richness and productivity of an area is commonly unimodal for marine ecosystems, with maximal species richness being found at intermediate levels of productivity.

In a survey of grazer abundance across several sites differing in productivity, a positive association between grazer density and eutrophication has been noted (Worm and Lotze, 2006). Such enhanced abundance is expected in early stages of eutrophication when habitats of the grazers are still able to maintain suitable living conditions. The species composition of grazers also changes in slightly eutrophied sites. The enhanced growth of periphyton, plankton and particulate organic matter (POM) favours gastropod grazers and various bivalves. For example, along a nutrient gradient from net-cage fish farms, the crustacean fauna on *F. vesiculosus* increased with increasing distance, while the sites near the nutrient source were dominated by molluscs (Korpinen et al., 2010). More eutrophied sites become characterised by heavy epiphyte biomass, strong grazing effects on perennial algae and finally loss of macroalgal beds and the associated fauna, resulting in rocky surfaces covered by only filamentous algae (Kautsky, 1991).

8.5.2 Impacts of eutrophication on aphotic hard substratum ecosystems: focus on mussel beds

Subtidal hard substrata below the photic zone are characterised by attached fauna, such as mussels, barnacles and polyps, as well as various cnidarians and especially cold-water coral reefs. The mussel beds and corals are important habitat-forming species, hosting a rich community of invertebrate species and offering an important feeding habitat for demersal fish and marine mammals.

Filter feeders generally benefit from slight eutrophication due to increased food abundance. In the Bay of Fundy, Canada, Worm (2000) observed an increase of 10% in cover of filter feeders as a result of higher nutrient availability. Westerbom (2006) argued, however, that the increase of mussels can only be seen in the early stages of eutrophication, while further eutrophication leads to decline of mussel beds and has severe consequences for the food web. In the northern Baltic Sea the abundance of blue mussels and roach (*Rutilus rutilus*) have increased as a result of eutrophication of the area. The increased roach populations migrate seasonally to offshore mussel banks to feed on blue mussels, a phenomenon not observed in the past (Lappalainen et al., 2005).

Like grazers, filter feeders can potentially buffer the eutrophication process by incorporating organic matter (phytoplankton and inorganic particles) into their biomass and into sediments. Benthic bivalves, such as oysters and blue mussels, reduce concentrations of phytoplankton and other suspended particles, thereby increasing light levels reaching the sediment surface (Newell et al., 2005). The increased light conditions enable benthic primary production and support seagrass and macroalgal habitats. While grazing by zooplankton is limited in temperate latitudes by low abundance in the winter time, bivalve filtering continues throughout the year. Filter feeders were estimated to filter the water volume of the upper and middle Chesapeake Bay in less than 4 days in the nineteenth century, whereas in the current eutrophicated condition, where oyster reefs are covered by organic matter, the filtering time may be several hundred days (Kemp et al., 2005). Similar estimates of lost filtering potential have been given for other coastal areas along the Atlantic temperate coast (e.g. Newell et al., 2005). The top-down control of phytoplankton by bivalves has been substantially reviewed (e.g. Dame, 1996) and as much as 44% of chlorophyll a and 26% of POM can be cleared from the water column in sheltered locations (Souchu et al., 2001).

Benthic filtration can also affect nutrient cycling processes; estimates of digestion and assimilation of N from all POM vary from 20 to 90% (Newell et al., 2005). While a part of the filtered phytoplankton and other particles is assimilated in bivalve biomass, most of the nutrients, in the form of biodeposits, settle initially on the seabed, where they support enhanced production of deposit-feeding animals and benthic microalgae. In oxygenated sediments, ammonium is nitrified to nitrate by bacterial nitrification processes and a proportion of the nitrates migrate to the boundary layer between oxic and anoxic sediment where other denitrifying bacteria reduce them to gaseous nitrogen, thus removing N from the aquatic system. Denitrification does not occur in anoxic seabeds as N remains in the form of ammonia. Based on conservative assumptions, Newell et al. (2005) estimated that oysters are able to remove annually 750 mg N and 170 mg P per 1 g DW oyster in chlorophyll concentrations of 5.5–16 µg L^{-1} if the denitrification and burial of biodeposits are included. According to their estimates for Chesapeake Bay, about half of the N inputs and 3.5 times more than the P inputs to the system would be removed by oysters at a moderate density (10 m^{-2}). At their current density of 1 g DW m^{-2}, oysters can only remove 0.6 and 8% of annual N and P inputs, respectively. Naturally, the strong filtering capacity of bivalves and the potential to remove nutrients from the water have raised great interest in the use of filter feeders in bioremediation of eutrophic sites. Cultivation of blue mussels for that purpose has been practiced in several areas with varying results and efforts continue to find optimal, cost-efficient conditions for the process. The farmed blue mussels are harvested in order to reduce the nutrient loading of the system.

8.5.3 Impacts of eutrophication on photic sandy and muddy ecosystems

Marine photic soft-bottom habitats include a wide range of habitat types, defined by sediment type, grain size, content of organic matter and degree of trophic state. While muddy, naturally mesotrophic or eutrophic photic habitats, rich in organic matter, may be more abundant in estuaries or sheltered or partly enclosed bays, photic sandy habitats may be more oligotrophic and are frequently found on open coastlines. The habitats are characterised by burrowing infauna and rooted plants, seagrasses (e.g. *Zostera* spp., *Posidonia* spp.) in brackish or marine environments and other vascular plants (e.g. *Chara* spp., *Ruppia* spp., *Potamogeton* spp.) in less saline environments. Seagrass meadows (as well as perennial macroalgal communities) harbour the highest biodiversity in temperate coastal ecosystems.

Coastal bays differ from deep estuaries in two fundamental ways (Cloern, 2001): (1) the seafloor is mainly photic, allowing the dominance of benthic primary production, and (2) the nutrient cycling driven by benthic primary producers results in strong benthic–pelagic coupling. As water column stratification typically does not occur in these shallow systems, anoxia occurs in eutrophic bays only as episodic events – driven by the collapse of autotrophic communities – rather than as seasonal anoxia.

The conceptual model of eutrophication in shallow, coastal bays describes a shift in dominance from seagrasses and perennial macroalgae to ephemeral, bloom-forming macroalgae and epiphytes and ultimately to phytoplankton dominance in the most heavily eutrophied systems (Dahlgren and Kautsky, 2004; McGlathery et al., 2007). In shallow bays, benthic primary producers maintain good light conditions on the bottom and buffer against the shift to a phytoplankton-dominated system. The large benthic macrophytes have relatively high light requirements, high biomass per unit area and low nutrient turnover rates, whereas opportunistic algae have lower light requirements, low biomass per unit area and high nutrient turnover rate. Thus, the shift from large phytobenthos to filamentous algae and/or phytoplankton means a loss of a benthic biogenic habitat, loss of nutrient transfer from water column to seabed and loss of retention of nutrients in sediment (Cloern, 2001). This leads very often to benthic hypoxic events and increased sulfide concentrations, which also inhibit the nitrification/denitrification process (McGlathery et al., 2007).

Eutrophication impacts do not appear to depend linearly on N loading in shallow bays but include a threshold (Howarth and Marino, 2006) and, therefore, potential restoration attempts require greater efforts to reverse the status. In macrophyte-dominated shallow bays, high productivity can also be achieved at low levels of external N loading due to high rates of N fixation by benthic and epiphytic cyanobacteria. Brackish-water coastal bays are found in different vegetative states according to their total P loading and concentration, water exchange rates, water transparency, chlorophyll concentrations and cover of benthic macrophytes. The shift from macrophyte domination to phytoplankton domination does not only depend on phosphorus loading but also on water exchange and the depth of the bay; longer water residence and deeper depth favour the shift to phytoplankton dominance (Dahlgren and Kautsky, 2004). However, dominance of dense charophyte vegetation seems to buffer against eutrophication even under diminished water renewal (Appelgren and Mattila, 2005). Charophytes, though, are sensitive to dredging which rapidly shifts the benthic species composition to sparsely growing vascular plants with lower buffering capability.

Eutrophication first increases the total production in shallow bays but after that the production stays stable and no change to pre-eutrophic conditions can be seen (McGlathery et al., 2007). Perennial macrophytes and, in particular, seagrass roots and rhizomes retain nutrients and organic matter for years, reducing nutrient cycling in the system (Duarte, 2000). Moreover, the seagrass canopies also reduce water motion and thus enable retention of organic matter and prevent erosion of sediments. Thus, macrophytes are a buffer against external nutrient loading, acting as so-called coastal filters. This capability arises from the active oxygenation of bottom sediments, retaining P as ferrophosphates and creating suitable conditions for nitrification and denitrification. Seagrasses transfer oxygen and dissolved organic matter (DOM) from the water to the roots where extra material leaks to the sediment around the roots, creating a rhizosphere, a special microhabitat for bacterial activity and oxidisation of iron and sulfur, fixing phosphorus within the sediment. The anoxic conditions are prevented by active O_2 pumping, the strength of which depends on the amount of photosynthesis (Duarte, 2000; Borum et al., 2005).

As a result of the trapping of DOM, seagrass sediments are rich in organic materials compared to adjacent non-vegetated sediments and thus sensitive to anoxic bacterial metabolism. Eutrophication-induced phytoplankton and macroalgal blooms increase the organic matter in the sediments of the seagrass meadows. Reduced photosynthetic oxygen production of seagrasses decreases oxygen translocation and release to the rhizosphere (Duarte, 2000). In addition, oxygen depletion and the consequent high ammonium (NH_4^+) concentrations can cause seagrass mortality, especially for new shoots (Borum et al., 2005). Thus, exceeding the tipping point of sediment oxygenation is the primary factor in the shift to a phytoplankton dominated state. However, recovery may be possible if external nutrient loading is reduced, water transparency increases and macrophytes can recolonise the seafloor and start oxygenating sediment, cycling nutrients and retaining them.

Large-scale declining trends of seagrass are documented from both temperate and tropical ecosystems (Duarte, 2000). Globally, anthropogenic eutrophication is the most common and significant cause of seagrass decline (Orth et al., 2006). Seagrasses are adapted to oligotrophic conditions as they require only little nutrients for their growth and they are efficient nutrient recyclers. Sewage and aquaculture are therefore the worst drivers for eutrophication impacts in seagrass meadows. A strong inverse relationship has been shown between eelgrass (*Zostera marina*) habitat and nitrogen concentration in the overlying water (Olesen, 1996). A comparison of historic and recent data sets on depth limits of eelgrass meadows from the same sites in the Baltic Sea showed that in 1900 the depth limit of eelgrass meadows averaged 5–6 m in estuaries and 7–8 m in open waters, while in the 1990s the depth limit of the meadows was reduced by about 50% to 2–3 m in estuaries and 4–5 m in open waters (HELCOM, 2009). A similar upward shift was found in other Baltic monitoring programmes, too, and a consequent loss of meadows of up to 85% (Schubert and Schories, 2008). Interestingly, Schubert and Schories (2008) report on a coastal area which does not have agriculture in the catchment area but good water exchange, and it has shown almost no shift in the macrophyte abundance or depth distribution since the 1930s.

As described above, eutrophication enhances the biomass of filamentous algae which grow not only on hard substrata but also as epiphytes on perennial plants. Epiphytes attenuate up to 90% of the light at the scale of the seagrass blade (Sand-Jensen, 1977; Howard and Short, 1986). The longer the filaments grow the easier they detach in wave motion. Bloom-forming macroalgae (e.g. *Ulva* spp., *Chaetomorpha* spp., *Enteromorpha* spp., *Gracilaria* sp., *Polysiphonia* sp.) continue to live as an unattached drifting mat. These form dense canopies up to 1 m thick over seagrass beds in eutrophic waters and can decrease light levels by similar magnitudes or higher as the epiphytes (Sundbäck and McGlathery, 2005). The drifting algal mats become entangled with seagrass meadows or cover large areas of seabed below the influence of waves and consume oxygen from the underlying seabed. High mortality of benthic fauna has been found from areas affected by the loose-lying algae (Norkko and Bonsdorff, 1996), although the algae can also provide an alternative habitat to a variety of invertebrate species (Salovius et al., 2005).

Numerous experimental and observational studies from coastal bays indicate the capacity of a variety of benthic grazers including filter-feeding bivalves, ascidians, sponges, polychaetes, amphipods, gastropods and small herbivorous fish to buffer the effects of nutrient enrichment (reviewed by McGlathery et al., 2007). The grazer potential to consume plant biomass will, however, saturate as productivity increases. Depending on local geomorphological characteristics accumulation of dead plant material on the seabed increases and grazer abundance will be negatively impacted by low oxygen, high sulfide and NH_4^+. In early stages of eutrophication, heavy grazing could buffer against loss of seagrass by removing the fast-growing epiphytes and phytoplankton in the deeper end of the meadows where seagrass is already stressed by the lower light availability (Duarte, 2000).

Nutrient enrichment has been attributed to the global loss of salt marshes, which are highly productive wetland areas providing several ecosystem services. Increased nitrogen levels have been shown to decrease the root:shoot biomass ratio of *Spartina* spp., the main species of the ecosystem. The decreased root biomass destabilises the marsh soil, leading to increased erosion and bare mud flats, and decreased *Spartina* coverage and carbon-storage potential of marshes (Deegan et al., 2012).

McGlathery et al. (2007) suggest the following generalisations, as eutrophication proceeds in coastal bays:

1. Long-term retention of recalcitrant, dissolved and particulate organic matter will decline as seagrasses are replaced by algae with less refractory material.
2. Benthic grazers buffer the early effects of nutrient enrichment but consumption rates will decline as physico-chemical conditions stress consumer populations.
3. Mass transport of plant-bound nutrients will increase because attached perennial macrophytes will be replaced by unattached, ephemeral algae that move with the water.

4. Denitrification will be an unimportant sink for N because primary producers typically outcompete bacteria for available N and partitioning of nitrate reduction will shift to DNRA in later stages of eutrophication.

8.5.4 Impacts of eutrophication on aphotic sandy and muddy ecosystems

Aphotic soft-bottom habitats are the dominant benthic habitat types in shallow seas worldwide. The habitats are characterised by diverse fauna, including burrowing infauna living at species-specific sediment depths, epifauna moving and feeding in the sediment-water interface, benthic-feeding zooplankton and nekton as well as demersal fish species.

Development of oxygen deficiency impacts benthic communities
Perhaps the single strongest factor influencing the biodiversity of benthic communities is the increased prevalence of oxygen-depleted bottom water since the 1960s (Díaz and Rosenberg, 2008). The increasing frequency and magnitude of oxygen (O_2) depletion from bottom waters is a phenomenon linked closely with anthropogenic nutrient enrichment. Hypoxia develops as the increased plant biomass on the seabed degrades in chemical and biological consumption processes and due to the increased metabolisation of the increased abundance of benthic consumers. Anoxic conditions result in the formation of hydrogen sulfide (H_2S), which is lethal to higher organisms.

Oxygen deficiency, i.e. hypoxia and in extreme cases anoxia, is a global phenomenon in coastal waters and shallow seas as a result of the eutrophication process (e.g. Craig *et al.*, 2001; Kemp *et al.*, 2005; HELCOM, 2009). When the phenomenon was first analysed on a global scale (up to 1969), the hotspots were centred around several estuaries along the coasts of the United States (primarily north-east Atlantic) and the North Sea/Baltic Sea (Díaz *et al.*, 2010; Figure 8.3). Twenty years later the problem had grown substantially in the same regions, and also in the Gulf of Mexico, the Mediterranean Sea, South-east Asia (primarily Japan) and a few sites in the southern hemisphere (Australia and south-east Brazil). Today the situation is even worse, with over 600 sites and cases of reported coastal hypoxia related to eutrophication, worldwide, including previously unaffected regions even in the Pacific (Díaz and Rosenberg, 2008; Conley *et al.*, 2009a, 2011; Figure 8.3). The prime hotspots are still the East Coast of the United States (and a growing number of estuaries, bays and harbours on the US West Coast) and north-west Europe.

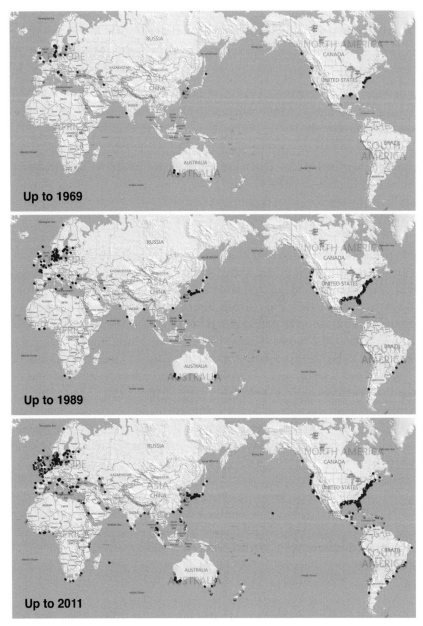

Figure 8.3 Global patterns in the development of coastal hypoxia. Each red dot represents a documented case of hypoxia related to human activities. Blue dots indicate previously hypoxic sites that have improved and yellow dots indicate eutrophic sites. Numbers of sites are cumulative through time. Based on Díaz *et al.* (2010), adapted with permission from http://www.wri.org/project/eutrophication. A black and white version of this figure will appear in some formats. For the colour version, please refer to the plate section.

The continental shelf of the north-western Gulf of Mexico has intense summer hypoxia because of freshwater inflow in the spring leading to salinity stratification, high nutrient levels and phytoplankton production in the upper water column and the deposition of this material to the seabed (Craig et al., 2001). Autumn storms break down the stratification and bring oxygenated water to the seabed. The Baltic is the largest coastal area in the world to suffer from eutrophication-induced hypoxia (Karlson et al., 2002; Conley et al., 2009a). Due to strong vertical and horizontal salinity gradients, large areas of the central basin are constantly anoxic/hypoxic and the size of the areas varies annually. Seasonal hypoxia is found in some of the shallower gulfs, while the northernmost sub-basins are well oxygenated due to lower nutrient inputs (HELCOM, 2009).

While advective and vertical mixing of the water column may prevent the formation of hypoxia even under high nutrient loads, climate change is predicted to increase the occurrence of hypoxia as a response to higher water temperatures (increased metabolisation, lower oxygen saturation), increased stratification due to higher freshwater runoff and increased runoff of organic matter and nutrients from catchment areas (Meier et al., 2011). Repeated hypoxic events enhance eutrophication with positive feedback mechanisms and make the system susceptible to further hypoxia (Conley et al., 2009b). Hypoxia is related to nutrient inputs in a nonlinear way with a threshold and hysteresis can arise, reducing the chance of a return to the original status (Conley et al., 2009b).

Hypoxia is an escalating problem with enormous impacts on coastal ecosystems and the services they provide. Hypoxia diminishes biodiversity (Rabalais, 2005), and with an increase in scale (time and system size) the potential for human activities to counteract the trend is reduced (Rabalais et al., 2010). As the increase in coastal hypoxia ('dead zones') has been exponential and growing for at least half a century, the problem will most likely prevail throughout the twenty-second century unless very strict measures are taken to reduce nutrient input from land on a global scale.

Responses of benthic fauna to hypoxia
Healthy benthic communities play an important role in the mineralisation of organic matter settling on the seafloor and generally respond to organic enrichment in a predictable manner. In contrast to macrophytes, the macrozoobenthos does not respond directly to increased levels of nutrients but to the organic matter.

Sediments and the benthic communities are considered as particularly sensitive parts of the coastal ecosystem to eutrophication and hypoxia. The Pearson–Rosenberg (P–R) model of the responses of benthic fauna to input of organic matter predicts that benthic species richness, biomass and abundance may first increase slightly with increasing enrichment and then the species richness decreases smoothly as sensitive species disappear first and then increasing hypoxia reduces species from the site. Biomass and abundance may, however, have a second peak at a moderate increase of organic matter as a result of increase of opportunistic species when hypoxia has not yet formed (Pearson and Rosenberg, 1978). The model has been shown to be valid for many coastal regions of the world, but may be limited to geomorphologic environments characterised by limited water exchanges in their bottom waters and with predominantly silt-clay sediments, typical of sedimentation basins (Gray et al., 2002). A similar response of species richness has been shown by the unimodal productivity model, where increasing productivity of the site first increases species richness and later reduces it, causing a unimodal response graph (e.g. Rosenzweig and Abramsky, 1993; Worm et al., 2002). The P–R model also resembles the intermediate disturbance hypothesis, if organic enrichment is considered as a disturbance gradient (Rosenberg et al., 2004). Nilsson and Rosenberg (2000) photographed the successional stages of the sediment profile and its fauna after hypoxic events. They noticed that the faunal communities follow successional stages suggested by the P–R model and developed a benthic habitat quality (BHQ) index to categorise the state of the infaunal community. This is further supported by the large-scale studies by Rumohr et al. (1996), Bonsdorff and Pearson (1999) and Villnäs and Norkko (2011).

After the formation of hypoxia, macrofauna disappear usually at oxygen concentrations below 1.4 mg l^{-1}, while meiofaunal taxa do not show significant declines in abundance at those concentrations (reviewed in Gray et al., 2002 and in Karlson et al., 2002). The concentration of 2 mg l^{-1} is often considered the critical threshold for decline of macrofauna and that seems to hold well as an average sensitivity threshold in global analyses (Vaquer-Sunyer and Duarte, 2008). However, the thresholds of oxygen deficiency for survival of benthic organisms are species specific and the variation in the sensitivity is high. The traditional threshold of 2 mg O_2 l^{-1} seems to be far too low for over half of the organisms (Vaquer-Sunyer and Duarte, 2008), which explains also the early decline of species richness in the P–R model. Crustaceans, echinoderms and fish have been found to be more sensitive to hypoxia than annelids, cnidarians, bivalves and gastropods. Median lethal concentrations were even 8.6 mg O_2 l^{-1} for a crayfish species (all taxa 2.05) and sublethal responses were found at 10.2 mg O_2 l^{-1} for cod (all taxa 2.61). Hence, eutrophication-induced hypoxia modifies the species composition after relatively small reductions in oxygen levels.

Benthic fauna play an important role in ecosystem resistance to the formation of hypoxia in sediments by their burrowing, irrigation and feeding activity. Bioturbation enhances the vertical penetration of oxygen, which alters the rates and pathways of benthic mineralisation and nutrient cycling, and is strongly influenced by species composition and abundance (Lohrer et al., 2004; Karlson et al., 2007). The increased surface area of oxic–anoxic interfaces and the excretion of ammonia by bioturbators facilitate the nitrification and denitrification processes leading to N removal from the system. Also the sediment P-retention capacity depends on ventilation and redox conditions, which are modified by bioturbation. In hypoxic conditions, benthic uptake and processing of organic matter may shift from macrofauna to meiofauna or bacteria, which are not as efficient at processing organic matter (Woulds et al., 2007). Thus, a reduction of bioturbation may decrease the natural purification capacity and increase the internal nutrient loading of sediments (Karlson et al., 2007). Benthic ecosystem functionality has been shown to change in threshold-like shifts which take place when species tolerance levels are exceeded (Villnäs et al., 2012).

Large deep-burrowing, actively bioturbating species are more prone to hypoxia because their long generation times prevent development of viable populations in an environment facing frequent hypoxic events (Solan et al., 2004). The decline of the macrozoobenthos may lead to functional extinction – i.e. loss of a specific function in the ecosystem – due to the low abundance of a species. Usually surface-deposit feeders are favoured over subsurface or suspension feeders. However, in locations where hypoxia is an aperiodic event, mobile species are able to re-occupy previously hypoxic areas fairly rapidly (days to weeks) (Craig et al., 2001). In certain conditions unwanted non-native invasive species may turn out to perform a positive function in this respect, i.e. irrigate the hypoxic bottom water and hence in the long run reduce the leaking of nutrients from the sediment into the water mass (Norkko et al., 2012). In the Baltic Sea, where anoxia/hypoxia is a more or less permanent feature at present, macrobenthic communities in deeper water are never fully developed due to low oxygen concentrations and are characterised by small shallow-dwelling species (Rumohr et al., 1996; Bonsdorff, 2006;

Villnäs and Norkko, 2011). Macrobenthic communities in deeper waters are severely degraded and below a 40-year average for the entire area.

Mobile species, such as turtles, fish and invertebrates, tend to move away from areas of low dissolved oxygen (reviewed by Craig et al., 2001). In the Baltic Sea, oxygen depletion can lead to increased catches of crustaceans and fish while these animals attempt to escape low oxygen conditions (HELCOM, 2009). In the Gulf of Mexico, benthic-feeding sea turtles did not occur in hypoxic areas, even though they were abundant in nearby areas. Lower trophic-level organisms may, however, decrease activity or change behaviour when exposed to low dissolved oxygen, possibly to conserve energy until levels increase again. The predatory isopod *Saduria entomon* was inhibited in its predatory activities on the Baltic Sea benthos (Sandberg et al., 1994) and the blue crab *Callinectes sapidus* altered its predatory behaviour (Taylor and Eggleston, 2000). Some species are able to use alternative breathing modes when subjected to low dissolved oxygen that may be ecologically costly or physiologically less efficient. Increased energy requirements for the maintenance of metabolism may impact growth, reproduction and ecological performance.

8.5.5 Impacts of eutrophication on planktonic communities

The general responses of pelagic ecosystems to nutrient enrichment are a gradual change towards:

1. increased planktonic primary production compared to benthic production;
2. a dominance of microbial food webs over the 'classic' planktonic food chain from small to large organisms;
3. a dominance of non-siliceous phytoplankton species over diatom species;
4. a dominance of gelatinous zooplankton (jellyfish) over crustacean zooplankton;
5. increased sedimentation of organic matter to the seafloor; and
6. loss of higher life forms owing to poor benthic oxygen conditions, including fish and invertebrates feeding on benthic prey.

Eutrophication increases the abundance of phytoplankton
Phytoplankton respond rapidly and accurately to changes in nutrient levels and, therefore, the biomass and species composition of phytoplankton can be used as indicators of eutrophication (Uusitalo et al., 2013).

Chlorophyll *a* concentration is a commonly used proxy of phytoplankton biomass and status of eutrophication, because all phytoplankton cells contain chlorophyll *a* as their main photosynthetic pigment (HELCOM, 2009; OSPAR, 2010; US EPA, 2012). For example, the shift of the Baltic Sea from an oligotrophic basin to a highly eutrophied one is exemplified by the clear increase in chlorophyll concentrations between early and late twentieth century from 48–55 mg C m^{-3} to 83–216 mg C m^{-3}, the increase occurring in the 1960–70s (reviewed in Wasmund and Siegel, 2008). The nutrient inputs to the Baltic Sea have declined since early 1990s, which have resulted in a decrease in phytoplankton abundance.

Marine primary production seems to depend on nitrogen as the proximate limiting nutrient, whereas phosphorus availability may set the level of productivity over longer time periods (Tyrrel, 1999). The two nutrients occur in aquatic systems in the molar ratio of 15–16 (N):1 (P). In freshwater systems, estuaries and fjords as well as in many subtropical and tropical systems, P may be more limiting than N (Howarth and Marino, 2006; Howarth, 2008) and some coastal marine areas can be co-limited by both the nutrients (HELCOM, 2009). In some coastal systems the limiting element may be silica (Si), as its input to the water has not increased along with N and P, whereas in some oceanic systems the limiting element can be also iron (Fe) or molybdenum (Mo) (e.g. Sarthou et al., 2005).

Impacts of eutrophication on phytoplankton species composition
Nutrient availability and the ratio of N and P (and possibly Si, Fe and Mo) are important factors affecting the species composition of phytoplankton communities. In eutrophicated marine areas, typical changes in phytoplankton species composition are a shift from Chrysophyta and diatoms to flagellated algae (e.g. cryptophytes) and a dominance of nano- and picophytoplankton (green algae and cyanobacteria). The shifts lead often to algal blooms, many of which are toxic or which may not be consumed effectively by aquatic grazers. As dinoflagellates are less nutritious than diatoms and Chrysophyta, this shift in dominance may have consequences for the trophic web.

Anthropogenic nutrient inputs provide N and P and other elements in ratios which are not natural to the system. The ratios of incoming nutrients cause shifts in species composition of phytoplankton communities as species have specific strategies in assimilating nutrients. A common feature is the decrease in the intensity of the diatom blooms and changes of species in the diatom community as a result of decreased proportions of

Si in comparison to N and P, especially in the coastal waters (e.g. Hällfors et al., 2013). Once Si is depleted, high levels of inorganic P may promote dinoflagellates, in particular when the N:P ratio is low (Howarth and Marino, 2006). Wasmund and Siegel (2008) reviewed changes in the phytoplankton spring bloom in the Baltic Sea and noticed that, although diatoms and dinoflagellates co-dominate the spring bloom, the proportion of diatoms has decreased in several sub-basins. The shift from diatoms to dinoflagellates in the spring can influence the nutrient dynamics in the summer and organic matter load to the sediment as the diatoms usually settle to the seabed at the end of the bloom, whereas the dinoflagellates are mostly remineralised in the upper water layers (Tamelander and Heiskanen, 2004).

The N:P ratio changes also as a result of different nutrient cycling processes in eutrophied systems. The P release from reduced sediments (see the section on nutrient cycles) decreases the N:P ratio and leads to greater blooms of nitrogen-fixing cyanobacteria, which are P limited but can produce their own N from atmospheric N_2 (Vahtera et al., 2007). Common species forming extensive cyanobacterial blooms are the N_2-fixing, filamentous *Nodularia spumigena* (hepatoxic), *Aphanizomenon flos-aquae* (non-toxic) and *Anabaena* spp. (potentially neurotoxic). In coastal areas hepatoxic *Microcystis* spp. are also dominant. Typically, cyanobacterial blooms start growing after the pool of DIN has been emptied, but up-welled phosphorus may also initiate the blooms. Variations in the intensity and occurrence of blooms are dependent upon winter nutrient conditions, surface layer salinity, temperature and solar irradiation, summer stratification conditions, upwelling, and frontal processes (Vahtera et al., 2007). Meteorological conditions also shape the spatial occurrence of blooms. Sedimentary and grazing losses of filamentous species are generally small, and most of the biomass is decomposed in the surface layer. Blooms are terminated by several factors, including nutrient limitation, mixing events, decreased solar irradiation, decreasing water temperature, and possibly viral lysis.

Eutrophication alters the seasonality of phytoplankton succession. In Atlantic North America, the time of phytoplankton blooms has changed from the 1930s to the 1990s: strong spring blooms of diatoms in the 1930s have been largely replaced by blooms of dinoflagellates in the summer and autumn (Lotze and Milewski, 2004). It has been shown that nutrient inputs – either from anthropogenic sources, upwelling of deep water or sediments – enhance summer blooms of diatoms, dinoflagellates or cyanobacteria (Wasmund and Siegel, 2008). The decline of the highly nutritious diatoms has, however, a profound effect on energy flow in the food web.

Many species of the cyanobacteria, dinoflagellates and nano- and picoplankton produce toxic blooms, known as harmful algal blooms (HAB). According to the review by Cloern (2001), there is clear evidence of increased frequency of dinoflagellate blooms producing toxins (so-called red tides) and resulting in fish kills and shellfish closures. This shift can also be reversed as shown, for example, in the oligohaline Potomac River estuary in Chesapeake Bay where the frequency of summer blooms of the toxic cyanobacterium *Microcystis aeruginosa* declined sharply in the early 1970s as a result of P removal from sewage waters. The eutrophication-generated microbial food has also increased mixotrophy which was recognised as a significant factor in maintaining HAB (Burkholder *et al.*, 2008).

Eutrophication-induced changes in phytoplankton community structure and O_2 conditions of estuaries can increase the significance of the microbial food web and the importance of gelatinous predators (e.g. Purcell *et al.*, 2001; Turner, 2001). The microbial food web, also called a microbial loop, operates where small-sized organisms dominate production throughout the year. In such systems, sedimentation is unimportant as the particles are remineralised by microorganisms within the surface layers (Kiørboe *et al.*, 1996). In such systems grazing control is highly efficient and the system is rarely food-limited. However, if the phytoplankton change is episodic there is considerable sedimentation and oxygen deficits can occur in deeper waters (Gray, 2002).

Impacts of eutrophication on zooplankton
Marine zooplankton biomass does not seem to increase as a response to increased productivity (Micheli, 1999). It seems that eutrophication increases zooplankton abundance and biomass only in estuarine areas, where smaller species have become dominant (Telesh, 2004). However, Kemp *et al.* (2005) noted in their review of the state of the Chesapeake Bay ecosystem that the zooplankton community responded only weakly if at all to increased phytoplankton biomass. As many marine food webs have shifted towards predominance of zooplanktivorous mesopredators as a result of overfishing, the increased predation pressure may also mask any bottom-up enhancement of zooplankton. However, a typical change in the zooplankton community in eutrophicated ecosystems is a shift from copepods to heterotrophic dinoflagellates, ciliates and rotifers (Park and Marshall, 1999; Pinto-Coelho *et al.*, 2005).

8.6 Combined effects of eutrophication and exploitation of trophic webs

The view that resource availability (e.g. nutrients) and top-down forces (i.e. grazing and predation) interact in community regulation has gained growing attention (e.g. Borer *et al.*, 2006; Burkepile and Hay, 2006; Schmitz *et al.*, 2006; Heck and Valentine, 2007; Marczak *et al.*, 2007; Svensson *et al.*, 2007). A standing hypothesis states that the productivity of the environment is reflected in higher trophic levels (Oksanen *et al.*, 1981), whereas a recent meta-analysis of manipulative experiments in terrestrial, marine and freshwater ecosystems suggested that top-down control is more efficient than bottom-up control in cascading through the food web (Borer *et al.*, 2006).

In general, increases of fish stocks and, in particular, the ratio of pelagic to demersal fish species have been associated with the early stages of eutrophication (de Leiva Moreno *et al.*, 2000; Rabalais *et al.*, 2009). In Limfjorden, Denmark, decreasing trends of most of the flatfishes, other demersal species and gadoids have been observed, while the biomass of small, pelagic and opportunistic species has increased (jellyfish, horse mackerel, black goby, pipefish and stickleback) (ICES, 2011). Similarly, Fortibuoni *et al.* (2010) describe the increase of small-bodied species in the Adriatic Sea and Ljunggren *et al.* (2010) report of the increased three-spined stickleback (*Gasterosteus aculeatus*) abundance in the northern Baltic Sea. The Baltic herring (*Clupea harengus membras*) has greatly increased in the Bothnian Sea, a small sub-basin of the Baltic Sea which has recently started to eutrophicate.

In Chesapeake Bay, the ratio of pelagic to demersal species increased from 1.90 in the 1960s to 2.66 in the 1990s and the relation of commercial fishery yield to net primary production declined from 0.54% to 0.45% between 1980s and mid-1990s, indicating declining trophic efficiency in the Bay (Kemp *et al.*, 2005). The conceptual model by Caddy (2000) suggests that eutrophication effects on fish communities follow a sequence of three stages: (1) nutrient-enhanced production of demersal and pelagic species as a response to more food, (2) declines of demersal fish due to increased hypoxic events but continued increase in pelagic fish species, and (3) a general decline in total fish production under conditions of broadly deteriorating water and habitat quality. Österblom *et al.* (2007) argued that the exploitation of fish and eutrophication have caused a series of anthropogenic impacts, such as population crashes of seals, increased primary production and overfishing of cod. According to this model, the increased primary production in 1950–70 enhanced the abundance of fish in the Baltic.

External subsidies, such as nutrient loading from adjacent sources, have been found to increase consumer abundance in terrestrial, freshwater and marine communities (Marczak *et al.*, 2007). An increase in resources may lead to oscillatory dynamics in consumer populations, leading in extreme cases to population extinctions (the 'paradox of enrichment'; see Rosenzweig, 1971). Since, however, these are not common in nature, Murdoch *et al.* (1998) suggest that inedible species, such as opportunistic algae, may act as nutrient sponges and thereby stabilise ecosystems. Furthermore, in multitrophic food chains the paradox of enrichment does not lead to grazer extinction if predators inhibit the population density of grazers (Oksanen and Oksanen, 2000; Vos *et al.*, 2004).

Heck and Valentine (2007) in their synthesis of top-down and bottom-up regulation challenge the view of bottom-up driven eutrophication process and argue that the well-known outcomes of eutrophication are substantially caused by reduced herbivore abundance rather than by nutrients alone. They base their claim on the strong result that grazing is a much stronger regulator of algae than nutrient supply, and that the harvesting of top predators (fish, seals or birds) at the fourth trophic level has led to dampened grazing intensity (see also Burkepile and Hay, 2006 and Figure 4.5). Historical hunting, persecution and fishing had already depleted the big predatory species in many marine areas a hundred years ago (e.g. Österblom *et al.*, 2007). The ongoing overexploitation of fish stocks has received much attention globally (e.g. Worm *et al.*, 2006) and the unintentional bycatch of turtles, marine mammals and seabirds in fishing gears severely reduces populations of big predators (Heithaus *et al.*, 2008).

8.7 An overview of the occurrence of eutrophication in the marine environment

There are large differences among world's coastal ecosystems in the magnitude and character of their responses to enrichment. As almost all of the anthropogenic phosphorus and most of the anthropogenic nitrogen enter marine systems from land-based sources, eutrophication problems are greatest in estuaries, fjords, bays, archipelagos and enclosed sea areas. Cloern (2001) lists estuarine coastal systems that appear to be very sensitive to change in nutrient inputs (e.g. Chesapeake Bay, Adriatic Sea, Baltic Sea, Black Sea, northern Gulf of Mexico) and others that appear to have system attributes that dampen the direct responses to enrichment (e.g. San Francisco Bay, Bay of Brest, Ythan Estuary, Moresby Estuary, Westerschelde Estuary). The filters are inherent physical and biological attributes (e.g. tidal energy, salinity gradient generating vertical stratification, horizontal transport and the residence times, optical properties and abundance of suspension feeders).

The Baltic Sea is one of the world's largest brackish-water basins (402 000 km^2) with a catchment area of 1.7 million km^2, and eutrophication is its main environmental problem (Vahtera et al., 2007; HELCOM, 2009). In 2008 the total inputs of P and N to the Baltic Sea were 29 000 metric tonnes and 859 600 metric tonnes, respectively, but the sediment-released nutrients are considered as a much higher annual source of nutrients (HELCOM, 2012). The P and N inputs per square kilometre varied between 9 and 62 kg and 200 and 1700 kg. Pollution of the sea basin had already started in medieval times, but the last century has seen very intensive industrialisation, population growth, agriculture and animal farming. As a consequence, the last century nutrient loads have increased by c. 2.5 times for nitrogen and 3.7 times for phosphorus (Savchuk et al., 2008). In modern times, most of the deep sub-basins are continuously hypoxic. During years with large areas of hypoxia, low oxygen zones migrate higher up into the water column and coastal areas, and the individual basins become connected to form one large hypoxic area (Karlson et al., 2002; Conley et al., 2009a). The hypoxic sediments continuously release sediment bound phosphates to the water column to foster intensive cyanobacterial blooms (toxic *Nodularia spumigena* and non-toxic *Aphanizomenon flos-aquae*) at seasons when nitrogen is limiting the phytoplankton community (HELCOM, 2009).

The US National Coastal Condition Report (US EPA, 2012) shows that the overall water quality – assessed as concentrations of O_2, chlorophyll *a*, DIN and DIP, and water transparency – is less than good in all the four coastal regions of the US continental marine coasts. The state of the coasts has not improved during the past 10 years. Chesapeake Bay is a very productive aquatic system on the US Atlantic coast. It has a long water residence time, stratified water column and several shallow channels alongside the main channel. The water area of the bay is 7200 km^2 and the catchment area is 103 000 km^2. The first signs of organic enrichment appeared in sediment profiles c. 200 years ago, whereas increased phytoplankton abundance and decreased water clarity became apparent 100 years ago when benthic autotrophs were estimated to have a major role in the Bay's primary production (Kemp et al., 2005). The eutrophication of the Bay caused a wide loss of vascular plants and oysters and deep-water hypoxia in 1950–60s. The estimates of TN and TP loads in Chesapeake Bay and its tributaries during 1990–2011 were on average 21 400 kg N km^{-2} y^{-1}, and 1200 kg P km^{-2} y^{-1} (www.chesapeakebay.net). Improved sewage treatment has decreased algal biomass, higher water column O_2 concentrations, extensive recolonisation of macrophytes and increased water clarity in some parts of the Bay.

In the Bay of Fundy, Atlantic Canada, increasing human population, fish processing, plants, aquaculture operations, as well as loss of wetlands, introduction of artificial fertilisers, and atmospheric deposition have all contributed to high nutrient inputs (Lotze and Milewski, 2004). At eutrophied sites, per cent cover of perennial rockweeds (*A. nodosum, F. vesiculosus*) was reduced and replaced by annual algae (*U. intestinalis, U. lactuca*) or filter feeders (*M. edulis, Balanus* spp.) and diatom dominance shifted to dinoflagellate dominance. Lotze and Milewski (2004) summarise that similar bottom-up impacts – concurring with the multiple top-down (exploitation) and 'side-in' (habitat destruction, pollution) impacts – have occurred in the estuaries and coastal waters of the entire Atlantic North America since the colonisation of the continent.

The Mediterranean Sea is characterised by low primary production and low phytoplankton biomass, resulting in high transparency and deep light penetration into the water column. Eutrophication affects the Mediterranean mainly along the northern coastline such as the Adriatic Sea and the Nile Delta (UNEP/MAP-Plan Bleu, 2009). Eutrophication has increased gradually over the last decades. Altered nutrient ratios have changed the phytoplankton species composition from diatoms to non-siliceous species. High chlorophyll concentrations are found adjacent to estuaries and large cities. In contrast, offshore waters are almost oligotrophic, except in deep waters which are rich in nutrients.

In the Black Sea, during the last three decades, eutrophication has been identified as a key ecological problem for the coastal regions and especially for its northwestern part where strong anthropogenic nutrient and pollution loads – starting in the early 1970s from the River Danube – resulted in dramatic alterations in chemical and biological regimes (BSC, 2008). The TN input from the River Danube catchments increased from about 400 kt y^{-1} in the 1950s to 900 kt y^{-1} in 1985–1990 and then reduced to 760 kt y^{-1} in 2000–2005. Phosphorus emission changed from 40 kt y^{-1} in the 1950s to its peak value of 115 kt y^{-1} during the first half of the 1990s and then to 70 kt y^{-1} in 2000–2005. The Sea of Azov – a partly enclosed part of the Black Sea – is highly eutrophied. Seasonal hypoxia is regular in the north-west Black Sea but that has decreased in recent years. The eutrophication has caused a shift in the phytoplankton community from diatoms to dinoflagellates, increase in the dinoflagellate *Noctiluca scintillans* and the jelly fish *Aurelia aurita*, shift from perennial macroalgae to ephemeral ones, reduction of the sublittoral *Cystoseira* belt due to reduced water transparency, severe changes in zoobenthos and decline of demersal fish or fish dependent on the seabed during their life cycle. In addition, food web changes due to eutrophication and the invasion of the comb jelly *Mnemiopsis leidyi* combined with overfishing caused mass mortalities of the Black Sea harbour porpoise, short-beaked common dolphin and common bottlenose dolphin. The invasive comb jelly *M. leidyi* had a major outbreak in the late 1980s and its dominance was likely partly enhanced by the eutrophic state of the ecosystem. After the decline of the comb jelly by an invasive predatory comb jelly *Beroe ovate* and slight decrease in eutrophication, the zooplankton diversity was observed to increase (BSC, 2008). The zoobenthos experienced a drastic decrease of the specific diversity, simplified zoobenthic community structures, decreasing abundance and biomass of benthic populations, reduction in biofiltering due to the loss of filter-feeder populations and flourishing of some opportunistic worms.

In the North Sea, eutrophication is mainly a problem in coastal areas, estuaries and embayments (OSPAR, 2010). During the period 2001–2005, the entire eastern North Sea, Kattegat, the coast of Brittany and estuaries of the Ireland and Great Britain were considered as eutrophied. In addition, potentially eutrophied areas were identified from estuaries of the Iberian coast. According to OSPAR (2010), die-offs of cultured mussels and benthic animals have been linked to the decay of massive algal blooms in some estuaries in the Netherlands, kills of fish and invertebrates due to extreme oxygen deficiency have occurred in fjords and estuaries of Sweden and Denmark, kills of benthic invertebrates have occurred in Norwegian fjords, toxic hydrogen sulfide has been released from rotting sea lettuce on Brittany's beaches and algal foam on beaches in Belgium has been estimated to cause an annual economic loss of around 0.5% of revenue to the tourism industry. However, in many coastal areas high natural turbidity prevents algal growth and, hence, eutrophication impacts despite high nutrient concentrations. There have been only local improvements in the eutrophication status in the North Sea.

In the Caribbean Sea, untreated sewage waters are the main reason for eutrophication impacts, which include overgrowth of coral reefs by macroalgae, fish mortalities, decline of seagrass meadows and red tides (UNEP/CEP, 2013).

The Northwest Pacific region includes parts of northeast China, Japan, Korea and south-east Russia. It is one of the most densely populated areas in the world, and its coastal systems are subject to significant human-induced nutrient inputs (NOWPAP CEARAC, 2011). The sea areas are subject to frequent hypoxic and red tide events and also the increased abundance of giant jelly fish and green tides, changes in phytoplankton community and a general loss of biodiversity have been attributed to eutrophication.

8.8 Socioeconomic impacts of eutrophication

Eutrophication is not only an environmental concern but its adverse impacts on people's everyday lives and economics around the world are an increasing problem (Carpenter *et al.*, 1997). Almost every step in the process of eutrophication has certain sociological or economic implications.

During the early stages of eutrophication, increased loose-lying macroalgae clog gillnets, reducing fish catch and increasing maintenance efforts of the nets, and accumulate on beaches, reducing beach value and increasing cleaning costs. Increased nutrient availability and altered N:P:Si ratios foster phytoplankton blooms and cause shifts in phytoplankton communities, favouring blooms of dinoflagellates and cyanobacteria, many of which are toxic to humans, pets, cattle, marine mammals, fish and invertebrates. The toxic blooms shut down oyster farms and prevent human consumption or household use of water. Eutrophicated water also causes problems in the taste and odour of drinking water and filtration problems in drinking water supplies.

In a more advanced stage of eutrophication, oxygen deficiency causes mass mortalities of fish and invertebrates or indirectly, via reduced prey abundance, reduces catches of commercial and recreational fisheries. Moreover, changes in habitat conditions alter fish species composition towards less valued species, such as cyprinids, and cause loss of suitable spawning grounds for commercially exploited fish species.

Thus, eutrophication brings along clear monetary costs, loss of income, loss of property value and decline in aesthetic and recreational value. As 63% of the value of ecosystem services is contributed by marine ecosystems (Costanza *et al.*, 1987), the well-being of the seas should be of high concern. Although the costs of mitigation and remediation can be high, people's willingness to pay for a healthy marine environment can clearly exceed those (Ericsdotter *et al.*, 2013). Intensive research efforts have been focused on understanding human-induced eutrophication and the mechanisms of the abiotic and biotic processes are now quite well known. This has allowed a shift to increased interdisciplinary research on the economic values of a healthy marine environment and management options based on several scenarios and economic analyses (Wulff *et al.*, 2007; Ericsdotter *et al.*, 2013). It has also become clear that reversing eutrophication requires reductions in both N and P (Paerl, 2009). As the scale of eutrophicated marine waters is too large for individual management measures, international regional conventions for the protection of the marine environment have agreed on political goals which further steer and support the scientific research and allow a tight linkage between science and policy (e.g. the ecological objectives of the Helsinki Convention, www.helcom.fi, and OSPAR Convention, www.ospar.org). However, marine systems change all the time as a result of climate change, post-glacial land up-lift, changes in the catchment area and the increased global population growth. Therefore, research questions, political decisions and management need to be dynamic and take into account other stressors on the marine environment (Chapter 4).

8.9 Key findings and recommendations for decision makers

- Eutrophication is a global anthropogenic pressure on marine ecosystems and no significant reduction of the problem can be observed; waterborne inputs of nutrients and organic matter seem to increase due to continuously increasing use of fertilisers, meat production and inadequate wastewater treatment. However, technology to limit discharges of nutrients and organic matter is relatively cheap and national and international policies (e.g. regional sea conventions) can guide towards better practices.
- Impacts of eutrophication are documented from all marine ecosystems, but coastal sheltered systems may be more vulnerable than systems with greater water exchange. Stratified sea areas, such as estuaries, coastal bays and partly enclosed seas, are more prone to hypoxic events or permanent hypoxia/anoxia as a result of long-lasting enrichment.
- Eutrophication in shallow, coastal bays causes a shift in dominance from seagrasses and perennial macroalgae to ephemeral, bloom-forming macroalgae and epiphytes, and ultimately to phytoplankton dominance in the most heavily eutrophied systems.
- Phytoplanktonic communities become dominated by non-siliceous forms, such as dinoflagellates, rather than diatoms. As such, eutrophication is associated with increased occurrence of HABs. In the

zooplankton, gelatinous forms become more prevalent, tending to replace crustacean species.
- In more advanced stages of eutrophication, oxygen deficiency causes mass mortalities of fish and invertebrates or affects them indirectly, via reduced prey abundance, reducing catches of commercial and recreational fisheries. Moreover, changes in habitat conditions alter fish species composition towards less valued species, such as cyprinids, and cause loss of suitable spawning grounds for commercially exploited fish species.
- Reversing the state of such ecosystems is slow, because of high amounts of accumulated phosphorus in sediments, and requires strong political commitment and often international co-operation. It is important to acknowledge that recovery of the ecosystem may be to a different state than the original or desired, and hence management strategies must look ahead, and not strive backwards.
- Eutrophication processes are enhanced by exploitation of big fish and other top predators, as well as the use of other ecosystem goods and services. Therefore mitigation of eutrophication impacts may not be possible without managing fisheries and other human use of marine resources.

Acknowledgements

The authors would like to thank Robert Díaz for the global map of hypoxia and HELCOM for the concept diagram of eutrophication. Shane O'Boyle is thanked for the comments on the text.

References

Appelgren, K. and Mattila, J. (2005). Variation in vegetation communities in shallow bays of the northern Baltic Sea. *Aquatic Botany,* 83, 1–13.

Banta, G. T., Pedersen, M. F., Nielsen, S. L. (2004). Decomposition of marine primary producers: consequences for nutrient recycling and retention in coastal ecosystems. In *Estuarine Nutrient Cycling: The Influence of Primary Producers,* ed. S. L. Nielsen, G. T. Banta and M. F. Pedersen. Dordrecht, The Netherlands: Kluwer Academic, pp. 187–216.

Bonsdorff, E. (2006). Zoobenthic diversity gradients in the Baltic Sea: continuous post-glacial succession in a stressed ecosystem. *Journal of Experimental Marine Biology and Ecology,* 330, 383–391.

Bonsdorff, E., Blomqvist, E. M., Mattila, J. and Norkko, A. (1997). Coastal eutrophication: causes, consequences and perspectives in the archipelago areas of the northern Baltic Sea. *Estuarine, Coastal and Shelf Science,* 44 (Supplement A), 63–72.

Bonsdorff, E. and Pearson, T. H. (1999). Variation in the sublittoral macrozoobenthos of the Baltic Sea along environmental gradients: a functional-group approach. *Australian Journal of Ecology*, 24, 312–326.

Borer, E. T., Halpern, B. S. and Seabloom, E. W. (2006). Asymmetry in community regulation: effects of predators and productivity. *Ecology*, 87, 2813–2820.

Borum, J. and Sand-Jensen, K. (1996). Is total primary production in shallow coastal marine waters stimulated by nitrogen loading? *Oikos*, 76, 406–410.

Borum, J., Pedersen, O., Greve, T. M., et al. (2005). The potential role of plant oxygen and sulphide dynamics in die-off events of the tropical seagrass, *Thalassia testudinum*. *Journal of Ecology*, 93, 148–158.

BSC (2008). State of the environment of the Black Sea (2001–2006/7), ed. T. Oguz. *Publications of the Commission on the Protection of the Black Sea Against Pollution* (BSC) 2008–3, Istanbul, Turkey.

Burkepile, D. E. and Hay, M. E. (2006). Herbivore vs. nutrient control of marine primary producers: context-dependent effects. *Ecology*, 87, 3128–3139.

Burkholder, J. M., Glibert, P. M. and Skelton, H. M. (2008). Mixotrophy, a major mode of nutrition for harmful algal species in eutrophic waters. *Harmful Algae*, 8, 77–93.

Caddy, J. F. (2000). Marine catchment basin versus impacts of fisheries on semi-encloses seas. *ICES Journal of Marine Science*, 57, 628–640.

Carpenter, S. R., Bolgrien, D., Lathrop, R. C. et al. (1997). Ecological and economic analysis of lake eutrophication by nonpoint pollution. *Australian Journal of Ecology*, 23, 68–79.

Cloern, J. E. (2001). Our evolving conceptual model of the coastal eutrophication problem. *Marine Ecology Progress Series*, 210, 223–253.

Conley, D. J., Humborg, C., Rahm, L., Savchuk O. P. and Wulff, F. (2002). Hypoxia in the Baltic Sea and basin-scale changes in phosphorus biogeochemistry. *Environmental Science and Technology*, 36, 5315–5320.

Conley, D. J., Björk, S., Bonsdorff, E. et al. (2009a). Hypoxia-related processes in the Baltic Sea. *Environmental Science and Technology*, 43, 3412–3420.

Conley, D. J., Carstensen, J., Vaquer-Sunyer, R. and Duarte, C. M. (2009b). Ecosystem thresholds with hypoxia. *Hydrobiologia*, 629, 21–29.

Conley, D. J., Carstensen, J., Aigars, J. et al. (2011). Hypoxia is increasing in the coastal zone of the Baltic Sea. *Environmental Science and Technology*, 45, 6777–6783.

Costanza, R., d'Arge, R., deGroot, R., Farber, S. et al. (1987). The value of the world's ecosystem services and natural capital. *Nature*, 387, 253–260.

Craig, J. K., Crowder, L. B., Gray, C. D. et al. (2001). Ecological effects of hypoxia on fish, sea turtles, and marine mammals in the northwestern Gulf of Mexico. In *Coastal Hypoxia: Consequences for Living Resources and Ecosystems*, ed. N. N. Rabalais and R. E. Turner. Washington DC: American Geophysical Union, pp. 269–292.

Dahlgren, S. and Kautsky, L. (2004). Can different vegetative states in shallow coastal bays of the Baltic Sea be linked to internal nutrient levels and external nutrient load? *Hydrobiologia*, 514, 249–258.

Dalsgaard, T., Canfield, D. E., Petersen, J., Thamdrup, B. and Acuña-Gonzalez, J. (2003). N_2 production by the anammox reaction in the anoxic water column of Golfo Dulce, Costa Rica. *Nature*, 422, 606–608.

Dame, R. F. (1996). *Ecology of Marine Bivalves: An Ecosystem Approach*. Boca Raton, FL: CRC Press.
de Leiva Moreno, J. I., Agostini, V. N., Caddy, J. F. and Carocci, F. (2000). Is the pelagic–demersal ratio from fishery landings a useful proxy for nutrient availability? A preliminary data exploration for the semi-enclosed seas around Europe. *ICES Journal of Marine Science*, 57, 1091–1102.
Deegan, L. A., Johnson, D. S., Warren, R. S. et al. (2012). Coastal eutrophication as a driver of salt marsh loss. *Nature*, 490, 388–391.
Deutsch, C., Sarmiento, J. L., Sigman, D. M., Gruber, N. and Dunne, J. P. (2007). Spatial coupling of nitrogen inputs and losses in the ocean. *Nature*, 445, 163–167.
Diáz, R. J. and Rosenberg, R. (2008). Spreading dead zone and consequences for marine ecosystems. *Science*, 321, 926–929.
Díaz, R., Selman, M. and Chique, C. (2010). Global eutrophic and hypoxic coastal systems. World Resources Institute, Washington DC. Available at: http://www.wri.org/project/eutrophication, accessed 27 Feb 2013.
Duarte, C. M. (2000). Marine biodiversity and ecosystem services: an elusive link. *Journal of Experimental Marine Biology and Ecology*, 250, 117–131.
Duce, R. A., LaRoche, J., Altieri, K. et al. (2008). Impacts of atmospheric anthropogenic nitrogen on the open ocean. *Science*, 320, 893–897.
Elser, J. J., Bracken, M. E. S., Cleland, E. E. et al. (2007). Global analysis of nitrogen and phosphorus limitation of primary producers in freshwater, marine and terrestrial ecosystems. *Ecology Letters*, 10, 1135–1142.
Ericsdotter, S., Nekoro, M. and Scharin, H. (2013). The Baltic Sea: our common treasure. Economics of saving the sea. Report 2013:4, Swedish Agency for Marine and Water Management, Gothenburg, Sweden.
FAO (2006). World agriculture: towards 2030/2050. Prospects for food, nutrition, agriculture and major commodity groups. Interim report, Food and Agriculture Organization of the United Nations, Rome.
FAO (2011). Current world fertilizer trends and outlook to 2015. Food and Agriculture Organization of the United Nations, Rome.
Fortibuoni, T., Libralato, S., Raicevich, S., Giovanardi, O. and Solidoro, C. (2010). Coding early naturalists' accounts into long-term fish community changes in the Adriatic Sea (1800–2000). *PLoS ONE* 5: e15502.
Gray, J. S., Shiu-sun Wu, R. and Or, Y. Y. (2002). Effects of hypoxia and organic enrichment on the coastal marine environment. *Marine Ecology Progress Series*, 238, 249–279.
Gruber, N. (2008). The marine nitrogen cycle: overview and challenges. In *Nitrogen in Marine Environment*, 2nd edn, ed. D. G. Capone, D. Bronk, M. R. Mulholland and E. Carpenter. Amsterdam: Elsevier Inc., pp. 1–50.
Hällfors, H., Backer, H., Leppänen, J-M. et al. (2013). The northern Baltic Sea phytoplankton communities in 1903–1911 and 1993–2005: a comparison of historical and modern species data. *Hydrobiologia*, 707, 109–133.
Hansell, D. A. and Follows, M. J. (2008). Nitrogen in Atlantic Ocean. In *Nitrogen in Marine Environment*, 2nd edn, ed. D. G. Capone, D. Bronk, M. R. Mulholland and E. Carpenter. Amsterdam: Elsevier Inc., pp. 597–630.
Heck, K. L. Jr., and Valentine, J. F. (2007). The primacy of top-down effects in shallow benthic ecosystems. *Estuaries and Coasts*, 30, 371–381.

Heithaus, M. R., Frid, A., Wirsing, A. J. and Worm, B. (2008). Predicting ecological consequences of marine top predator declines. *Trends in Ecology and Evolution*, 23, 202–210.
HELCOM (2009). Eutrophication of the Baltic Sea: an integrated thematic assessment of the effects of nutrient enrichment in the Baltic Sea region. *Baltic Sea Environment Proceedings*, 115.
HELCOM (2010). Ecosystem Health of the Baltic Sea 2003–2007: HELCOM initial holistic assessment. *Baltic Sea Environment Proceedings*, 122.
HELCOM (2012). Fifth Pollution Load Compilation (PLC-5). *Baltic Sea Environment Proceedings*, 128.
Hillebrand, H., Worm, B. and Lotze, H. K. (2000). Marine microbenthic community structure regulated by nitrogen loading and herbivore pressure. *Marine Ecology Progress Series*, 204, 27–38.
Howard, R. K. and Short, F. T. (1986). Seagrass growth and survivorship under the influence of epiphyte grazers. *Aquatic Biology*, 24, 287–302.
Howarth, R. W. (1993). The role of nutrients in coastal waters. In *Managing Wastewater in Coastal Urban Areas*. Report from the National Research Council on Wastewater Management for Coastal Urban Areas. Washington DC: National Academy Press, pp. 177–202.
Howarth, R. W. and Marino, R. M. (2006). Nitrogen as the limiting nutrient for eutrophication in coastal marine ecosystems: evolving views over three decades. *Limnology and Oceanography*, 51, 364–376.
Howarth, R. W. (2008). Coastal nitrogen pollution: a review of sources and trends globally and regionally. *Harmful Algae*, 8, 14–20.
ICES (2011). Report of the ICES/HELCOM Working Group for Integrated Assessment of the Baltic Sea Ecosystem (WGIAB), ICES C.M. 2011/SSGRSP:03, 4–8 April 2011, Mallorca, Spain.
Karlson, K., Rosenberg, R. and Bonsdorf, E. (2002). Temporal and spatial large-scale effects of eutrophication and oxygen deficiency on benthic fauna in Scandinavian and Baltic waters: a review. *Oceanography and Marine Biology*, 40, 427–489.
Karlson, K., Bonsdorff, E. and Rosenberg, R. (2007). The impact of benthic macrofauna for nutrient fluxes from Baltic Sea sediments. *Ambio*, 36, 161–167.
Kautsky, H. (1991). Influence of eutrophication on the distribution of phytobenthic plant and animal communities. *Internationale Revue der Gesamten Hydrobiologie und Hydrographie*, 76, 423–432.
Kemp, W. M., Boynton, W. R., Adolf, J. E. et al. (2005). Eutrophication of Chesapeake Bay: historical trends and ecological interactions. *Marine Ecology Progress Series*, 303, 1–29.
Kiørboe, T., Hansen, J. L. S. and Alldredge, A. L. (1996). Sedimentation of phytoplankton during a diatom bloom: rates and mechanisms. *Journal of Marine Research*, 54, 1123–1148.
Korpinen, S., Jormalainen, V. and Honkanen, T. (2007a). Effects of nutrients, herbivory, and depth on the macroalgal community in the rocky sublittoral. *Ecology*, 88, 839–852.
Korpinen, S., Honkanen, T., Vesakoski, O. et al. (2007b). Macroalgal communities face the challenge of changing biotic interactions: review with focus on the Baltic Sea. *Ambio*, 36, 203–211.

Korpinen, S., Jormalainen, V. and Pettay, E. (2010). Nutrient availability modifies species abundance and community structure of *Fucus*-associated littoral benthic fauna. *Marine Environment Research*, 70, 283–292.

Kuenen, J. G. (2008). Anammox bacteria: from discovery to application. *Nature Reviews Microbiology*, 6, 320–326.

Lapointe, B. E. (1997). Nutrient thresholds for bottom-up control of macroalgal blooms on coral reefs in Jamaica and Southeast Florida. *Limnology and Oceanography*, 42, 1119–1131.

Lappalainen, A., Westerbom, M. and Heikinheimo, O. (2005). Roach (*Rutilus rutilus*) as an important predator on blue mussel (*Mytilus edulis*) populations in a brackish water environment, the northern Baltic Sea. *Marine Biology*, 147, 323–330.

Ljunggren, L., Sandström, A., Bergström, U. et al. (2010). Recruitment failure of coastal predatory fish in the Baltic Sea coincident with an offshore ecosystem regime shift. *Journal of Marine Science*, 67, 1587–1595.

Lohrer, A. M., Thrush, S. F. and Gibbs, M. M. (2004). Bioturbators enhance ecosystem function through complex biogeochemical interactions. *Nature*, 431, 1092–1095.

Lotze, H. and Milewski, I. (2004). Two centuries of multiple human impacts and successive changes in a North Atlantic food web. *Ecological Applications*, 14, 1428–1447.

Marczak, L. B., Thompson, R. M. and Richardson, J. S. (2007). Meta-analysis: trophic level, habitat, and productivity shape the food web effects of resource subsidies. *Ecology*, 88, 140–148.

Matson, P. A., Parton, W. J., Power, A. G. and Swift, M. J. (1997). Agricultural intensification and ecosystem properties. *Science*, 277, 504–509.

McCook, L. J., Jompa, J. and Díaz-Pulido, G. (2001). Competition between corals and algae on coral reefs: a review of evidence and mechanisms. *Coral Reefs*, 19, 400–417.

McGlathery, K. J., Sundbäck, K. and Anderson, I. C. (2007). Eutrophication in shallow coastal bays and lagoons: the role of plants in the coastal filter. *Marine Ecology Progress Series*, 348, 1–18.

Meier, H. E. M., Anderson, H. C., Eilola, K. et al. (2011). Hypoxia in future climates: a model ensemble study for the Baltic Sea. *Geophysical Research Letters*, 38, L24608.

Micheli, F. (1999). Eutrophication, fisheries, and consumer-resource dynamics in marine pelagic ecosystems. *Science*, 285, 1396–1398.

Middelburg, J. J. and Levin, L. A. (2009). Coastal hypoxia and sediment biogeochemistry. *Biogeosciences*, 6, 1273–1293.

Murdoch, W. W., Nisbet, R. M., McCauley, E., DeRoos, A. M. and Gurney, W. S. C. (1998). Plankton abundance and dynamics across nutrient levels: tests of hypotheses. *Ecology*, 79, 1339–1356.

Newell, R. I. E., Fisher, T. R., Holyoke, R. R. and Cornwell, J. C. (2005). Influence of eastern oysters on nitrogen and phosphorus regeneration in Chesapeake Bay, USA. In *Comparative Roles of Suspension Feeders in Ecosystems*, ed. R. Dame and S. Olenin. NATO Science Series. IV: Earth and Environmental Sciences Vol. 47. Berlin: Springer.

Nilsson, H. C. and Rosenberg, R. (1997). Benthic habitat quality assessment of an oxygen stressed fjord by surface and sediment profile images. *Journal of Marine Systems*, 11, 249–264.

Nilsson, H. C. and Rosenberg, R. (2000). Succession in marine benthic habitats and fauna in response to oxygen deficiency: analysed by sediment profile-imaging and by grab samples. *Marine Ecology Progress Series*, 197, 139–149.

Norkko, A. and Bonsdorff, E. (1996). Rapid zoobenthic community responses to accumulations of drifting algae. *Marine Ecology Progress Series*, 131, 143–157.

Norkko, J., Bonsdorff, E. and Norkko, A. (2000). Drifting algal mats as an alternative habitat for benthic invertebrates: Species specific responses to a transient resource. *Journal of Experimental Marine Biology and Ecology*, 248, 79–104.

Norkko, J., Reed, D. C., Timmermann, K., Norkko, A. et al. (2012). A welcome can of worms? Hypoxia mitigation by an invasive species. *Global Change Biology*, 18, 422–434.

NOWPAP CEARAC (2011). Integrated report on eutrophication assessment in selected sea areas in the NOWPAP region: evaluation of the NOWPAP Common Procedure. ISBN 978–4–9902809–5–6.

Oksanen, L., Fretwell, S. D., Arruda, J. and Niemela, P. (1981). Exploitation ecosystems in gradients of primary productivity. *American Naturalist*, 118, 240–261.

Oksanen, L. and Oksanen, T. (2000). The logic and realism of the hypothesis of exploitation ecosystems. *American Naturalist*, 155, 703–723.

Olesen, B. (1996). Regulation of light attenuation and eelgrass *Zostera marina* depth distribution in a Danish embayment. *Marine Ecology Progress Series*, 134, 187–194.

Orth, R. J., Carruthers, T. J. B., Dennison, W. C. et al. (2006). A global crisis for seagrass ecosystems. *BioScience*, 56, 987–996.

OSPAR (2010). Quality Status Report 2010. London, OSPAR Commission.

Österblom, H., Hansson, S., Larsson, U. et al. (2007). Human-induced trophic cascades and ecological regime shifts in the Baltic Sea. *Ecosystems*, 10, 877–889.

Paerl, H. W. (2009). Controlling eutrophication along the freshwater–marine continuum: dual nutrient (N and P) reductions are essential. *Estuaries and Coasts*, 32, 593–601.

Park, G. S. and Marshall, H. G. (1999). Estuarine relationships between zooplankton community structure and trophic gradients. *Journal of Plankton Research*, 22, 121–136.

Pearson, T. H. and Rosenberg, R. (1978). Macrobenthic succession in relation to organic enrichment and pollution of the marine environment. *Oceanography and Marine Biology: An Annual Review*, 16, 229–311.

Pinto-Coelho, R. M., Bezerra-Neto, J. F. and Morais-Jr., C. A. (2005). Effects of eutrophication on size and biomass of crustacean zooplankton in a tropical reservoir. *Brazilian Journal of Biology*, 65, 325–338.

Purcell, J. E., Breitburg, D. L., Decker, M. B. et al. (2001). Pelagic cnidarians and ctenophores in low dissolved oxygen environments: a review. In *Coastal Hypoxia: Consequences for Living Resources and Ecosystems*, ed. N. N. Rabalais and R. E. Turner. Washington DC: American Geophysical Union, pp. 77–100.

Rabalais, N. N. (2005). The potential for nutrient overenrichment to diminish marine biodiversity. In *Marine Conservation Biology: The Science of Maintaining the Sea's Biodiversity*, ed. E. A. Norse and L. B. Crowder. Washington DC: Island Press, pp. 109–122.

Rabalais, N. N., Turner, R. E., Díaz, R. J., and Justić, D. (2009). Global change and eutrophication of coastal waters. *ICES Journal of Marine Science*, 66, 1528–1537.

Rabalais, N. N., Díaz, R. J., Levin, L. A., Turner, R.E., Gilbert, D. and Zhang, J. (2010). Dynamics and distribution of natural and human-caused hypoxia. *Biogeosciences*, 7, 585–619.

Rask, N., Pedersen, S. T. and Jensen, M. H. (1999). Response to lowered nutrient discharges in the coastal waters around the island of Funen, Denmark. *Hydrobiologia*, 393, 69–81.

Rosenberg, R., Blomqvist, M., Nilsson, H. C., Cederwall, H. and Dimming, A. (2004). Marine quality assessment by use of benthic species–abundance distributions: a proposed new protocol within the European Union Water Framework Directive. *Marine Pollution Bulletin*, 49, 728–739.

Rosenzweig, M. L. (1971). Paradox of enrichment: destabilization of exploitation ecosystems in ecological time. *Science*, 171, 385–387.

Rosenzweig, M. L. and Abramsky, Z. (1993). How are diversity and productivity related. In *Species Diversity in Biological Communities*, ed. R.E. Ricklefs and D. Schlutee. Chicago, IL: University of Chicago Press, pp. 52–65.

Rumohr, H., Bonsdorff, E. and Pearson, T. H. (1996). Zoobenthic succession in Baltic sedimentary habitats. *Archive of Fishery And Marine Research*, 44, 179–214.

Salovius, S., Nyqvist, M. and Bonsdorff, E. (2005). Life in the fast lane: macrobenthos use temporary drifting algal habitats. *Journal of Sea Research*, 53, 169–180.

Sand-Jensen, K. (1977). Effects of epiphytes on eelgrass photosynthesis. *Aquatic Botany*, 3, 55–63.

Sandberg, E. (1994). Does short-term oxygen depletion affect predator–prey relationships in zoobenthos? Experiments with the isopod *Saduria entomon*. *Marine Ecology Progress Series*, 103, 73–80.

Sarthou, G., Timmermans, K. R., Blain, S. and Tréguer, P. (2005). Growth physiology and fate of diatoms in the ocean: a review. *Journal of Sea Research*, 53, 25–42.

Savchuk, O. P., Wulff, F., Hille, S., Humborg, C. and Pollehne, F. (2008). The Baltic Sea a century ago: a reconstruction from model simulations, verified by observations. *Journal of Marine Systems*, 74, 485–494.

Schmitz, O. J., Kalies, E. L. and Booth, M. G. (2006). Alternative dynamic regimes and trophic control of plant succession. *Ecosystems*, 9, 459–472.

Schubert, H. and Schories, D. (2008). Macrophytobenthos. In *State and Evolution of the Baltic Sea, 1952–2005. A Detailed 50-year Survey of Morphology and Climate, Physics, Chemistry, Biology and Marine Environment*, ed. R. Feistel, G. Nausch and N. Wasmund. Hoboken, NJ: John Wiley and Sons, Inc., pp. 413–516.

Smith, V. H., Tilman, G. D. and Nekola, J. C. (1999). Eutrophication: impacts of excess nutrient inputs on freshwater, marine, and terrestrial ecosystems. *Environmental Pollution*, 100, 179–196.

Smith, V. H. (2003). Eutrophication of freshwater and coastal marine ecosystems: a global problem. *Environmental Science and Pollution Research*, 10, 126–139.

Solan, M., Cardinale, B. J., Downing, A. L. et al. (2004). Extinction and ecosystem function in the marine benthos. *Science*, 306, 1177–1180.

Souchu, P., Vaquer, A., Collos, Y. et al. (2001). Influence of shellfish farming activities on the biogeochemical composition of the water column in Thau lagoon. *Marine Ecology Progress Series*, 218, 141–152.

Suding, K. N., Collins, S. L., Gough, L. et al. (2005). Functional- and abundance-based mechanisms explain diversity loss due to N fertilization. *Proceedings of the National Academy of Sciences of the United States of America*, 102, 4387–4392.

Sundbäck, K. and McGlathery, K. (2005). Interactions between benthic macroalgal and microalgal mats. In *Interactions Between Macro- and Microorganisms in Marine Sediments. Coastal and Estuarine Studies* 60, ed. E. Kristensen, R. R. Haese and J. E. Kostka. Washington DC: American Geophysical Union, pp. 7–29.

Sundbäck, K., Miles, A., Hulth, S. et al. (2003). Importance of benthic nutrient regeneration during initiation of macroalgal blooms in shallow bays. *Marine Ecology Progress Series*, 246, 115–126.

Svensson, J. R., Lindegarth, M., Siccha, M. et al. (2007). Maximum species richness at intermediate frequencies of disturbance: consistency among levels of productivity. *Ecology*, 88, 830–838.

Tamelander, T. and Heiskanen, A-S. (2004). Effects of spring bloom phytoplankton dynamics and hydrography on the composition of settling material in the coastal northern Baltic Sea. *Journal of Marine Systems*, 52, 217–234.

Telesh, I. V. (2004). Plankton of the Baltic estuarine ecosystems with emphasis on Neva Estuary: a review of present knowledge and research perspectives. *Marine Pollution Bulletin*, 49, 206–219.

Taylor, D. L. and Eggleston, D. B. (2000). Effects of hypoxia on an estuarine predator-prey interaction: foraging behavior and mutual interference in the blue crab *Callinectes sapidus* and the infaunal clam prey *Mya arenaria*. *Marine Ecology Progress Series*, 196, 221–237.

Turner, R. E. (2001). Some effects of eutrophication on pelagic and demersal marine food webs. In *Coastal Hypoxia: Consequences for Living Resources and Ecosystems*, ed. N. N. Rabalais and R. E. Turner. Washington DC: American Geophysical Union, pp. 371–393.

Turner, R. E. and Rabalais, N. N. (1994). Coastal eutrophication near the Mississippi river delta. *Nature*, 368, 619–621.

Tyrrel, T. (1999). The relative influences of nitrogen and phosphorus on oceanic primary production. *Nature*, 400, 525–535.

UNEP/CEP (2013). Waste water, sewage and sanitation. The UNEP Caribbean Environment Programme. Available at: http://www.cep.unep.org/publications-and-resources/marine-and-coastal-issues-links/wastewater-sewage-and-sanitation, accessed 8 January 2013.

UNEP/MAP-Plan Bleu (2009). State of the Environment and Development in the Mediterranean. UNEP/MAP-Plan Bleu, Athens.

US EPA (2012). National Coastal Condition Report IV. United States Environmental Protection Agency, EPA-842-R-10-003. Washington, DC.

Uusitalo, L., Fleming-Lehtinen, V., Hällfors, H. et al. (2013). A novel approach for estimating phytoplankton biodiversity. *ICES Journal of Marine Science*, 70, 408–417.

Vahtera, E., Conley, D. J., Gustafsson, B. G. et al. (2007). Internal ecosystem feedbacks enhance nitrogen-fixing cyanobacteria blooms and complicate management in the Baltic Sea. *Ambio*, 36, 1–10.

Valiela, I., McClelland, J., Hauxwell, J. et al. (1997). Macroalgal blooms in shallow estuaries: controls and ecophysiological and ecosystem consequences. *Limnology and Oceanography*, 42, 1105–1118.

Vaquer-Sunyer, R. and Duarte C. M. (2008). Thresholds of hypoxia for marine biodiversity. *Proceedings of the National Academy of Sciences*, 105, 15452–15457.

Wasmund, N. and Siegel, H. (2008). Phytoplankton. In *State and Evolution of the Baltic Sea, 1952–2005. A Detailed 50-year Survey of Morphology and Climate, Physics, Chemistry, Biology and Marine Environment.*, ed. R., Feistel, G. Nausch and N. Wasmund. Hoboken, NJ: John Wiley and Sons, Inc., pp. 441–482.

Westerbom, M. (2006). Population dynamics of blue mussels in a variable environment at the edge of their range. Doctoral thesis, Faculty of Biosciences, Department of Biological and Environmental Sciences, University of Helsinki.

Villnäs, A. and Norkko, A. (2011). Benthic diversity gradients and shifting baselines: implications for assessing environmental status. *Ecological Applications*, 21, 2172–2186.

Villnäs, A., Norkko, J., Lukkari, K., Hewitt, J. and Norkko, A. (2012). Consequences of increasing hypoxic disturbance on benthic communities and ecosystem functioning. *PLOS One*, 7, e44920.

Vitousek, P. M., Aber, J., Howarth, R. W. et al. (1997). Human alteration of the global nitrogen cycle: causes and consequences. *Ecological Applications*, 7, 737–750.

Vos, M., Kooi, B. W., DeAngelis, D. L. and Mooij, W. M. (2004). Inducible defences and the paradox of enrichment. *Oikos*, 105, 471–480.

Worm, B. (2000). Consumer versus resource control in rocky shore food webs: Baltic Sea and NW Atlantic Ocean. *Berichte aus dem Institut fuer Meereskunde Kiel*, 316, 1–147.

Worm, B., Barbier, E. B., Beaumont, N. et al. (2006). Impacts of biodiversity loss on ocean ecosystem services. *Science*, 314, 787–760.

Worm, B. and Lotze, H. K. (2006). Effects of eutrophication, grazing, and algal blooms on rocky shores. *Limnology and Oceanography*, 51, 569–579.

Worm, B., Lotze, H. K., Hillebrand, H. and Sommer, U. (2002). Consumer versus resource control of species diversity and ecosystem functioning. *Nature*, 417, 848–851.

Woulds, C., Cowie, G. L., Levin, L. A. et al. (2007). Oxygen as a control on sea floor biological communities and their roles in sedimentary carbon cycling. *Limnology and Oceanography*, 52, 1698–1709.

Wulff, F., Savchuk, O. P., Sokolov, A., Humborg, C. and Mörth, C.-M. (2007). Management options and effects on a marine ecosystem: assessing the future of the Baltic. *Ambio*, 36, 243–249.

9 · Pollution: effects of chemical contaminants and debris

EMMA L. JOHNSTON AND
MARIANA MAYER-PINTO

9.1 Introduction

Coastal marine ecosystems are both diverse and productive (Suchanek, 1994; Gray, 1997). They are ecologically significant and socioeconomically important providing an array of critical services and benefits to humans (see Chapters 1 and 2). The delivery of services depends, however, on the efficient functioning of ecosystems, which in turn, is influenced by biological diversity. Marine ecosystems are subject to a range of threats that may affect both diversity and function. In coastal and estuarine systems, contaminants can be a major problem (Thompson et al., 2002). Pollution has been associated with reductions in the densities, biomass and number of species in marine systems (e.g. Johnston and Roberts, 2009) and reductions in ecosystem function (Breitburg et al., 1999; Gonzalez et al., 2009; Johnston et al., 2015; Mayer-Pinto et al., in review). Declining water quality and contamination have been identified as critical factors contributing to the degradation and loss of ecologically and economically important marine biogenic habitats, such as seagrass, kelp beds and coral reefs (see e.g. Benedetti-Cecchi et al., 2001; Duarte, 2002; Bellwood et al., 2004; Foster and Schiel, 2010). Understanding the effect of contaminants on both biodiversity and ecosystem function is critical if we are to effectively prioritise conservation actions and ensure the security of ecosystem services.

In this chapter we review the effects of chemical contaminants on the biological diversity and functioning of marine systems. Most researchers have studied contaminant effects on diversity or function independently and this review is the first to consider the two sets of studies together. By taking a holistic approach, we have been able to identify common patterns in the effects of contaminants and critical areas that require further study.

Marine Ecosystems: Human Impacts on Biodiversity, Functioning and Services, eds T. P. Crowe and C. L. J. Frid. Published by Cambridge University Press. © Cambridge University Press 2015.

9.2 What are contaminants and where do they come from?

In its most generalised form, contamination is an increase in the concentration of any constituent in the water, soil, sediment and/or organism above the natural background level for that area or those organisms (Clark, 2001). When contaminants have a biological and/or ecological effect on organisms, populations and/or assemblages they are called pollutants and, consequently, they cause pollution (GESAMP, 1982).

There are many types of contamination that emanate from human activities. Chemical contaminants are among the most diverse stressors and the extent of their threat is large and increasing (Crain et al., 2008). In the European Union and United States alone, there are more than 100 000 registered chemicals (EU, 2001), each with different chemical properties that have the potential to dramatically affect several physiological traits of a wide range of organisms (e.g. Bryan, 1971; McCahon and Pascoe, 1990). Below we introduce the main sources of polluting contaminants in marine ecosystems which are summarised in Table 9.1.

Extensive growth of human populations and urban centres in close proximity to waterways has led to an increase in contaminants entering coastal habitats through runoff and deliberate release. In urbanised or industrialised areas, there are several primary pathways by which contaminants enter coastal environments, such as non-point-source runoff from developments and waste dumping into rivers, estuaries or nearshore environments (see e.g. GESAMP, 1982; Kennish, 2002). Runoff in urbanised coastal areas is a significant pathway for organic and inorganic contaminants entering the environment. In addition, 2 million tons of sewage and industrial and agricultural waste are discharged into the oceans daily (United Nations, 2003). In past decades, the amount of coastal runoff has significantly increased, becoming a primary cause of degradation of water quality (e.g. Vitousek, 1994; Birch and Rochford, 2010). Climate change is predicted to increase the frequency, duration and strength of storms and rainfall, which, in turn, is expected to increase the number and concentration of contaminants washed into coastal ecosystems (Schiedek et al., 2007).

Agriculture is another important source of pollution of natural waters, due to the application of millions of tons of fertilisers and pesticides each year (FAO and Nations, 2006; Schwarzenbach et al., 2006). These contaminants are washed or blown away from target areas and eventually discharged into rivers, groundwater and marine systems. For example,

Table 9.1 Summary of the main types of contaminants that might affect the diversity and functioning of marine systems.

Types of contaminant	Constitution	Bioaccumulation	Persistence in environment	Main effects	Examples	Sources	Comments
Metals	Metallic chemical elements naturally present in rocks and soils.	Y	Varies according to the metal (e.g. Hg is highly persistent)	Binds with important enzymes and proteins, changing their ability to function properly, causing malfunctioning or death of cells	Mercury, Lead, Cadmium, etc.	Industrial and mining wastes	Some metals are necessary for normal growth and development of marine organisms, but, at some concentrations, all metals are toxic
Hydrocarbons	Simplest organic compounds, containing only atoms of carbon and hydrogen.	Y	Moderate	Toxic contamination and/or physical contact	PAHs; oil	Sources of hydrocarbons can be natural, such as volcanoes and forests fires and anthropogenic such as petroleum and internal combustion engines	
Organic compounds	Manmade toxicant chemical substances	Y	High	Disrupt the endocrine, reproductive and immune systems	PCBs and some pesticides	Anthropogenic and agricultural activities, etc.	Travel long distances from their origin via wind and ocean currents

Herbicides or pesticides	Manmade toxicant chemical substances	N	Varies according to the type of herbicide	Varies according to the type of pesticide or herbicide. Examples are: disruption of photosystem II systems and basic cellular functions	Permethrin, Argarol, etc.	Run-offs from agriculture	Although some herbicides are classified as organic compounds, we separated them in this chapter
Tributyltin (TBT)	TBT is an organotin: a compound consisting of one to four organic components attached to a tin atom via carbon–tin covalent bonds	Y	Moderate	Disruption of endocrine system. Causes imposex (the development of male characteristics) in female gastropods	TBT	Antifouling paints	Mainly used as marine antifoulant.
Radionuclides	Atoms with unstable nucleus	Y	Varies. Some radionuclides are highly persistent	Radiation	Uranium and caesium	Industries' activities	While radionuclides occur naturally in the environment, those categorised as harmful are typically of *(cont.)*

Table 9.1 (cont.)

Types of contaminant	Constitution	Bioaccumulation	Persistence in environment	Main effects	Examples	Sources	Comments
							anthropogenic origin, released through industrial processes. Not reviewed in this chapter
Nutrients	Compounds found in the environment that plants and animals need to grow and survive	N	N/A	Hypoxia (anoxia); algal blooms which can lead to altered food webs	Phosphorus and nitrogen	Sewage, run-offs, etc.	Discussed in Chapter 8
Marine debris	Typically defined as any manmade object discarded, disposed of, or abandoned that enters the coastal or marine environment	Y	Highly	Can be physically or chemically harmful to organisms – either because they are themselves potentially toxic or because they absorb other pollutants	Plastic, glass, derelict fishing gear, etc.	Anthropogenic source; unplanned disposal	Although extremely common and harmful to environment, marine debris are not classified as hazards by environmental agencies

concentrations of the common herbicide Diuron have been measured at ecologically relevant concentrations in waters of the Great Barrier Reef up to 90 km offshore (Lewis et al., 2009; Shaw et al., 2010; Smith et al., 2012).

Vessels and their associated coastal development are additional sources of pollution in marine systems. Shipping has been identified as one of the major causes of pollution in places often assumed to be protected (e.g. Antarctica) due to their great distance from urbanised areas (Negri and Marshall, 2009). Vessel-associated contamination is often in the form of antifouling paint released or abraded from ship hulls. Vessel mooring facilities such as ports and marinas are associated with increased concentrations of metals and poly-aromatic hydrocarbons (PAHs) (Dafforn et al., 2009). In addition, accidental spills of oil and gasoline represent an important source of pollution (Kennish, 2002; Schwarzenbach et al., 2006).

Mining is another major, if localised, source of marine pollution. The exploration and production of oil and gas reservoirs around the world has resulted in large quantities of contaminants, mainly in the form of drill cuttings, i.e. a combination of drilling mud, speciality chemicals and fragments of reservoir rock, being deposited onto the seafloor (e.g. Breuer et al., 2004). Ocean drilling can sometimes go wrong and result in major contamination events such as in 2010, when the Deepwater Horizon oil rig exploded and sunk, releasing the equivalent of 4.9 million barrels of oil into the Gulf of Mexico. Coastal mines that discharge cuttings to rivers may also deposit large quantities of metals and other contaminants in nearshore and estuarine systems (e.g. Stauber, 1998).

9.3 Effects of contaminants

Several types of contaminants, derived from the activities and sources mentioned above, such as metals, PAHs, pesticides and antifouling compounds (e.g. tributyltin (TBT)), can be found in coastal and estuarine waters near urbanised areas. Each of these contaminants cause different effects on organisms, such as changes to their physiology or behaviour (e.g. Gray et al., 1988; Qiu et al., 2005; Culbertson et al., 2007), which can lead to local extinctions and the loss of biodiversity and ecosystem services (Chapters 2, 3, 5). Direct effects of contaminants vary with the intensity and duration of exposure to contaminant(s) (Chapter 4), but often reduce the abundance of organisms (Fleeger et al., 2003).

Metals, for instance, in trace amounts are natural constituents of the biosphere and are essential elements for growth, however, at high enough concentrations, all metals are toxic (Bryan, 1971). The diversity of assemblages tends to decrease with increased concentrations of metals and there are differences in the structure of the assemblages in the presence of high concentrations of metals (e.g. Rygg, 1985; Bryan and Langston, 1992; Stark, 1998). Moreover, metals have been shown to cause a decrease in growth and settlement of many benthic organisms and/or an increase in larval mortality (e.g. Wisely and Blick, 1967; Peters et al., 1997; Bellas et al., 2004).

Pesticides and herbicides are common contaminants that can threaten the natural functioning of assemblages (Rohr and Crumrine, 2005). The application of insecticides, for example, can reduce the diversity and abundance of invertebrates, decrease decomposition rates and increase primary productivity (Relyea, 2005; Relyea and Hoverman, 2008). Photosystem II herbicides penetrate coral tissues and, within minutes, reduce the photochemical efficiency of the intracellular algal symbionts (Jones, 2005). Many pesticides and herbicides can also be toxic to non-target organisms which can, consequently, affect whole assemblages in terms of composition, biomass and number of species.

Oil spills can result in both immediate and long-term damage to the environment. Some of the effects caused by oil spills can last for decades after the spill and can be due to the physical contact of the oil with organisms or due to their toxic effects (Peterson et al., 2003). Some of the most well-known effects of oil spills are the death of sea birds and marine mammals. Oil coats the feather of birds, preventing them from flying and damaging their natural waterproofing and insulation. It may also clog the blowholes of whales and dolphins, preventing them from breathing or communicating. Less obvious are the effects of hydrocarbons from oil spills that are absorbed by marine organisms in the form of fat-soluble compounds. These can be extremely prejudicial to the environment and can accumulate through food webs, contaminating organisms at the top of marine food chains. Oil also contains ring-structured PAH compounds, which can be particularly toxic to animals. Furthermore, oil spills can limit the amount of light absorbed by planktonic algae, causing their eventual death and consequently reducing the amount of food available for other organisms, such as zooplankton. This may result in knock-on effects reflected in other parts of the food web. The clean-up following spills may sometimes be more damaging to organisms and assemblages than the oil itself. The chemical dispersants used in the clean-up may be toxic to several organisms, including important grazers such as molluscs, the death of which may result in cascading effects such as the proliferation of opportunistic macroalgae (Peterson, 2001; Thompson et al., 2002).

Extraction of crude oil and gas is often accompanied by the production of great quantities of water, called produced water. This water generally contains traces of drilling lubricants (e.g. barite and bentonite) and naturally occurring radionuclides (Hosseini et al., 2012). Barite and bentonite are both toxic and can cause impacts on marine invertebrates (Strachan and Kingston, 2012). Radionuclides are also released in the marine environment via discharges of nuclear power plants or nuclear accidents, either in small or large amounts (Dallas et al., 2012). Very little is known about the potential effects of radionuclides on the marine environment. A decreased survival in aquatic organisms after irradiation has been reported (Dallas et al., 2012) as well as high levels of polonium in plankton communities, indicating the potential for significant bioaccumulation (Hosseini et al., 2008).

Marine debris, although not yet classified as a hazard by government and protective agencies, can also cause significant damage to marine organisms. Plastics are synthetic organic polymers and make up most of the marine litter worldwide (Derraik, 2002). Although plastics have only existed for just over a century, their annual production has reached 280 million tonnes in 2011, reaching the marine environment via accidental release and indiscriminate discard (Derraik, 2002; Wright et al., 2013). Plastic debris can harm organisms physically and/or chemically, by releasing toxic substances (that they either absorb or contain) (Rochman and Browne, 2013). Large pieces of plastic can kill and injure several marine species, such as marine mammals and sea birds via ingestion or entanglement (Rochman and Browne, 2013). Although the effects of microplastic on marine and estuarine organisms remain poorly understood, studies have shown that such pollutants also have the potential to cause considerable harm to organisms (Browne et al., 2007).

Contaminants may affect organisms through ecological interactions. These are generally called indirect effects and may increase or decrease the number of species and/or abundances of populations (Fleeger et al., 2003). For example, contaminants may affect competitive dominants, favouring inferior competitors, which are better colonisers or more tolerant to the contaminant. Short-term copper exposures, for example, by causing a reduction in the survival of longer lived solitary ascidians, have been shown to increase the abundance of rapid colonisers such as serpulid polychaetes (e.g. Johnston and Keough, 2002; Johnston et al., 2002). Another common indirect effect of contaminants occurs when the abundance of primary producers is increased in the presence of certain contaminants because the numbers and/or activities of grazers are decreased (Fleeger *et al.*, 2003, for review). Herbicides and fungicides can have bottom-up effects because they reduce primary productivity and fungi, the major food sources for higher trophic levels (Rohr *et al.*, 2006).

When contaminants affect the abundance of ecologically influential organisms, such as keystone species (Paine, 1969, 1995) and ecosystem engineers, including habitat-forming or foundation species (Bruno and Bertness, 2001), their influence on communities and ecosystem processes can be particularly pervasive. Kelp forests, in temperate systems and coral reefs, in tropical areas, provide complex habitat and food for a great diversity of fish and invertebrates, including many economically important species (Jackson, 2001; Steneck *et al.*, 2002). The elimination of kelp plants and some reef-building corals would have, therefore, dramatic impacts on diversity and functioning of systems through a series of direct and indirect effects. Also, some physiological functions of foundation species, such as clearance rates or photosynthetic activity, contribute directly to the functioning of ecosystems. Habitat-forming invertebrates such as oysters and mussels, for example, are the dominant filter feeders in the habitat they form and their clearance and growth rates can usually be as approximated to the total clearance and productivity of these same habitats. Similarly, the photosynthetic activity of habitat-forming macroalgae can be directly indicative of the primary productivity of the ecosystem. Impacts caused by contaminants on such functional groups would be expected to have dramatic consequences, not only for associated biota, but also for ecosystem functioning.

9.4 Scope of the chapter

Here we synthesise the results of three separate systematic literature reviews that each used a consistent process. The first review, Johnston and Roberts (2009) focused on the impacts of contaminants on marine biodiversity. The second and third reviews (Johnston *et al.*, 2015, and Mayer-Pinto *et al.*, in review) focused on the impacts of contaminants on ecosystem function for communities and habitat-forming species, respectively. In this chapter we bring the findings of these three reviews together to enable a synthesis and comparison of contaminant impacts on biodiversity and ecosystem functioning.

Table 9.2 *List of the keywords used in the structured search through the online resources for each review.*

Review	Diversity	Ecosystem functioning	Habitat-forming organisms	Keywords common to all reviews
Diversity	Biodiversity, diversity			
Ecosystem functioning	N/A	System*, ecosystem, ecosystem function*, resilience, tolerance, resistance, recovery, stability, energy cycling, food web, nutrient flux*, carbon or nitrogen cycle		contamina*, pollut*, marine, estuar*, hydrocarbon*, PAH, metal*
Habitat-forming organisms	N/A	productivity, production, respiration, respir*, biomass, photosynthesis	brown alga*, kelp*, mangrove*, salt marsh*, seagrass*, bivalve*, clearance rate*, filter*, growth, purification	

The review methodology is outlined in detail within each of the separate papers, but for completeness we describe it briefly below. The literature reviews used structured searches through online search engines, assessing published studies of the effects of contaminants on biological diversity, ecosystem functioning and functional traits of habitat-forming organisms. Several keywords and their combinations were used in the searches (Table 9.2). The titles of all papers that appeared in these searches were read and those with a marine focus and which reported direct comparisons of contaminated areas or treatments with controls were selected. If the title was not sufficient to gather the necessary information, abstracts were read. For papers that met the selection criteria, their citation lists were reviewed in order to capture studies that had not been included in the initial searches or that had been published in journals not indexed in the databases searched.

Studies were classified according to the type of contaminant used (metals, hydrocarbons, organic compounds, herbicides, TBT, combined contaminants and others), habitat type or organisms studied (in case of diversity studies and habitat-formers, respectively) or component of the system studied (in case of studies on ecosystem functioning, e.g. phytoplankton, benthic invertebrates, phyto- and zooplankton, etc.) and function or diversity measured (e.g. gross primary productivity or Shannon index, respectively).

Data were then collected from each paper, including the overall finding of the research (reduced, increased or no effects on diversity, functioning or relevant physiological traits) as concluded by the authors of the paper. Where concentrations of the same contaminant were varied within a single study, the outcome from the contrast between the greatest concentration and the controls was used.

9.4.1 Inclusion criteria

Only studies on the effects of chemical contaminants are discussed in this chapter (contamination by radionuclides were excluded). No other stressors such as habitat modification, aquaculture, overfishing, climatic changes, etc. were included. In addition, we excluded studies of excess nutrients (eutrophication) since these represent a vast field of research that warrants their own chapter (Chapter 8). Multiple stressor effects (Chapter 4) were considered beyond the scope of this chapter unless both stressors belonged to the same category of stressor (e.g. combined effects of two different metals) or where the effects of contaminants were clearly separated from the other stressors. Some of the possible effects of contaminants and other stressors are, however, briefly discussed in Section 9.6.3.

The term biodiversity per se has many accepted definitions and its use may be strongly context dependent (DeLong, 1996). In this chapter, we focused on biodiversity in terms of the number of species per unit area either as species richness or with a measure of relative abundance (e.g. Shannon–Wiener index (H') and Pielou evenness (J) (Johnston and Roberts, 2009).

For the purpose of this chapter, ecosystem functioning was defined as changes in stocks of energy and materials, such as carbon, water, mineral

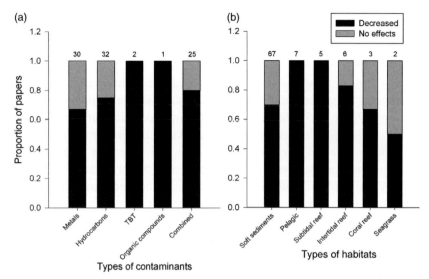

Figure 9.1 Summary of a qualitative literature review on studies of the effects of contaminants on marine biodiversity. Studies are characterized by (a) type of contaminants studied and (b) study systems in which research is conducted. The number of research papers in each category is shown above each bar.

nutrients and the rates of processes involving fluxes of energy and matter (e.g. productivity and decomposition) between trophic levels and to or from the environment (e.g. Lamont, 1995; Hooper et al., 2005; Srivastava and Vellend, 2005). Studies that evaluated the effects of contaminants on at least one aspect of physiological function of habitat-forming organisms that could be directly translated to ecosystem functioning (such as clearance rate or photosynthetic efficiency) were included.

9.5 Effects of contaminants on biodiversity and ecosystem functioning

Contaminants were associated with reductions in biodiversity within every marine habitat (Johnston and Roberts, 2009). The vast majority of published studies found a negative effect of contaminants on diversity of marine and estuarine systems, irrespective of type of contaminant or habitat studied (Figure 9.1). Pollution by contaminants was correlated with consistent negative impacts on the species richness (approximately 30–50% reduction) and evenness of recipient communities, with effect sizes upon species richness (S) and Shannon–Wiener diversity (H′)

tending to be greater than reductions in Pielou evenness (J) (Johnston and Roberts, 2009).

Metals, for instance, have been shown to decrease abundance and diversity of sessile invertebrates (e.g. Medina et al., 2005; Perrett et al., 2006). Even remote habitats, once considered to be pristine such as Antarctica, have been affected by anthropogenic contamination (Lenihan and Oliver, 1995; Stark et al., 2003; Stark et al., 2005; Negri et al., 2006). In a comprehensive survey of Antarctic sedimentary marine communities in relation to contaminants, such as hydrocarbons, metals and organic compounds, Lenihan and Oliver (1995) found that the most heavily contaminated site had low infaunal and epifaunal abundance and was dominated by a few opportunistic species of polychaetes. The uncontaminated sites, on the other hand, had a more diverse community, including crustaceans and a large population of bivalves. A meta-analysis done on the effects of contaminants on fish also found an overall negative effect of contaminants on the abundance of fish (McKinley and Johnston, 2010). The authors found, however, weak links with species richness, which suggests that contaminants do not have as strong an effect on the diversity of these higher order and highly mobile organisms (McKinley and Johnston, 2010).

Contaminants caused a decrease in the majority of ecosystem's functional endpoints assessed, such as net and gross primary production (NPP and GPP) and bacterial production of systems (Figure 9.2). The main exception to this finding was when considering total respiration of systems. Contaminants caused an overall increase in respiration, which is expected, since a decrease in primary production is often associated with an increase in respiration. These effects were generally consistent regardless of the type of contaminant studied (i.e. metals, hydrocarbons, pesticides, etc.).

The negative effects of contaminants on ecosystem functioning were also consistent among different ecosystem components (e.g. pelagic producers or benthic microorganisms) when each was studied individually. However, if more than one component of the system was examined simultaneously, for instance, phyto- *and* zooplankton or benthic invertebrates *and* fish, effects of contamination were less likely to be found. This was true for different functional endpoints such as GPP and chlorophyll *a*. Kuiper (1981), for instance, did an experiment to evaluate the effects of a metal on a coastal planktonic community (i.e. phyto- and zooplankton combined) and found no effects of the contaminant on the GPP of the system. On the other hand, in a study on phytoplankton and

Pollution: effects of chemical contaminants and debris · 257

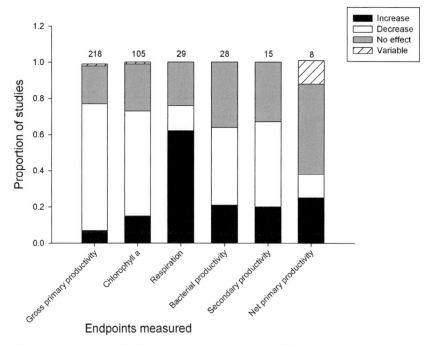

Figure 9.2 Summary of a literature review on studies of the effects of contaminants on functioning endpoints of marine systems. The number of studies in each category is shown above each bar.

metals, Maraldo and Dahllof (2004) found a decrease in the GPP of systems caused by the contaminant. These apparently contrasting results are probably the consequence of differential sensitivity of interacting components of the system (e.g. if consumers are affected by contaminants then impacts on primary producers may be ameliorated). This highlights the importance of having integrated and multidisciplinary studies.

Habitat-formers, such as kelps, seagrasses and mussels, are also negatively affected by contaminants. Such pollutants caused a decrease in most functional variables of habitat-forming organisms (Figure 9.3). Remarkably, however, a combination of contaminants, i.e. contaminants from different categories, such as metals and organic compounds, seems to have no effects on primary productivity of habitat-formers. This might be due to the occurrence of antagonistic effects of different types of contaminants (Chapter 4).

Between the two different types of invertebrate habitat-formers, mussels (Mytiloida) and oysters (Ostreoida), there were not many differences

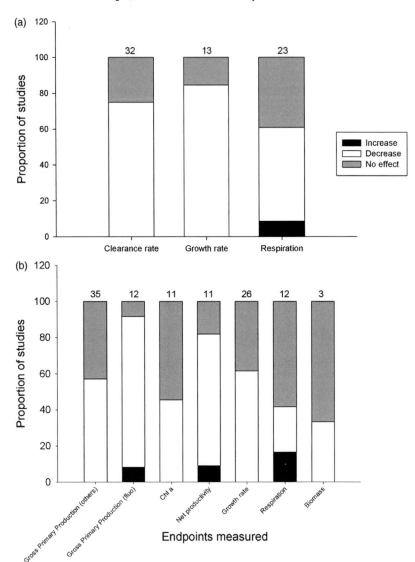

Figure 9.3 Proportion of studies that found an increase, decrease or no effects of contaminants on all of the reviewed functional traits of habitat-forming (a) invertebrates and (b) primary producers. The number of research papers in each category is shown above each bar.

in the responses of clearance and growth rates to contaminants. This may be due to their similarity regarding physiological and morphological features. Effects of contaminants did seem to vary, however, amongst the habitat-forming producers. It appears that the photosynthetic activity (measured by fluorescence) of fucoids and seagrass is more resistant to contaminants than that of kelps and saltmarsh plants. Furthermore, contaminants did not always affect gross primary productivity of fucoids or seagrasses, whereas the GPP of kelps and saltmarsh plants was always reduced by exposure to contaminants. Huovinen et al. (2010), for example, found that metals decreased the primary productivity of three different species of kelp. It is important to note, however, that there is a great discrepancy in the number of studies done on these different groups (14 and 17 studies on fucoids and seagrasses, respectively, and only 3 studies on kelps and 1 on saltmarsh plants). Interestingly, when considering growth, respiration and chlorophyll *a*, kelps appear to be more resistant to contaminants, with the great majority of studies showing no effects of contaminants on those variables. Overall, contaminants appear to increase the respiration rate of fucoid algae. There were no clear trends for seagrass for any of the response variables analysed, with approximately half of the studies showing a negative effect of toxicants and the other half showing no effects. These differences observed in the general effects of contaminants on different types of producers are probably due, not only to the different physiological characteristics of the organisms, but also to differences in potential exposure. Kelps, for instance, require hard substrate to attach and live, whereas seagrasses are found in soft sediments, which have a greater potential to absorb contaminants.

Marine debris can also cause a variety of deleterious effects in the biodiversity and ecosystem functioning worldwide. Their threats to the marine environment are not only mechanical, due to ingestion of plastic debris and entanglement in synthetic ropes and lines, drift nets, etc. (Derraik, 2002), but also chemical. Marine debris is not classified as a chemical contaminant, but can leach constituent contaminants such as monomers and plastic additives (Wright *et al.*, 2013), and can adsorb organic pollutants such as PAHs (Fisner *et al.*, 2013). Microplastics – a pervasive component of marine debris – are found in considerable amounts in benthic and sedimentary habitats (Wright *et al.*, 2013), making some benthic suspension and deposit feeders especially likely to be exposed to this type of contamination (Wright *et al.*, 2013). In addition, these pollutants have the potential to modify the structure of populations, by creating, for example, substrata for rafting communities, as well as increasing biogeographic connectivity via long distance transport mechanisms (Wright et al., 2013). There is also a potential threat to marine ecosystems due to the accumulation of plastics debris on the seafloor (Derraik, 2002). This accumulation can prevent the gas exchange between the overlying waters and the pore waters of the sediments, leading to hypoxia and anoxia (Derraik, 2002). Marine debris has, therefore, the capacity to greatly affect diversity and functioning of marine and estuarine systems via a variety of physical and chemical processes.

9.6 Limitations of existing research

9.6.1 Differences in types of communities and contaminants

Most research on contaminants and diversity has been done in soft-sediment systems with studies in hard-substrata habitats, such as intertidal rocky shores and subtidal reefs, being rare (Johnston and Roberts, 2009; Mayer-Pinto et al., 2010). Johnston and Roberts (2009) also identified a substantial lack of studies on coral reefs and seagrass meadows, with only five studies having evaluated the effects of contaminants on the biodiversity of these systems. A search carried out in February 2013 revealed that not much had changed since the 2009 review. Only four new studies have been published on contaminant impacts on coral reef diversity and none was found for seagrass.

On the other hand, to date, most studies on ecosystem functioning and contaminants have been done in coastal pelagic systems, mainly plankton. Furthermore, most research so far has focused on a single trophic level, with only 25% of the studies addressing two or more trophic levels. To fully understand the effects of contaminants on ecosystem functioning, however, studies analysing several components of the system are essential due to the many chemical and ecological interactions occurring among species.

Of the studies on habitat-formers, approximately 57% were done on producers, i.e. macroalgae and kelps, while 43% were on invertebrates, such as mussels and oysters. Seagrasses were the most studied taxa of the producers, followed by fucoids and kelps (i.e. Laminarians), respectively. Only species from the orders Mytiloida and Ostreioda were identified in the studies of invertebrate habitat-formers, with the great majority of studies having evaluated effects of contaminants on mussels (i.e. Mytiloida). This bias in the type of model organisms and habitats studied is likely to affect our understanding of the real effects of contaminants on marine ecosystems.

Studies have generally focused either on effects of contaminants on biodiversity (Reish et al., 1999; Johnston and Roberts, 2009; Clements et al., 2012), or on how biodiversity links to functioning (e.g. Loreau et al., 2001; Hooper et al., 2005; Hooper et al., 2012). The relationship among toxicants, biodiversity and ecosystem functioning is, however, usually dependent on the composition of species, the type of toxicant and abiotic factors that vary across studies, space and time (Hooper et al., 2005; McMahon et al., 2012). Contaminants can affect organisms that do or do not contribute substantially to the functioning of systems. Thus, contaminants' effects on diversity may not necessarily affect functioning and vice versa (McMahon et al., 2012). So, whether contaminants affect functionally redundant species or species that contribute little to the functioning of particular ecosystems, or whether they affect influential species, such as habitat-formers, will determine the extent to which contaminants affect ecosystem functioning directly (McMahon et al., 2012). Translating structural measures such as abundance and diversity of organisms into their functional equivalents is prone to error, leading to a patchy understanding of how toxicants transform ecosystem function. Consequently, the differences found between studies on diversity and functional studies regarding the focal ecosystems, do not allow the formulation of strong causal links between diversity and function and the full understanding of the role that contaminants may play in mediating this relationship. There is substantial opportunity to add value to existing toxicant studies by incorporating structural and functional endpoints. For instance, a well-designed and relevant study done *in situ* on copper mine tailings and meiofauna assemblages in Chile (Lee and Correa, 2005), although very comprehensive regarding structure and diversity of assemblages, could have greatly contributed to our understanding of contaminants and functioning of systems, if the authors had included some functional endpoints such as the primary productivity of the sediment.

Furthermore, although hydrocarbons and metals overall were the most studied type of contaminants in both biodiversity and functional studies, we have observed some striking differences in the toxicant focus between these two types of studies. Herbicides, for instance, were the third most common contaminant evaluated in toxicity tests of ecosystem function, however, they have not yet been used as model contaminant in any of the marine biodiversity studies (see Johnston and Roberts, 2009). This focus on herbicides in functional studies is likely to reflect the predicted impact of these toxicants on photosynthetic organisms and therefore primary productivity. The lack of diversity measurements as part of such studies might be a reflection of the difficulty in confidently identifying many photosynthetic plankton and therefore the difficulty in establishing impacts on biological diversity in that system. Advances in molecular sequencing technology will enable further studies of microscopic organisms that combine both diversity and functional measures of stressor impact (e.g. Sun *et al.*, 2012).

9.6.2 Research approach and rigour of study design

The vast majority of studies reviewed here were either done in laboratory conditions or were purely field-based surveys. Johnston and Roberts (2009) showed, however, that manipulative field experiments tended to have lower effect sizes associated with contaminant impact on Shannon–Weiner diversity than did survey studies. There were similar strong trends towards smaller effect sizes in field and laboratory experiments with respect to species richness and evenness. In addition, when considering field experiments versus field surveys, a greater proportion of the experimental studies found no effects of contaminants than did field surveys. Field studies done with habitat-formers were, however, more likely to find no effects of contaminants than those done in the laboratory, irrespective of the type of the habitat-forming organism.

Laboratory experiments do not reflect the complex conditions to which the organisms are naturally subjected, such as the interactions among and within species and the physical dynamics of the system (Connell, 1974; Underwood and Peterson, 1988). Laboratory toxicological studies also lack the ability to predict indirect effects of toxicants on organisms mediated through other processes (Underwood and Peterson, 1988), which are crucial to fully understand ecosystem functioning. For instance, interactions between contamination and ecological processes such as competition can modify responses of organisms to a toxicant (Johnston and Keough, 2003), which can alter their effects on the diversity and, consequently, the functioning of a system. Also, when a contaminant is released in the field, there are many other factors influencing its toxicity (see Section 9.6.3), which are lacking in the laboratory. This might be one of the reasons for the greater proportion of studies that found no effects of contaminants in the field relative to the laboratory, however exposure mechanisms and scale of exposure are alternative explanations.

Although field surveys give a better idea of what is actually happening in a natural system, they are unable to demonstrate a causal relationship between the contamination and the changes in diversity and/or function of systems. Manipulative experiments in the field are necessary to allow tests of hypotheses about effects of pollution under relevant environmental conditions and to understand how effects of contaminants vary spatially and temporally. Since every approach has its own strengths and weaknesses, it is important that we gather multiple lines of evidence in order to fully comprehend the impact of contaminants under a range of scenarios (Chapter 3).

Unfortunately, almost half of the studies on the functioning of ecosystems and habitat-formers had problems with their design, such as spatially confounded controls or a lack of replication of control and contaminated sites. This is likely to influence the outcome of results and/or affect the interpretation of their findings (see e.g. Mayer-Pinto et al., 2010 for review). Unless researchers and journal editors restrict publishing to well-designed studies, which allow for unambiguous tests of hypotheses, we will continue to have difficulty interpreting and understanding the effects of contaminants on ecosystems.

9.6.3 Contaminants and other stressors

Although this chapter deals with one main type of stressor – contamination – natural systems are often simultaneously subjected to multiple human-derived stressors (Crain et al., 2008; Chapter 4). Abiotic and biotic stressors (i.e. temperature, contaminants and disruptions and changes in biological interactions) usually interact to produce combined impacts on biodiversity and ecosystem functioning (Vinebrooke et al., 2004).

Climatic changes, for instance, are expected to affect contaminant exposures and their toxic effects (Schiedek et al., 2007). Altered atmospheric conditions are predicted to increase the frequency, duration and strength of storms and rainfall (IPCC, 2007). Greater rainfall will not only increase freshwater inputs in coastal systems, but also the amount of contaminants (e.g. pesticides, metals, etc.) being released via runoff. Changes in atmospheric conditions also affect current circulation and upwelling regimes, which, in turn, affect the availability of nutrients, as well as contaminants, in the system.

Increased temperatures, predicted to occur with climate change, are also likely to influence the effects of contaminants on marine systems, since it is well known that temperatures modify the chemistry of several contaminants, altering their toxicities. The toxicity of metals, for example, varies according to the salinity, pH and temperature of the water, etc. (Bryan, 1971; McLusky et al., 1986). In addition, greater temperatures are likely to increase the rate of uptake of contaminants due to responses to increased metabolic rates and decreased oxygen solubility (Kennedy et al., 1989).

Pollution also 'interacts' with stressors such as habitat modification (Schiff et al., 2007), overfishing (Jackson, 2001) and invasive species (Piola and Johnston, 2008), often having synergistic effects on diversity and functioning. Understanding the effects of contaminants, separately and in conjunction with other common stressors, is therefore crucial to being able to devise sensible and efficient management and conservation policies.

9.6.4 Temporal and spatial scales

One of the main challenges for environmental managers is to determine the extent and duration of a possible impact due to contamination. Spatial and temporal scales of impacts are usually not known, making it difficult to establish efficient protocols for restoration and management policies. Species respond to a myriad of environmental factors and their combinations on different spatial and temporal scales (Wiens et al., 1993). In marine systems, contaminants are diluted with seawater and dissipate according to local environmental factors, such as waves, currents, temperature, etc., which can lead to a 'patchy' distribution. Amounts of contaminants derived from a point-source disturbance, for instance, such as mining wastes and sewage outfalls, tend to decrease with increasing distance from the point source, creating a gradient (Bishop et al., 2002). Therefore, the distribution of contaminants interacts with the natural distribution of organisms. Gradients of contaminants are, however, often poorly matched with ecological gradients of abundance of organisms or structure of assemblages (e.g. Raimondi and Reed, 1996; Bishop et al., 2002). The perception of ecological patterns and processes is greatly determined by the scale at which they are measured, with studies done on different spatial scales usually resulting in the definition of different ecological patterns (Wiens, 1989; Wiens et al., 1993). Determining the proper spatial scale at which an impact might be occurring is, therefore, crucial if managers and regulators are to be able to successfully manage the impact. Many studies tend, however, to ignore different spatial scales on their assessment of impacts by contaminants on diversity or functioning of systems and whether impacts are consistent across spatial scales. This gap in our knowledge may critically affect our perception of the extent of effects of contaminants on systems.

Temporal scales should also be considered in studies of contaminants. Subtle changes in organisms or assemblages in the short term can have further consequences in the long term through a series of direct and indirect interactions. It is believed, for instance, that long-term studies help the understanding of the complex interactions among organisms and the effect(s) that stressors may have on these interactions (Hawkins et al., 2002; Steinbeck et al., 2005; Hobbs et al., 2007). It can be difficult, however, to separate the effects of long-term contamination from other environmental variation.

9.7 Conclusions

Contaminants, in general, cause a decrease in the biological diversity and functioning of marine and estuarine systems. There are, however, clear differences in the toxicant focus for functional studies relative to studies of biodiversity. We have observed a strong bias in the ecotoxicological literature towards the study of ecosystem function in pelagic systems and biodiversity in benthic systems. Plankton are the most investigated organisms in relation to ecosystem function while benthic communities, particularly those in soft sediments, are the focus of biodiversity studies. Contaminants can affect organisms that do or do not contribute substantially to the functioning of systems. Thus, contaminants effects on diversity may not necessarily affect function and vice versa (McMahon et al., 2012). Different species usually influence different functions. Therefore, risk assessments and management strategies focusing on isolated individual processes underestimate the diversity needed to maintain multifunctional ecosystems (Hector and Bagchi, 2007). Because most existing research has only measured diversity *or* function and because each type of study has focused on a different set of habitats and contaminants – it is not currently possible to make strong causal links between diversity and function or the role that contaminants may play in this relationship. Identifying critical function endpoints that are directly linked to biological diversity is therefore important to establish new – more holistic and integrated – ecological risk assessments (Apitz, 2013).

We therefore conclude that unless we improve our integrated understanding of how contaminants influence multiple aspects of marine systems, we will not be able to provide sound and efficient managerial solutions for conservation purposes.

9.8 Key points and recommendations

- Pollution reduces both the diversity and function of marine ecosystems.
- Contaminants pose a serious threat to the diversity and functioning of systems and should be considered a priority for managers and regulators.
- Effects of contaminants were generally similar across types of contaminants, but varied according to the type of habitat-forming organism.
- Studies of the effects of contaminants on diversity have generally been done on benthic communities, whereas studies of function have been done on planktonic communities or keystone species. There is an urgent need, therefore, for more integrated research and risk assessment.
- Systems and/or habitats formed by one or few foundations species should be considered by environmental managers as especially susceptible systems to the effects of contaminants and prone to collapse.
- Marine debris should be considered as an ubiquitous hazard by managers and regulators and should be included in risk assessments.
- Benthic suspension and deposit-feeder organisms are likely to be more susceptible to ingestion of marine debris, such as microplastics.
- Other stressors – from invasive species and artificial structures to stressors associated with climatic changes – interact with contaminants, having the capacity to exacerbate their effects via synergistic processes. Risk assessments should, therefore, take into consideration a variety of probable concomitant stressors.
- The complexity of experimental designs needs to be increased if researchers, managers and regulators are to fully understand the effects of contaminants on biodiversity and function:
  - Assessments should include studies with more than one trophic component.
  - Assessments should take into account relevant temporal and spatial scales.
  - Assessments should measure structure *and* function wherever possible.
- Environmental and regulators should include functional endpoints in future ecosystem risk assessments.

References

Apitz, S. E. (2013). Ecosystem services and environmental decision making: seeking order in complexity. *Integrated Environmental Assessment and Management*, 9, 214–230.
Bellas, J., Beiras, R. and Vazquez, E. (2004). Sublethal effects of trace metals (Cd, Cr, Cu, Hg) on embryogenesis and larval settlement of the ascidian *Ciona intestinalis*. *Archives of Environmental Contamination and Toxicology*, 46, 61–66.
Bellwood, D. R., Hughes, T. P., Folke, C. and Nyström, M. (2004). Confronting the coral reef crisis. *Nature*, 429, 827–833.
Benedetti-Cecchi, L., Pannacciulli, F., Bulleri, F. *et al.* (2001). Predicting the consequences of anthropogenic disturbance: large-scale effects of loss of canopy algae on rocky shores. *Marine Ecology Progress Series*, 214, 137–150.
Birch, G. F. and Rochford, L. (2010). Stormwater metal loading to a well-mixed/stratified estuary (Sydney Estuary, Australia) and management implications. *Environmental Monitoring and Assessment*, 169, 531–551.
Bishop, M. J., Underwood, A. J. and Archambault, P. (2002). Sewage and environmental impacts on rocky shores: necessity of identifying relevant spatial scales. *Marine Ecology Progress Series*, 236, 121–128.
Breitburg, D. L., Sanders, J. G., Gilmour, C. C. *et al.* (1999). Variability in responses to nutrients and trace elements, and transmission of stressor effects through an estuarine food web. *Limnology and Oceanography*, 44, 837–863.
Breuer, E., Stevenson, A. G., Howe, J. A., Carroll, J. and Shimmield, G. B. (2004). Drill cutting accumulations in the Northern and Central North Sea: a review of environmental interactions and chemical fate. *Marine Pollution Bulletin*, 48, 12–25.
Browne, M. A., Galloway, T. and Thompson, R. (2007). Microplastic: an emerging contaminant of potential concern? *Integrated Environmental Assessment and Management*, 3, 559–561.
Bruno J. F. and Bertness, M. D. (2001). Habitat modification and facilitation in benthic marine communities. In *Marine Community Ecology*, ed. M. D. Bertness, S. D. Gaines and M. E. Hay. Sunderland, MA: Sinauer Associates, Inc., pp. 201–218.
Bryan, G. W. (1971). Effects of heavy metals (other than mercury) on marine and estuarine organisms. *Proceedings of the Royal Society of London Series B-Biological Sciences*, 177, 389–410.
Bryan, G. W. and Langston, W. J. (1992). Bioavailability, accumulation and effects of heavy metals in sediments with special reference to United Kingdom estuaries: a review. *Environmental Pollution*, 76, 89–131.
Clark, R. B. (2001). *Marine Pollution*. New York: Oxford University Press.
Clements, W. H., Hickey, C. W. and Kidd, K. A. (2012). How do aquatic communities respond to contaminants? It depends on the ecological context. *Environmental Toxicology and Chemistry*, 31, 1932–1940.
Connell, J. (1974). Ecology: field experiments in marine ecology. In *Experimental Marine Biology*, ed. R. Mariscal. New York: Academic Press, pp. 21–54.

Crain C. M., Kroeker, K. and Halpern, B. S. (2008). Interactive and cumulative effects of multiple human stressors in marine systems. *Ecology Letters*, 11, 1304–1315.

Culbertson, J. B., Valiela, I., Peacock, E. E. *et al.* (2007). Long-term biological effects of petroleum residues on fiddler crabs in salt marshes. *Marine Pollution Bulletin*, 54, 955–962.

Dafforn, K. A., Johnston, E. L. and Glasby T. M. (2009). Shallow moving structures promote marine invader dominance. *Biofouling*, 25, 277–287.

Dallas, L. J., Keith-Roach, M., Lyons, B. P. and Jha, A. N. (2012). Assessing the impact of ionizing radiation on aquatic invertebrates: a critical review. *Radiation Research*, 177, 693–716.

DeLong, D. C. (1996). Defining biodiversity. *Wildlife Society Bulletin*, 24, 738–749.

Derraik J. G. B. (2002). The pollution of the marine environment by plastic debris: a review. *Marine Pollution Bulletin*, 44, 842–852.

Duarte, C. M. (2002). The future of seagrass meadows. *Environmental Conservation*, 29, 192–206.

EU (2001). White Paper. Strategy for a future chemicals policy. Commission of the European Communities B, Belgium. Available at: http://eur-lex.europa.eu/LexUriServ/LexUriServ.do?uri=COM:2001:0088:FIN:EN:PDF.

FAO (2006). Statistical database. FAO, Rome.

Fisner, M., Taniguchi, S., Moreira, F., Bicego, M. C. and Turra, A. (2013). Polycyclic aromatic hydrocarbons (PAHs) in plastic pellets: variability in the concentration and composition at different sediment depths in a sandy beach. *Marine Pollution Bulletin*, 70, 219–226.

Fleeger, J. W., Carman, K. R. and Nisbet, R. M. (2003). Indirect effects of contaminants in aquatic ecosystems. *Science of the Total Environment*, 317, 207–233.

Foster, M. S. and Schiel, D. R. (2010). Loss of predators and the collapse of southern California kelp forests: alternatives, explanations and generalizations. *Journal of Experimental Marine Biology and Ecology*, 393, 59–70.

GESAMP (1982). *Scientific Criteria for the Selection of Waste Disposal Sites at Sea*. London: Inter-Governmental Maritime Consultative Organization.

Gonzalez, J., Figueiras, F. G., Aranguren-Gassis, M. *et al.* (2009). Effect of a simulated oil spill on natural assemblages of marine phytoplankton enclosed in microcosms. *Estuarine Coastal and Shelf Science*, 83, 265–276.

Gray, J. S. (1997). Marine biodiversity: patterns, threats and conservation needs. *Biodiversity and Conservation*, 6, 153–175.

Gray, J. S., Aschan, M., Carr, M. R. *et al.* (1988). Analysis of community attributes of the benthic macrofauna of Frierfjord-Langesundfjord and in a mesocosm experiment. *Marine Ecology Progress Series*, 46, 151–165.

Hawkins, S. J., Gibbs, P. E., Pope, N. D. *et al.* (2002). Recovery of polluted ecosystems: the case for long-term studies. *Marine Environmental Research*, 54, 215–222.

Hector, A. and Bagchi, R. (2007). Biodiversity and ecosystem multifunctionality. *Nature*, 448, 188–U186.

Hobbs, R. J., Yates, S. and Mooney, H. A. (2007). Long-term data reveal complex dynamics in grassland in relation to climate and disturbance. *Ecological Monographs*, 77, 545–568.

Hooper, D. U., Adair, E. C., Cardinale, B. J. et al. (2012). A global synthesis reveals biodiversity loss as a major driver of ecosystem change. *Nature*, 486, 105–U129.

Hooper, D. U., Chapin, F. S., Ewel, J. J. et al. (2005). Effects of biodiversity on ecosystem functioning: a consensus of current knowledge. *Ecological Monographs*, 75, 3–35.

Hosseini, A., Brown, J. E., Gwynn, J. P. and Dowdall, M. (2012). Review of research on impacts to biota of discharges of naturally occurring radionuclides in produced water to the marine environment. *Science of the Total Environment*, 438, 325–333.

Hosseini, A., Thorring, H., Brown, J. E., Saxen, R. and Ilus, E. (2008). Transfer of radionuclides in aquatic ecosystems: default concentration ratios for aquatic biota in the Erica Tool. *Journal of Environmental Radioactivity*, 99, 1408–1429.

Huovinen, P., Leal, P. and Gomez, I. (2010). Interacting effects of copper, nitrogen and ultraviolet radiation on the physiology of three South Pacific kelps. *Marine and Freshwater Research*, 61, 330–341.

IPCC (2007). Climate change 2007. The physical science basis: Working group I contribution to the fourth assessment report of the IPCC. Cambridge: Cambridge University Press.

Jackson, J. B. C. (2001). What was natural in the coastal oceans? *Proceedings of the National Academy of Sciences of the United States of America*, 98, 5411–5418.

Johnston, E. L. and Keough, M. J. (2002). Direct and indirect effects of repeated pollution events on marine hard-substrate assemblages. *Ecological Applications*, 12, 1212–1228.

Johnston, E. L. and Keough, M. J. (2003). Competition modifies the response of organisms to toxic disturbance. *Marine Ecology Progress Series*, 251, 15–26.

Johnston, E. L. and Roberts D. A. (2009). Contaminants reduce the richness and evenness of marine communities: a review and meta-analysis. *Environmental Pollution*, 157, 1745–1752.

Johnston, E. L., Keough, M. J. and Qian, P. Y. (2002). Maintenance of species dominance through pulse disturbances to a sessile marine invertebrate assemblage in Port Shelter, Hong Kong. *Marine Ecology Progress Series*, 226, 103–114.

Johnston, E. L., Mayer-Pinto, M. and Crowe, T. P. (2015). Chemical contaminant effects on marine ecosystem functioning. *Journal of Applied Ecology*, 52, 140–149.

Jones, R. (2005). The ecotoxicological effects of photosystem II herbicides on corals. *Marine Pollution Bulletin*, 51, 495–506.

Kennedy, C. J., Gill, K. A. and Walsh, P. J. (1989). Thermal modulation of benzoapyrene uptake in the Gulf toadfish, *Opsanus beta*. *Environmental Toxicology and Chemistry*, 8, 863–869.

Kennish, M. J. (2002). Environmental threats and environmental future of estuaries. *Environmental Conservation*, 29, 78–107.

Kuiper, J. (1981). Fate and effects of cadmium in marine plankton communities in experimental enclosures. *Marine Ecology Progress Series*, 6, 161–174.

Lamont, B. B. (1995). Testing the effect of ecosystem composition structure on its functioning. *Oikos*, 74, 283–295.

Lee, M. R. and Correa, J. A. (2005). Effects of copper mine tailings disposal on littoral meiofaunal assemblages in the Atacama region of northern Chile. *Marine Environmental Research*, 59, 1–18.

Lenihan, H. S. and Oliver, J. S. (1995). Anthropogenic and natural disturbances to marine benthic communities in Antarctica. *Ecological Applications*, 5, 311–326.

Lewis, S. E., Brodie, J. E., Bainbridge, Z. T. et al. (2009). Herbicides: a new threat to the Great Barrier Reef. *Environmental Pollution*, 157, 2470–2484.

Loreau, M., Naeem, S., Inchausti, P. et al. (2001). Ecology–biodiversity and ecosystem functioning: current knowledge and future challenges. *Science*, 294, 804–808.

Maraldo, K. and Dahllof, I. (2004). Seasonal variations in the effect of zinc pyrithione and copper pyrithione on pelagic phytoplankton communities. *Aquatic Toxicology*, 69, 189–198.

Mayer-Pinto, M., Underwood, A. J., Coleman, R. A. and Tolhurst, T. (2010). Effects of metals on assemblages: what do we know? *Journal of Experimental Marine Biology and Ecology*, 391, 1–9.

Mayer-Pinto, M., Crowe, T. P. and Johnston, E. L. (in review). Impacts of contaminants on ecosystem functioning through sub-lethal effects on habitat-forming species: a review. *Environmental Pollution*.

McCahon C. P. and Pascoe, D. (1990). Episodic pollution: causes, toxicological effects and ecological significance. *Functional Ecology*, 4, 375–383.

McKinley, A. and Johnston, E. L. (2010). Impacts of contaminant sources on marine fish abundance and species richness: a review and meta-analysis of evidence from the field. *Marine Ecology Progress Series*, 420, 175–191.

McLusky, D. S., Bryant, V. and Campbell, R. (1986). The effects of temperature and salinity on the toxicity of heavy metals to marine and estuarine invertebrates. *Oceanography and Marine Biology*, 24, 481–520.

McMahon, T. A., Halstead, N. T., Johnson, S. et al. (2012). Fungicide-induced declines of freshwater biodiversity modify ecosystem functions and services. *Ecology Letters*, 15, 714–722.

Medina, M., Andrade, S., Faugeron, S. et al. (2005). Biodiversity of rocky intertidal benthic communities associated with copper mine tailing discharges in northern Chile. *Marine Pollution Bulletin*, 50, 396–409.

Negri, A., Burns, K., Boyle, S., Brinkman, D. and Webster, N. (2006). Contamination in sediments, bivalves and sponges of McMurdo Sound, Antarctica. *Environmental Pollution*, 143, 456–467.

Negri, A. and Marshall, P. (2009). TBT contamination of remote marine environments: ship groundings and ice-breakers as sources of organotins in the Great Barrier Reef and Antarctica. *Journal of Environmental Management*, 90, S31–S40.

Paine, R. T. (1969). *Pisaster–Tegul* interaction: prey patches, predator food preference and intertidal community structure. *Ecology*, 50, 950.

Paine, R. T. (1995). A conversation on refining the concept of keystone species. *Conservation Biology*, 9, 962–964.

Perrett, L. A., Johnston, E. L. and Poore, A. G. B. (2006). Impact by association: direct and indirect effects of copper exposure on mobile invertebrate fauna. *Marine Ecology Progress Series*, 326, 195–205.

Peters, E. C., Gassman, N. J., Firman, J. C., Richmond, R. H. and Power, E. A. (1997). Ecotoxicology of tropical marine ecosystems. *Environmental Toxicology and Chemistry*, 16, 12–40.

Peterson, C. H. (2001). The *Exxon Valdez* oil spill in Alaska: acute, indirect and chronic effects on the ecosystem. *Advances in Marine Biology*, 39, 1–103.

Peterson, C. H., Rice, S. D., Short, J. W. et al. (2003). Long-term ecosystem response to the Exxon Valdez oil spill. *Science*, 302, 2082–2086.

Piola R. F. and Johnston, E. L. (2008). Pollution reduces native diversity and increases invader dominance in marine hard-substrate communities. *Diversity and Distributions*, 14, 329–342.

Qiu, J. W., Thiyagarajan, V., Cheung, S. and Qian, P. Y. (2005). Toxic effects of copper on larval development of the barnacle *Balanus amphitrite*. *Marine Pollution Bulletin*, 51, 688–693.

Raimondi, P. T. and Reed, D. C. (1996). Determining the spatial extent of ecological impacts caused by local anthropogenic disturbances in coastal marine habitats. In *Detecting Ecological Impacts: Concepts and Applications in Coastal Habitats*, ed. R. J. Schmitt and C. W. Osenberg. New York: Academic Press, pp. 179–198.

Reish D. J., Oshida, P. S., Mearns, A. J., Ginn, T. C. and Buchman, M. (1999). Effects of pollution on marine organisms. *Water Environment Research*, 71, 1100–1115.

Relyea, R. A. (2005). The impact of insecticides and herbicides on the biodiversity and productivity of aquatic communities. *Ecological Applications*, 15, 618–627.

Relyea, R. A. and Hoverman, J. T. (2008). Interactive effects of predators and a pesticide on aquatic communities. *Oikos*, 117, 1647–1658.

Rochman, C. M. and Browne, M. A. (2013). Classify plastic waste as hazardous. *Nature*, 494, 169–171.

Rohr, J. R. and Crumrine, P. W. (2005). Effects of an herbicide and an insecticide on pond community structure and processes. *Ecological Applications*, 15, 1135–1147.

Rohr, J. R., Kerby, J. L. and Sih, A. (2006). Community ecology as a framework for predicting contaminant effects. *Trends in Ecology and Evolution*, 21, 606–613.

Rygg, B. (1985). Distribution of species along pollution-induced diversity gradients in benthic communities in Norwegian Fjords. *Marine Pollution Bulletin*, 16, 469–474.

Schiedek, D., Sundelin, B., Readman, J. W. and Macdonald, R. W. (2007). Interactions between climate change and contaminants. *Marine Pollution Bulletin*, 54, 1845–1856.

Schiff, K., Brown, J., Diehl, D. and Greenstein, D. (2007). Extent and magnitude of copper contamination in marinas of the San Diego region, California, USA. *Marine Pollution Bulletin*, 54, 322–328.

Schwarzenbach, R. P., Escher, B. I., Fenner, K. et al. (2006). The challenge of micropollutants in aquatic systems. *Science*, 313, 1072–1077.

Shaw, M., Furnas, M. J., Fabricius, K. et al. (2010). Monitoring pesticides in the Great Barrier Reef. *Marine Pollution Bulletin*, 60, 113–122.

Smith, R., Middlebrook, R., Turner, R. et al. (2012). Large-scale pesticide monitoring across Great Barrier Reef catchments: Paddock to Reef Integrated Monitoring, Modelling and Reporting Program. *Marine Pollution Bulletin*, 65, 117–127.

Srivastava, D. S. and Vellend, M. (2005). Biodiversity–ecosystem function research: is it relevant to conservation? *Annual Review of Ecology Evolution and Systematics*, 36, 267–294.

Stark, J. S. (1998). Heavy metal pollution and macrobenthic assemblages in soft sediments in two Sydney estuaries, Australia. *Marine and Freshwater Research*, 49, 533–540.

Stark, J. S., Snape, I. and Riddle, M. J. (2003). The effects of petroleum hydrocarbon and heavy metal contamination of marine sediments on recruitment of Antarctic soft-sediment assemblages: a field experimental investigation. *Journal of Experimental Marine Biology and Ecology*, 283, 21–50.

Stark, J. S., Snape, I., Riddle, M. J. and Stark, S. C. (2005). Constraints on spatial variability in soft-sediment communities affected by contamination from an Antarctic waste disposal site. *Marine Pollution Bulletin*, 50, 276–290.

Stauber, J. L. (1998). Toxicity of chlorate to marine microalgae. *Aquatic Toxicology*, 41, 213–227.

Steinbeck, J. R., Schiel, D. R. and Foster, M. S. (2005). Detecting long-term change in complex communities: a case study from the rocky intertidal zone. *Ecological Applications*, 15, 1813–1832.

Steneck, R. S., Graham, M. H., Bourque, B. J. et al. (2002). Kelp forest ecosystems: biodiversity, stability, resilience and future. *Environmental Conservation*, 29, 436–459.

Strachan, M. F. and Kingston, P. F. (2012). A comparative study on the effects of barite, ilmenite and bentonite on four suspension feeding bivalves. *Marine Pollution Bulletin*, 64, 2029–2038.

Suchanek, T. H. (1994). Temperate coastal marine communities: biodiversity and threats. *American Zoologist*, 34, 100–114.

Sun, M. Y., Dafforn, K. A., Brown, M. V. and Johnston, E. L. (2012). Bacterial communities are sensitive indicators of contaminant stress. *Marine Pollution Bulletin*, 64, 1029–1038.

Thompson, R. C., Crowe, T. P. and Hawkins, S. J. (2002). Rocky intertidal communities: past environmental changes, present status and predictions for the next 25 years. *Environmental Conservation*, 29, 168–191.

Underwood, A. J. and Peterson, C. H. (1988). Towards an ecological framework for investigating pollution. *Marine Ecology Progress Series*, 46, 227–234.

United Nations Educational, Scientific and Cultural Organization (2003). Water for people, water for life: the United Nations World Water Development Report. UNESCO, Barcelona.

Vinebrooke, R. D., Cottingham, K. L., Norberg, J. et al. (2004). Impacts of multiple stressors on biodiversity and ecosystem functioning: the role of species co-tolerance. *Oikos*, 104, 451–457.

Vitousek, P. M. (1994). Beyond global warming: ecology and global change. *Ecology*, 75, 1861–1876.

Wiens, J. A. (1989). Spatial scaling in ecology. *Functional Ecology*, 3, 385–397.

Wiens, J. A., Patten, D. T., Botkin, D. B. (1993). Assessing ecological impact assessment: lessons from Mono Lake, California. *Ecological Applications,* 3, 595–609.
Wisely, B. and Blick, R. A. P. (1967). Mortality of marine invertebrate larvae in mercury, copper and zinc solutions. *Australian Journal of Marine and Freshwater Research,* 18, 63–72.
Wright, S. L., Thompson, R. C. and Galloway, T. S. (2013). The physical impacts of microplastics on marine organisms: a review. *Environmental Pollution,* 178, 483–492.

10 · Invasions by non-indigenous species

MADS SOLGAARD THOMSEN, THOMAS WERNBERG AND DAVID SCHIEL

10.1 Marine invaders: background

Invasions by marine non-indigenous species (NIS, see Box 10.1 for definitions) have long been recognised. For example, Carl Emil Hansen Ostenfeld described the invasion of the planktonic diatom *Biddulphia* (*Odontella*) *sinensis* Grev. into the North Sea in 1903, probably transported by ships (Ostenfeld, 1908), and over 50 years ago Charles Elton provided the first overview with narrative accounts of impacts associated with high profile marine invasions by oysters and cordgrass (Elton, 1958). However, marine invasions were first approached with a systematic research effort following J. Carlton's seminal work in San Francisco Bay (Carlton, 1979). Today it is recognised that marine NIS comprise a diverse group of organisms found in most marine systems. In a recent review, Hewitt and Campbell (2010) tallied almost 1800 marine NIS and cryptogenic species (Box 10.1) worldwide, dominated by arthropods (444), molluscs (350), fish (166), red algae (153), annelids (104, mainly polychaetes), cnidarians (100), heterokonts (73), bryozoans (73) and green algae (51). Although marine NIS have been introduced around the world, some bioregions are more invaded than others; the most heavily invaded regions are the Mediterranean Sea (467 NIS), Australia and New Zealand (429), the South Pacific (289), the north-east Pacific (284) and the north-east Atlantic (216) (Hewitt and Campbell, 2010). More locally, estuaries centred around metropolitan areas and with extensive shipping traffic are typically the most invaded systems. For example, *c.* 230, 200 and 100 marine NIS are found in San Francisco Bay, Chesapeake Bay and Port Phillip Bay, respectively (Hewitt *et al.*, 2004; Ruiz *et al.*, 2011). The most important marine vectors are transportation on ships (mainly hull fouling), inside ships (in ballast water, ballast sediments and sea-chests), with aquaculture (in particular associated with oyster transplantations) and via human-mediated removal of physical barriers (e.g. the Suez Canal which connects biota between the Red and the Mediterranean Seas).

No biological species live in an ecological vacuum (Agosta and Klemens, 2008) and each NIS that establishes a permanent population will therefore impact many local species via direct (e.g. consumer–resource and competitive–facilitative interactions) (Byers, 2000; Eastwood et al., 2007) and indirect (e.g. enemy/facilitation cascades) (Thomsen et al., 2010a) biotic interactions. These interactions will typically also influence local community structures and patterns of biodiversity and ecosystem functioning (BEF, see Chapters 1–5 for overview). For example, early in the invasion phase, local species richness automatically increases by one (the new NIS) and global homogenisation automatically increases simply because the invaded region now contains one more species from the global species pool (Olden and Rooney, 2006). Furthermore, each NIS will, via its demography (e.g. population growth), its interactions (e.g. competition), and its impact on community structures (e.g., adding a new species to the system) also modify ecosystem functioning and therefore the provision of ecosystem services (Crooks, 2002a, 2009).

The most pressing question about marine invasion impact is no longer *if* a NIS has an impact or not, but more precisely:

1. What is the direction of impact (is it positive or negative on local species, ecosystem function and biodiversity – see also Box 10.1)?
2. What is the pathway of impact (are effects direct or mediated via indirect pathways)?
3. What is the magnitude of impact (is it too weak to be detected in a variable world)?
4. How do direction, pathway and magnitude vary over space and time within and between invaders, taxa, life stages, invaded habitats and with co-occurring stressors? (how are impacts context dependent)?
5. Are there rules that can be used to predict context-dependent variability in impact?
6. How do NIS modify ecosystem functioning and its relationship with biodiversity?
7. What can be done – if anything – to minimise the impacts of NIS?

The remainder of this chapter will focus on these questions, summarising findings from reviews and case studies.

Box 10.1 *Definitions*

NIS. Non-indigenous species (= alien, exotic, non-native, introduced) are species living outside their native distributional range, which have arrived there by human activity, either deliberate or accidental (Wikipedia). We do not include new species that have arrived by their own dispersal mechanisms across natural barriers, but only survive due to climate changes, as NIS.

Invasive species. NIS are typically coined invasive if they are highly successful (e.g. rapid spread and high abundance) and/or have a strong impact on local ecosystems. In invasion biology all invaders are NIS, but in biogeographical research invaders can also be native that arrive to new regions by natural dispersal mechanisms or 'natural' breakdown of physical barriers.

Cryptogenic species. Of unknown origin; detailed taxonomic, biogeographical, and molecular analysis is required to determine if the species is a native or NIS.

Transport vectors. Pathway whereby NIS arrive to a new system associated with human activity. In marine systems the most important vectors are hull fouling (outside ships), ballast water and sediments (inside ships) and aquaculture (intentional, escapees, and 'blind passengers').

Invasion impact studies. Studies that focus on how invaders affect a property of an invaded habitat/system. The invader is assumed to be the causal agent of change and is considered the independent variable that, typically, is portrayed graphically on the x-axis (in invasion success studies the invader is considered the dependent variable). Impact can be larger or smaller than a reference value (often defined as 0) and impact is then either positive (>0) or negative (<0). Positive or negative impact is a statistical measure that differs from whether the impact is 'good' or 'bad' (the human interpretation of the statistical measure). Impact can be reported on cultural (economics, health, cultures) or natural properties. Natural properties can be divided into biotic and abiotic properties. Biotic properties can be divided into impact reported on or above the species level (the fundamental unit in biology); negative impact on the species level is simple to interpret whereas negative impact reported above the species level may hide opposing effects (i.e. that certain species benefit whereas others are harmed, by the invader).

Mensurative experiments. The researcher has no control over the abundance of the invader, but compares responses between sample sites and sample periods that (hopefully) reflect different levels of invaders. Mensurative experiments are important because they can be conducted on large and long space–time scales, with a wide range of invasive taxa and are without artefacts associated with cages, tethers or experimental disturbances.

Manipulative experiments. The researcher has control over the abundance (and other attributes) of the invader either adding invaders to uninvaded plots/locations or removing invaders from already invaded plots/locations. Manipulative experiments are important because causality can be inferred between the invader and the responses and because invader densities and the spatio-temporal context can be controlled to form part of advanced hypothesis testing.

Keystone and cascades terminology. 'Keystone' and 'cascade' describe 'a disproportionally large ecological effect' and 'a disproportional large ecological effect effectuated via repeated ecological interactions', respectively. 'Cascade' is the more precise term because the first and second interaction in the chain are the same, and therefore takes precedence over 'keystone'. Effects on focal species are always positive in three-level cascades, either via 'a friend of my friend is my friend' (cascading habitat formation, mutualism) or 'an enemy of my enemy is my friend' (cascading consumption, competition). By contrast, effects can be either positive or negative if the cascade contain more than three levels. We differentiate different keystone interactions based on the first interaction in a chain reaction; e.g. keystone consumption and keystone mutualism imply that the first interaction is consumption or mutualism, respectively. The effect on focal organisms from keystone interaction chains with more than three levels can be positive or negative depending on the direction of the two chained interactions (see Figure 10.1 for contrasting examples).

10.2 How do marine invaders impact the local biota?

In this section we exemplify common processes whereby marine NIS impact native organisms, classified according to direction (positive vs.

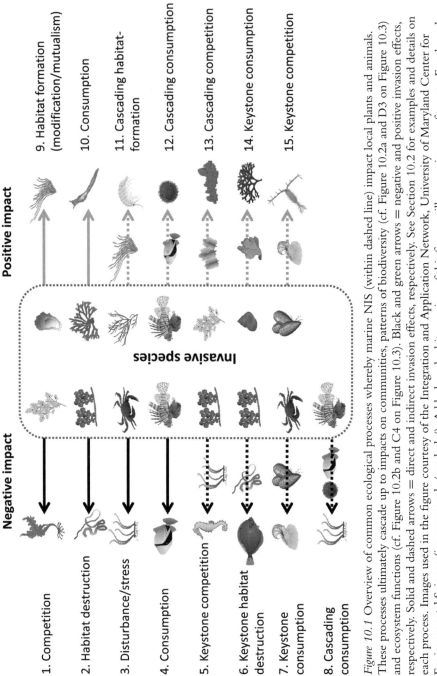

Figure 10.1 Overview of common ecological processes whereby marine NIS (within dashed line) impact local plants and animals. These processes ultimately cascade up to impacts on communities, patterns of biodiversity (cf. Figure 10.2a and D3 on Figure 10.3) and ecosystem functions (cf. Figure 10.2b and C4 on Figure 10.3). Black and green arrows = negative and positive invasion effects, respectively. Solid and dashed arrows = direct and indirect invasion effects, respectively. See Section 10.2 for examples and details on each process. Images used in the figure courtesy of the Integration and Application Network, University of Maryland Center for Environmental Science (ian.umces.edu/symbols/). A black and white version of this figure will appear in some formats. For the colour version, please refer to the plate section.

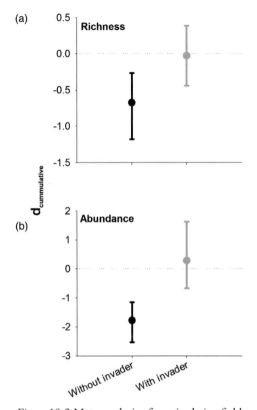

Figure 10.2 Meta-analysis of manipulative field experiments that quantify impact of marine plant invaders on local plant richness (a) and total plant abundance (b). Hedges' effect size ($d_{cumulative}$) was calculated as a traditional impact analysis, i.e. without inclusion of the invasive species itself (black circles) and contrasted to the same data set where we added the invaders own 'richness' (=1 species more per plot in invasion treatments) and abundance (a relatively high per cent cover or biomass in invasion treatments) (grey circles). Replication for richness = 30 experiments from 11 research papers. Replication for abundance = 23 experiments from eight research papers. The graphs show strong negative impact of marine plant invaders on richness and total abundance from traditional impact analysis, but that net community effects changes to zero (error bars overlap zero) when the taxonomic status and abundance of the invader itself was added to the community data.

negative) and pathway (direct vs. indirect effects). Note that 'direction' simply refers to whether the quantified effect is larger or smaller than a reference value (often defined as 0). Invasion impact is, therefore, here considered a statistical measure that can be either positive (>0) or negative (<0), and this measure gives no direct indication as to whether the impact should be considered 'good' or 'bad' (which represents the human interpretation of the statistical measure, see Chapter 1). Understanding the complexity of ecological interactions between invasive and native species is a prerequisite to understanding how biodiversity and ecosystem functioning also are impacted.

10.2.1 Negative impacts

Invasion impact studies have traditionally focused on how native species have been negatively affected by invaders (Rodriguez, 2006) via direct (Figure 10.1, 1–4) and indirect (Figure 10.1, 5–8) mechanisms. Most of these studies have focused on the direct effects (White et al., 2006) although with an increasing number of species in a community the number of possible indirect effects increases exponentially and much faster than the number of direct interactions (Wootton, 1994). In the following sections, we first describe direct effects followed by simple cases of indirect effects that involve 3–4 species interacting in 'chain of events', whereby a 'primary invader' has an effect on a 'focal native species', but mediated by 1–2 'intermediate species'.

Competition

Marine invaders often compete with native species for limited resources (reviewed in Byers, 2009). For example, the invasive *Sargassum muticum* competes with native canopy-forming seaweeds for light (Figure 10.1, 1) (Staehr et al., 2000; Britton-Simmons, 2004) – although the impact sometimes is minor, especially in the intertidal zone where light limitation is less important (Sánchez and Fernández, 2005; Olabarria et al., 2009). Similarly, invasive *Caulerpa* species compete with native seagrasses for space, light and nutrients, with some studies suggesting strong negative impacts (de Villele and Verlaque, 1995; Ceccherelli and Cinelli, 1997), whereas others do not (Jaubert et al., 1999; Ceccherelli and Sechi, 2002; Jaubert et al., 2003; Thomsen et al., 2012b). These examples suggest that invasion impact is context dependent, varying from dramatic to negligible, depending on specific invader–habitat interactions (Strayer et al., 2006; Thomsen et al., 2011b). Invasive seaweeds may also compete with sessile animals for limited resources. For example, accumulations of drifting mats of *Gracilaria vermiculophylla* can reduce recruitment of sessile organisms that require hard substrates (including the reef-forming oyster themselves), thereby altering sessile communities probably intercepting propagules from the water column and physical smothering (Thomsen and McGlathery, 2006). In this study, the impact on community structure was similar across hydrodynamic regimes (non-significant interaction in multivariate analysis), suggesting that at least in some cases impacts can be constant along environmental gradients. Other examples of competitive effects between plants and animals include invasive turf alga that competes with sessile coral animals for space (Linares et al., 2012; Cebrian et al., 2012) or encrusting fouling bryozoa (*Membranipora membranacea*) that live as epiphytes on kelp blades (Levin et al., 2002; Saunders and Metaxas, 2008), thereby shading the kelp and increasing susceptibility to frond breakage (note that the invasive bryozoa is positively affected by the kelp, using the kelp as habitat).

Habitat destruction
Habitat destruction by invaders can be common, particularly where unvegetated sedimentary habitats are converted to vegetated meadows, a process that may result in displacement of obligate infauna and other organisms that depend on sediments. However, there are surprisingly few cases documenting strong negative impacts on infaunal species from invasions by marine plants (Wright et al., 2007; Gribben et al., 2009b; Tsai et al., 2010), perhaps because many infaunal species are facultative mud-flat inhabitants that still survive in the vegetated meadows. Destruction of mud flats has been reported for invasive *Caulerpa* species (Figure 10.1, 2) (Gribben and Wright, 2006; McKinnon et al., 2009; Byers et al., 2010; Klein and Verlaque, 2011; Pacciardi et al., 2011), *Gracilaria vermiculphylla* (Thomsen et al., 2007, 2010a; Byers et al., 2012), *Sargassum muticum* (Strong et al., 2006), invasive seagrasses (Posey, 1988; Baldwin and Lovvorn, 1994; Berkenbusch et al., 2007; Willette and Ambrose, 2009; Ruesink et al., 2010; Willette and Ambrose, 2012), invasive salt marshes and mangroves (Neira et al., 2007; Thomsen et al., 2009a; Wu et al., 2009; Demopoulos and Smith, 2010) and several invasive sessile animals, such as oysters and mussels (Kochmann et al., 2008). Furthermore, habitat destruction could also occur if invasive consumers wipe out a habitat-former, as has been observed following outbreaks of native consumers (e.g. urchins converting kelps to barrens). However, we are not aware of examples documenting this process with invasive consumers (although disease outbreaks on habitat-formers may involve human-mediated introductions).

Disturbances
Marine invaders can cause negative impacts by physical disturbances. For example, native seagrasses (*Zostera marina* and *Halophila ovalis*) can be uprooted and their seeds buried by invasive crabs (*Carcinus maenas*)

(Figure 10.1, 3) (Davis et al., 1998), snails (*Batillaria australis*) (Hoeffle et al., 2012) and polychaetes (*Marenzelleria viridis*) (Kristensen et al., 2011). These disturbances are probably most important in sparse seagrasses, at seagrass edges, on individual seedlings and around newly established small patches where interconnected rhizomes are few and physical integration low. Furthermore marine invaders may also, via their activities, make the habitat more hostile for native species. For example, the invasive polychaete *Marenzelleria viridis* can decrease the oxic zone in sediments (Kristensen et al., 2011; Norkko et al., 2012), potentially having a negative impact on native seagrasses.

Consumption (by the invader)
Marine invasive animals can have a direct negative impact on native species via consumptive processes (reviewed in Rilov, 2009). For example, invasive lionfish exert strong top-down control in invaded Caribbean reefs, particularly on juvenile tropical reef fish, in part because the native prey appear to be naive to this new enemy (Figure 10.1, 4) (Albins and Hixon, 2008; Albins, 2013). Similarly, invasive crabs, such as *Carcinus maenas* or *Hemigrapsus sanguineus*, consume large quantities of native mussels (DeGraaf and Tyrrell, 2004) and the jellyfish *Mnemiopsis leidyi* has, via its predation on fish larvae, been implicated in the collapse of the anchovy fishery in the Black Sea (Kideys, 2002) (but see also Bilio and Niermann, 2004)). Another well-documented example is how the invasive snail *Littorina littorea* causes dramatic grazing effects on rocky intertidal shores and tidepools, denuding *Fucus* recruits (Lubchenco, 1983) and ephemeral algae from rocky tidepools (Lubchenco and Menge, 1978). Still, such top-down effects that are evaluated from manipulative experiments, may be overestimated, because invader-driven niche shifts or local reductions/extinctions of functionally similar native species, can be extremely difficult to detect. For example, *Littorina littorea* may have displaced the native sibling *Littorina saxatilis* (Eastwood et al., 2007) that otherwise can carry out similar ecosystem functioning in the absence of the invasive grazer. This highlights an important problem when evaluating impact from experiments conducted in already invaded locations.

Keystone competition
When invasive plants have negative impacts on native plants, this can lead to indirect negative impact on animals that prefer to forage around and inhabit the native plant. For example, syngnathid and monacanthid fish

have been found to be more abundant around native seagrass compared to invasive *Caulerpa* species (Figure 10.1, 5) (York *et al.*, 2006), juvenile fish can be more abundant in native seagrass beds compared to invasive *Halophila stipulacea* (Willette and Ambrose, 2012) and gastropods and seastars are typically more common on native kelp compared to invasive *Codium fragile* (Schmidt and Scheibling, 2006). Still, many of these indirect effects appear to be relatively minor, probably because marine animals are often generalists that can survive on different vegetation (Bell, 1991), a finding supported by many 'epibiota comparisons' that typically find larger spatio-temporal differences in epibiota community structures, compared to between an invasive and native plant competitor (Viejo, 1999; Wernberg *et al.*, 2004; Guerra-García *et al.*, 2012; Janiak and Whitlatch, 2012).

Keystone habitat destruction
Indirect negative effects associated with keystone habitat destruction is conceptually similar to keystone competition; i.e. if invasive plants destroy habitats (e.g. converting mudflats to meadows), this can lead to indirect negative impact on animals that prefer to inhabit and forage in sediments (Figure 10.1, 6). Organisms that are likely to be indirectly negatively influenced by mudflat–vegetation conversions include burying fish such as flounders and stargazers, as well as many wading birds that feed on mudflats, particularly during migrations (although few studies have documented this for invasive marine plants; Parks, 2006). Whether these types of negative effects arise through direct interactions with the invader or indirect effects, for example, mediated through the disappearance of a food source, may be difficult to evaluate, but can be important to know, particularly from a conservation perspective, to ensure that management action works according to intentions. For example, if a management objective is to conserve a species whose decline is possibly caused by habitat destruction by an invader, then 'encouraging' the establishment of alternative resources (in habitat and/or food) for the declining native species could be an alternative to control or eradication of the invader.

Keystone consumption
Consumers may also have large indirect effects on native organisms via keystone consumption (see Box 10.1), in particular where a consumer preys on important habitat-formers. For example, invasive crabs, such as *Carcinus maenas* or *Hemigrapsus sanguineus*, consume large quantities of sessile animals, in particular mussels (Grosholz and Ruiz, 1996; DeGraaf

and Tyrrell, 2004) that are important biogenic habitat-formers for sessile animals and plants (Figure 10.1, 7) (Altieri *et al.*, 2007; Lang and Buschbaum, 2010). Keystone consumption is likely to be particularly important in estuarine soft-sediment systems where biogenic habitat formation is particularly important (Buschbaum *et al.*, 2009; Thomsen *et al.*, 2010a).

Cascading consumption
Possibly the best studied indirect interaction is cascading consumption (= 'trophic cascades') (Hairston *et al.*, 1960; for a review of multitrophic effects and marine invaders, see Grosholz and Ruiz, 2009). It is well established that second order consumers have positive indirect effects on plants (Shurin *et al.*, 2005), and we can therefore extend this insight to speculate that third order consumers have negative impacts on plants (Daskalov *et al.*, 2007; Casini *et al.*, 2008). However, we are unaware of studies that have documented four-level cascading consumption involving marine invasive top predators (but see Tronstad *et al.*, 2010, for an example with a freshwater invader). Still, a hypothetical case can illustrate the process. For example, it has been documented that invasive lionfish *Pterois volitans* consume trigger fish (Albins, 2013), and that triggerfish consume urchins (McClanahan *et al.*, 1996; O'Leary and McClanahan, 2010) and that urchins can denude tropical seaweeds (McClanahan *et al.*, 1996; O'Leary and McClanahan, 2010) and seagrasses (Alcoverro and Mariani, 2004). Hence, we speculate that lionfish could have indirect negative impact on tropical seaweeds via four-level cascading consumption (Figure 10.1, 8).

10.2.2 Positive impacts

Although there has been a traditional research focus on negative effects of invasions, positive effects have gained increasing scientific interest (Rodriguez, 2006) and more and more examples are documented, both from direct (Figure 10.1, 9–10) and indirect (Figure 10.1, 11–15) mechanisms, which we describe below.

Habitat formation, modification and mutualism
Many marine invaders can have direct positive effects on native species by creating and modifying habitats. Thus, many invaders create three-dimensional structures that are inhabited by native species, including invasive oysters (Figure 10.1, 9) (Lang and Buschbaum, 2010; Markert

et al., 2010; Padilla, 2010), mussels (Sousa *et al.*, 2009), snails (Wonham *et al.*, 2005; Thomsen *et al.*, 2010b), tunicates (Castilla *et al.*, 2004), bryozoa (Wilson, 2011), seaweeds (Thomsen *et al.*, 2006b; Nyberg *et al.*, 2009; Thomsen *et al.*, 2010a; Klein and Verlaque, 2011; Byers *et al.*, 2012), and seagrasses (Posey, 1988; Berkenbusch and Rowden, 2007; Willette and Ambrose, 2012). Of these examples, oyster invasions are particularly important, because they have affected enormous areas of soft-bottom estuaries worldwide (Padilla, 2010), thereby creating hard substratum (a limiting resource for many sessile organisms in sedimentary systems) for native sessile organisms. Many of these direct positive effects are rather obvious as physical structures provide a place to escape enemies and abiotic stress and find 'friends' and resources, processes that are particular important for interstitial organisms (Huston 1994). Positive invasion effects may not only occur on barren sediments, but may also occur in already vegetated habitats, as invaders can still 'add' structure (Roscher *et al.*, 2005) to support 'more' native species. For example, the invasion of *Sargassum muticum* in Denmark has probably resulted in a system-wide increase in macroscopic plant biomass and thereby (additional) habitat for epiphytic plants (Thomsen *et al.*, 2006c) and invertebrates (Wernberg *et al.*, 2004). Positive effects are, like negative effects, typically context dependent. For example, experimental data suggest that positive habitat formation effects of the invasive seaweed *Gracilaria vermiculophylla* can vary between depth levels, invader density, native taxa and life stage of native taxa (e.g. threshold vs. continuous positive impact on snails vs. bivalves; Thomsen, 2010; Thomsen *et al.*, 2013). Density-dependent thresholds have also been documented experimentally with invasive *Sargassum muticum*, where positive habitat-formation effects changed to negative competitive effects at high densities (White and Shurin, 2011). In addition to creating habitat (autogenic engineering) for sessile plants and animals and mobile interstitial invertebrates, many invaders can also modify the environment (habitat modification; allogenic engineering) to make it more benign for native species. For examples, invasive seaweeds, oysters, mussels and tunicates may reduce desiccation stress on native species (Castilla *et al.*, 2004; Bulleri *et al.*, 2006), increase sediment organic matter via faecal production and increase water clarity by filtering out plankton thereby increasing light levels for native benthic plants (Petersen *et al.*, 2008; Thomsen *et al.*, 2010b). Invasive bioturbaters may also increase sediment oxygen levels and potentially thereby reduce stress on native seagrass and infauna. In contrast to common studies documenting direct positive effects from invasive habitat-formers, there has been less emphasis on how impacted native species also can influence the invader. It is possible that many reciprocal interactions can be classified as mutualisms, for example, if native grazers consume epiphytes on invasive plants, thereby increasing light levels, or if native animal excretions provide nutrients to invasive plants.

Consumption (of the invader)
The second common type of direct positive effect is through consumption, i.e. where the invader is a food source for native species. Thus, if invasive plant and animals increase system-wide standing stocks there should be more food available to native grazers and higher order consumers. For example, siphonalian invaders (*Codium* and *Caulerpa* spp.) can have positive impacts on specialist saccoglossan grazers (Figure 10.1, 10) (Trowbridge and Todd, 2001; Trowbridge, 2002; Harris and Jones, 2005). Similarly, juvenile invasive seaweeds (Thornber *et al.*, 2004; Sjotun *et al.*, 2007) and seagrass (Reynolds *et al.*, 2012) can provide an important seasonal food supply for grazers, and invasive *Codium fragile* and seagrass can be an important food source for periwinkles (Scheibling *et al.*, 2008) and waterfowl. Still, positive effects associated with consumption are probably less important than effects due to habitat formation, as many invasive plants are relatively poor food sources (Scheibling and Anthony, 2001; Britton-Simmons, 2004; Thomsen and McGlathery, 2007; Monteiro *et al.*, 2009; Nejrup and Pedersen, 2010; Cebrian *et al.*, 2011; Engelen *et al.*, 2011; Tomas *et al.*, 2011; Nejrup *et al.*, 2012), some of them having deterring toxins (Boudouresque *et al.*, 1996; Gollan and Wright, 2006; Nylund *et al.*, 2011). Invasive animals can also provide new food sources (Rilov, 2009). For example, invasive *Littorina littorea* snails are consumed by crabs (Trussell *et al.*, 2002, 2004; Eastwood *et al.*, 2007), invasive *Carcinus maenas* crabs are consumed by blue crabs (*Callinectes sapidus*) (DeRivera *et al.*, 2005), the invasive clam *Nuttallia obscurata* is consumed by native crabs (Byers, 2002), invasive copepods are consumed by native fish (Bollens *et al.*, 2002), and invasive lionfish may be consumed by groupers (Maljković *et al.*, 2008). Note, however, as shown for invasive seaweeds, many animal invaders are also relatively resistant to enemies, e.g. invasive *Batillaria australis* snails are rarely consumed by fish or crabs (Thomsen *et al.*, 2010b) and invasive lionfish are toxic to many native species (Whitfield *et al.*, 2002).

Cascading habitat formation and modification
Marine invaders can have many types of indirect positive effects. A common indirect facilitation mechanism is cascading habitat formation. For

example, the invasive snail *Batillaria australis* provides habitats for sessile plants and animals (in particular native seaweeds) (Thrring et al., 2014), that again provide additional habitat for many smaller hydrozoa, bryozoa and mobile invertebrates (Figure 10.1, 11) (Thomsen et al., 2010a). Therefore *Batillaria australis* has indirect positive impacts on invertebrates mediated through its epiphytes. Similar forms of cascading indirect effects may occur following all the direct positive habitat formation processes described above. That is, invasive habitat-formers like oysters, mussels, snails, seaweeds and angiosperms may have indirect positive effects on native organisms by being the 'primary habitat-former' that provides habitat for 'intermediate habitat-formers'. These types of indirect positive effects can also occur through cascading habitat modification. For example, invasive seaweeds may reduce sediment oxygen levels thereby forcing native bivalves to move to the sediment surface and be exposed to fouling by sessile species (Gribben et al., 2009a; Hoeffle et al., 2012). The invasive seaweed thereby has indirect positive effects on fouling species by modifying the environment for bivalves.

Cascading consumption
The most frequently studied indirect positive interaction involves three-level consumption cascades (Hairston et al., 1960; Estes and Palmisano, 1974) (for a review of multitrophic effects and marine invaders, see Grosholz and Ruiz, 2009). For example (as also outlined in Figure 10.1, 8), invasive lionfish can consume trigger fish (Albins, 2013) and herbivorous fish, thereby releasing predation pressure on urchins (Figure 10.1, 12) (McClanahan et al., 1996; O'Leary and McClanahan, 2010) and seaweeds (Lesser and Slattery, 2011), respectively. Another example is when invasive crabs consume or alter the behaviours of grazing snails (Trussell et al., 2002, 2004; Eastwood et al., 2007), thereby indirectly facilitating seaweeds in a three-level density or trait-mediated consumption cascade.

Cascading competition
Positive indirect effects can also occur via cascading competition, in particular when the basal and intermediate competitors compete for a different resource compared to the intermediate and focal competitors (Levine, 1999). We are not familiar with examples documenting cascading competition from marine invasive basal competitors, but we suggest that invasive canopy-forming *Sargassum muticum* mainly compete with smaller understory algae for light, whereas understory algae likely compete with encrusting plants and sessile animals for space (Figure 10.1, 13)

(Staehr et al., 2000; Thomsen et al., 2006c) (for an example of cascading competition, but with native canopy-formers, turfs and crusts, see Wernberg et al., 2012b). Thus, a marine invader may have an indirect positive impact on a con-trophic species by reducing the abundance of another con-trophic (strong) competitor.

Keystone consumption
Consumers can have indirect positive effects on lower trophic levels by preferentially consuming a competitively dominant species (Paine, 1966). A classic example is invasive *Littorina* snails that consume fast growing seaweeds like *Ulva* species thereby reducing competition for slower-growing less palatable species like *Chondrus crispus* (Lubchenco, 1978, 1983). Thus, in this case the consumer provides a similar function as the canopy former in Figure 10.1, 14, or a physical disturbance that opens up space, in each case indirectly facilitating a competitively inferior native species.

Keystone competition
Our final example involves indirect positive effects associated with keystone competition, such as when a basal competitor has a negative impact on an intermediate competitor, but the two competitors use different resources. On rocky shores, invasive mussels compete with native limpets and barnacles for space (Steffani and Branch, 2005), but whereas the mussels consume phytoplankton, the limpets and barnacles consume seaweeds and zooplankton, respectively. The mussel can therefore have indirect positive effects on seaweeds and zooplankton by outcompeting limpets and barnacles (Figure 10.1, 15). Similarly, invasive crabs compete with native crustaceans (Rossong et al., 2006; Williams et al., 2006; MacDonald et al., 2007), but consume slightly different prey, and invasive oysters compete with native mussels (Kochmann et al., 2008), but filter slightly different plankton components. In these examples, the native prey that is preferentially consumed by the native competitor may be indirectly facilitated by the invasive consumer.

The examples above represent common mechanisms whereby marine invaders affect local species and functional groups. These different types of effects eventually scale up to the entire community, as different species in the community are influenced by positive and/or negative impacts, to modify local patterns of biodiversity and ecosystem functioning. In the next sections we describe common processes whereby invaders modify the abiotic environment, and review studies of marine invasions and their impacts on biodiversity.

10.3 How do marine invaders affect the local abiotic environment?

All NIS modify abiotic conditions and elemental cycling, either directly by their metabolic and behavioural activities and/or indirectly by modifying the invaded biotic community. Invasion impacts on the abiotic environment can be classified as effects on *abiotic conditions* that are not utilised by the invader, such as temperature or salinity, and *abiotic resources* that are utilised by invader. Resources can become *depleted* (e.g. macronutrients) *or not* (calcium carbonate, used in shells). Impacts on resource levels can be quantified in terms of compartment themselves (e.g. total nitrogen in sediments) or as fluxes between compartments (e.g. the flux of nitrogen between water and sediment).

In marine systems, abiotic conditions like temperature, salinity or waves can in rare cases be modified by NIS. For example, invasive species dominating the intertidal zone can, at low tides, reduce *temperature* fluctuations and *evaporation* and thereby increase *moisture* levels within and beneath the NIS (Thomsen, unpubl.; Critchley et al., 1990). Similarly, structurally dominant NIS can reduce *hydrodynamic forces*, such as *Spartina* marshes reducing flows compared to unvegetated mudflats (Neira et al., 2006). However, a counter-example is that in a small-scale removal experiment *Sargassum muticum* did not affect flow rate (Britton-Simmons, 2004). Flow reductions are most likely to occur when the invader is structurally large and covers large areas. These types of invasion effects are probably of relatively minor importance in marine habitats.

Of greater ecological importance are invasion impacts on resource levels, particularly those that occasionally become depleted or where 'by-products' can be toxic. The most important abiotic resources impacted by marine NIS involve elemental fluxes of *nutrients such as nitrogen and phosphorous*. For example, invasive plant invaders can take up nitrogen and phosphorous from the water column and sometimes also sediments from soft-bottom systems (Tyler et al., 2005; Tyler and McGlathery, 2006), store it in live tissue, which can be transported to adjacent ecosystems (Thomsen et al., 2009b), and the nutrients released again through decomposition (Thomsen et al., 2009b; Byers et al., 2012). Fluxes and storage of nitrogen and phosphorous can be controlled by invaders, particularly when NIS dominate in habitats with low abundances of native structural species, e.g. *Crassostrea gigas*, *Gracilaria vermiculophylla* and *Caulerpa* species invading mudflats. However, nutrient fluxes can also change if a NIS replaces a native species that appears to be 'similar'. For example, invasive *Sargassum muticum* is morphologically similar to, and phylogenetic related to, the native *Halidrys siliquosa*, but has significantly faster nutrient uptake rate, growth and decomposition and only the invader sheds its laterals, a mechanism that locally can remove nutrients stored in tissue (Wernberg *et al.*, 2001; Pedersen *et al.*, 2005). Partial replacement of *Halidrys* with *Sargassum* (Staehr *et al.*, 2000) may therefore have relatively low impact on the general physiognomy of the invaded system, but a larger 'hidden' impact on elemental cycling and nutrient transfer. Sessile animals also modify fluxes of nutrients, for example by increasing pore-water ammonia in sediments as done by the invasive bivalves *Musculista senhousia* (Reusch and Williams, 1998) and *Crassostrea gigas* (Green *et al.*, 2013).

Another important abiotic resource that can be locally depleted is *light*. Large species that produce dense canopies intercept light (often >90% of incident light) and shade understory species. Such shading is typical for canopy-forming invasive macrophytes such as *Eucheuma denticulatum* in seagrass beds (Ekloef *et al.*, 2006), *Sargassum muticum* on rocky reefs (Critchley *et al.*, 1990; Britton-Simmons, 2004) and *Gracilaria vermiculophylla* morphologies on mudflats. By contrast, many invasive filter feeders (oysters, mussels, polychaetes) can potentially increase light levels by removing large amounts of organic materials from the water column (Davies *et al.*, 1989; Padilla, 2010).

Marine invaders can modify *oxygen* levels through photosynthesis, respiration and indirectly by bacterial oxygen consumption of decaying/lost organic material (Green *et al.*, 2012). For example, *Caulerpa taxifolia* reduced oxygen in the water column compared to unvegetated sandflats (often to hypoxic levels) (Wright *et al.*, 2010) and the invasive snail *Batillaria australis* reduced the depth to sulfide horizon (i.e. increased anoxia) (Hoeffle *et al.*, 2012). The latter result appears counterintuitive because the snail is a bioturbator that moves sediments. However, Hoeffle *et al.* hypothesised that the snail deposits organic matter within sediment and that bacterial decomposition caused reduced oxygen levels. Indeed, sediments are often reported to be modified by marine invaders, including organic matter content, grain size, sedimentation rates, and sediment stability/erosion rates. Thus, several studies have shown increases in *sediment organic matter* when unvegetated mudflats are invaded by NIS, such as saltmarsh plants (*Spartina altiniflora*) (Neira *et al.*, 2006) and sessile bivalves (*Musculista senhousia* and *Crassostrea gigas*) (Reusch and Williams, 1998; Kochmann et al., 2008). Note, however, that if oysters invaded systems already dominated by native bivalves, sediment organic matter was of similar magnitude (= a 'substitution' process, see Section 10.5 for meta-analytical example) (Kochmann et al., 2008). Increased organic matter in sediment comes from organic material deposited by the invader itself and because the NIS reduces hydrodynamic forces, which increases deposition and reduces resuspension. Still, not all studies have found increases in organic matter; there were no increases in organic matter on sand flats invaded by *Caulerpa racemosa* (Pacciardi et al., 2011) or in seagrass beds invaded by *Euchema denticulatum* (Ekloef et al., 2006). Marine NIS can also change the *inorganic sediment texture*. For example, mudflats invaded by *Spartina alterniflora* had a higher proportion of fine particles (Neira et al., 2006), whereas *Caulerpa* had the opposite impact, with a higher gravel content and less sand at invaded than non-invaded sites (but only for one of two invaded areas) (Pacciardi et al., 2011). Related to texture and organics are *sedimentation rates*, which also have been shown to be modified by invaders. For example, *Undaria pinnatifida* increased sedimentation rates but only in the absence of bulldozing urchins (Valentine and Johnson, 2005), as did *Spartina alterniflora* compared to adjacent mudflats (Neira et al., 2006). By contrast, invasive drift mats of *Gracilaria vermiculophylla* reduced sedimentation on oyster reefs (Thomsen and McGlathery, 2006) whereas *Sargassum muticum* had no detectable impact on shallow subtidal reefs (Britton-Simmons, 2004). Finally, mobile invaders can change *sediment stability and erosion rates*, such as through bioturbation and burrowing. For example, the Chinese mitten crab *Eriocheir sinensis* and the isopod *Sphaeroma quoyanum* make burrows in banks, potentially causing bank erosion or converting marsh to mud (Rudnick et al., 2005; Davidson and de Rivera, 2010), and at the same time changing invertebrate communities living in and around the holes (Davidson et al., 2010).

Finally, we present a case study to illustrate how a single invasive species (the mollusc *Batillaria australis*) could impact abiotic conditions in an entire ecosystem. A primary assumption is that the behaviour and activity of *Batillaria* can be scaled up from individuals to the entire population in the system (Parker et al., 1999; Ruesink et al., 2006). A secondary assumption is that *Batillaria* has not dramatically altered populations of native molluscs. This assumption is supported by circumstantial evidence; correlative survey data and field experiments suggest low impact of *Batillaria* on native molluscs because the invader can be found in extreme high densities together with native molluscs, and because few shell deposits have been found in sediment cores suggesting that no 'bivalve mass extinction' has occurred (Thomsen et al., 2010b, 2012a; Hoeffle et al., 2012). *Batillaria* was introduced to the Swan River in Perth with imported oysters about 50 years ago. It has been conservatively estimated that 3.6 billion *Batillaria* exist in the Swan River today (Thomsen et al., 2010b). Except for a minor population outside of the mouth of Swan River, the nearest population is more than 3000 km away. *Batillaria* is the dominant gastropod in the estuary, occupying most habitats and depth strata. It is on average 2.4 cm long and typically lives 3–4 years.

Here we restate the impact questions to 'what do these 3.6 billion snails do?' Most importantly, *Batillaria* produces hard substratum within the estuarine 'sea of soft sediments', corresponding to *c.* 1.7 km^2 of *Batillaria* substratum or 3 111 243 kg ash free dry weight (primarily $CaCO_3$) that move around on a daily basis in the Swan River. Assuming the population is not growing anymore and that the majority of empty shells eventually are buried in sediments, >3000 tonnes $CaCO_3$ are buried over a 3–4 year period in this single estuary. *Batillaria* moves around on the sediment surface or just below and we calculated that up to 450 000 m^3 sediments could be in physical contact, and thus potentially moved, with the snail daily in the estuary. An unknown proportion of this sediment will be ingested and oxygenated, and diatoms, bacteria, organic material and invertebrate eggs and larvae possibly consumed. Similarly, *Batillaria* can affect sediment properties through faecal production and re-mineralisation of nitrogen. We calculated that more than 2 kg nitrogen may be released daily in the system by this single snail. Finally, if *Batillaria* is a dual feeder like its sibling species *Batillaria zonalis* and *Batillaria flectosiphonata*, then, in addition to deposit feeding on sediments it may also filter the water column for particulate organic matter. Assuming that *B. australis* has similar traits to its native congeners, then, theoretically, up to 3 billion litres of Swan River water could be cleared of suspended particles each day. More specifically, assuming a mixed water column, *Batillaria* in a seagrass bed at 1 m depth could potentially clear the entire water column every few hours. Similar simple ballpark estimates could be made for most marine invaders to provide first-order approximation of which species influence specific biogeochemical cycles the most (for analogous ballpark calculations that emphasise productivity changes, see Ruesink et al., 2006). Ultimately, all of the above types of impact on local abiotic conditions and resources will interact with effects on local species and on BEF relationships.

10.4 How do invaders impact local communities?

The previous sections exemplified how marine invaders affect local abiotic conditions and biota, but with little consideration of how entire communities and patterns of biodiversity might be altered. Here we review and analyse the effects of marine invaders on entire communities.

10.4.1 Review of reviews

Several authors have already reviewed impacts of invasive marine seaweeds using vote-counting (counting studies that find a particular type of effect, Schaffelke and Hewitt, 2007; Williams and Smith, 2007) and quantitative meta-analysis (calulate average impact and confidence limits, Thomsen et al., 2009c). Schaffelke and Hewitt (SH) and Williams and Smith (WS) reviewed 69 and 68 impact papers, respectively, including both mensurative and manipulative experiments (Box 10.1). We revisited these studies to evaluate what impacts have been reported on local communities, as measured by univariate diversity metrics and/or multivariate community structures. For each reviewed study (Table 1 in both SH and WS) we tallied impact direction (positive, negative, zero, nondirectional), experimental method (mensurative vs. manipulative), and impacted organisms (plant or animal diversity, multivariate community structure) (Table 10.1). This table highlights that little is known about marine plant invasion impacts; for example, there were only one (SH) or three (WS) examples of documented negative or zero impacts on local plant diversity from manipulative experiments, and not a single experimental study documented positive effects of marine plant invaders on local plant diversity. Evidence was only slightly stronger for mensurative data; SW reported mostly negative impacts on plant diversity, whereas WS reported almost equal numbers of negative, positive or zero impacts. There were even fewer studies reporting impacts on local animal diversity than those on plant diversity; not a single manipulative experiment was identified, whereas a few mensurative experiments documented both positive and negative effects. Finally, no manipulative studies were found to alter multivariate community structures significantly (WS; 1 reviewed study with no effect).

Table 10.1 *Community impacts from invasive seaweeds tallied from Table 1 in Schaffelke and Hewitt (2007; SH) and Table 1 in Williams and Smith (2007; WS).*

Impact on →	Plant diversity				Animal diversity				Community structure			
Method →	Man.		Men.		Man.		Men.		Man.		Men.	
Reviewer →	SH	WS	SB	WS	SB	WS	SH	WS	SH	WS	SH	WS
Effect ↓												
Negative	1	2	11	4	0	0	5	5	0	0	4	10
Zero	0	1	1	3	0	0	0	4	0	1	5	2
Positive	0	0	1	3	0	0	4	2	NA	NA	NA	NA

Some studies documented both negative, positive and zero impact from the same single experiment (e.g. on plants, animals, and community structure, respectively). Impact on community structure was classified as 'negative' or 'zero' if the multivariate tests were significant or non-significant, respectively. The majority of studies reviewed by SH and WS documented negative impact of invasive seaweeds on diversity. Man. and Men. = Manipulative and Mensurative experiments (see Box 10.1). Diversity = univariate diversity metrics; Community structure = multivariate metric.

The meta-analytical review (Thomsen et al., 2009c) tested more specifically if marine plants have different effects within versus across trophic levels and on different levels in the ecological hierarchy. Evaluated from manipulative experiments only, it was shown that impact on plant abundances and plant diversity was negative, probably caused by competition processes. However, a net positive impact was reported on the diversity of animals, probably because the invasive plants created habitat and food for the animals (but this result was calculated from only two primary studies).

Several other reviews have evaluated invasion impacts in marine systems, focusing on ecological – not taxonomic – comparison, including competition, predator/prey, ecosystem-engineering and multitrophic effects (qualitative reviews/vote-counting; Byers, 2009; Crooks, 2009; Grosholz and Ruiz, 2009; Rilov, 2009) and how density, identity or origin of the invader may modify impact (quantitative meta-analysis; Thomsen et al., 2011a, 2011b, 2012b). However, none of these reviews emphasised impacts on local communities or biodiversity; a keyword search for 'diversity' or 'richness' in these seven papers resulted in only 10 hits and several of these hits used 'diversity' in a different context.

More telling, only two specific diversity-impact examples were provided in these reviews; *Phragmites australis* and *Spartina alterniflora* (intertidal marsh plants) can reduce resident plant richness, probably via competition for space, light and nutrients (reviewed in Byers, 2009), and infaunal invertebrate richness and diversity, probably by converting mud habitat to vegetated habitat (reviewed in Grosholz and Ruiz, 2009). Overall, our examination of invasion impact reviews found that only a few case studies reported 'large' impacts on community metrics, and that major research gaps exist concerning community impacts. These reviews also emphasise that the mechanisms that cause large or small community impacts are poorly known, but that there is a tendency for negative impacts to occur via plant–plant competition, and positive impacts to occur via habitat formation (on sessile epibiota and mobile animals), and food provision (on local grazers).

10.4.2 New meta-analysis

As indicated in the previous section, little is known about how marine invaders affect local community structures and ecosystem functions, a knowledge gap also highlighted for terrestrial plant invaders (Levine et al., 2003; Powell et al., 2011; Vilà et al., 2011; Pyšek et al., 2012), and more generally in invasion biology (Parker et al., 1999; Byers et al., 2002; Thomsen et al., 2011a). Importantly, invasion impacts on community metrics are traditionally analysed separately from the invader itself (because it is the independent test variable) – but the invader itself is at the same time an integral part of the new community. To better understand BEF relationships, we suggest evaluating invasion impacts on communities using both the traditional type of analysis, but also by adding the invader to the community/ecosystem variable that is quantified. We therefore conducted a meta-analysis that compared community effects with versus without inclusion of the invader (i.e. by adding the independent invader to the dependent community data set). More specifically, we tested if invasion impact, on *species richness* and *total standing stock* would differ if the invader itself was included as part of the total community. We focused on these two metrics because they are most commonly reported; far fewer field impact experiments have quantified impacts on ecosystem functions (but total plant abundance can be interpreted as a surrogate, as abundance often correlates with function).

We used seaweed invaders as model organisms because this literature has been summarised in detail (see previous section), and focus on experimental studies to be able to infer causality. We reviewed experimental studies that report impacts on local plant communities to ensure commensurability in units between the invader and response community (typically biomass or percentage cover of both the invasive seaweed and impacted local plants). Thus, studies that reported invader abundance as biomass, but impacted plant community as per cent cover, could not be included (Thomsen and McGlathery, 2006). Tests were evaluated from published replicated field experiments that compared invaded and non-invaded plots using invader-addition or -removal manipulations. Literature search, data extraction and meta-analyses generally followed methods described in recent reviews (Thomsen et al., 2009c; Thomsen et al., 2011a, 2011b). We found eight papers (describing 23 experiments) that reported both the abundance of the invader and total abundance of all local plant species, and 11 papers (describing 30 experiments) that reported impacts on local plant richness (Table 10.2).

More specifically, we extracted the last data-point from repeated-measure experiments and compared controls to the highest invader density from multi-density experiments. Nested, orthogonal and repeated experiments and different experiments within a paper were treated as 'independent' experiments. We calculated Hedges' effect size d, corrected for small sample sizes (= $d_{individual}$), where $d_{individual}$ is <0 if the invader reduces biodiversity or total abundance. We subsequently calculated 'combined-$d_{individual}$' values to represent invasion effects on the entire community that included the invader itself (as it is part of the 'new' plant community).

It is straightforward to compare richness effects with versus without inclusion of the invader, simply by adding '1' to the mean effect per plot (there is on average one more species in each invaded plot when the invader itself is included). Adding a constant (= '1') to all samples does not influence the standard deviation. We calculated the new total plant abundance in invasion treatments by adding the abundance of the invader to the total abundance of all the local species (see Table 10.2 for details). The new total standard deviations were averaged from individual SD-values. Multiple $d_{individual}$ from individual experiments were averaged into a single effect size using equal weighting (Borenstein et al., 2009). We calculated unweighted cumulative effect sizes and 95% bias-corrected CL in Metawin 2.1 from the independent $d_{experiment}$ values (Rosenberg et al., 2000; Thomsen et al., 2009c). Finally, we evaluated if a test-factor was significantly different from zero or another test-factor if 95% CL did not overlap zero or each other, respectively.

Table 10.2 *List of manipulative field experiments reporting impacts of invasive seaweeds on local plant richness and total plant abundances (and also reporting the abundance of the invader). Total community richness was calculated by adding one species (the invader itself) to reported impact on 'local species richness'. Total community abundance was calculated by adding the abundance of the invader itself to reported abundances of all local species combined (see footnotes for details).*

Reference	Invader	Impact on plant...
Britton-Simmons (2004)	*Sargassum muticum*	Richness
Bulleri *et al.* (2010)	*Caulerpa racemosa*	Richness, Abundance[7]
Casas *et al.* (2004)	*Undaria pinnatifida*	Richness, Abundance[3]
Gennaro and Piazzi (2011)	*Caulerpa racemosa*	Richness
Gribben *et al.* (2009a)	*Caulerpa taxifolia*	Richness
Klein and Verlaque (2011)	*Caulerpa racemosa*	Richness, Abundance[6]
Olabarria *et al.* (2009)	*Sargassum muticum*	Richness
Piazzi and Ceccherelli (2006)	*Caulerpa racemosa*	Richness
Piazzi *et al.* (2005)	*Caulerpa racemosa*	Richness, Abundance[5]
Sánchez and Fernández (2005)	*Sargassum muticum*	Richness, Abundance[1]
Thomsen and McGlathery (2006)	*Gracilaria vermiculophylla*	Richness
Valentine *et al.* (2007)	*Undaria pinnatifida*	Abundance[8]
Viejo (1997)	*Sargassum muticum*	Abundance[4]
White and Shurin (2011)	*Sargassum muticum*	Richness, Abundance[2]

Superscript numbers in the column 'Impact on plant...' refer to how total community abundance was calculated – by adding...

[1] Cover values for *Sargassum muticum* to cover values of *Bifurcaria bifurcata*, *Gelidium spinosum* and 'rest of species' (extracted from their Figure 1).
[2] Biomass of *Sargassum muticum* (high density treatment) to biomass of native seaweed (extracted from their Figure 4).
[3] Biomass of *Undaria pinnatifida* to biomass of all species (extracted from their Table 1).
[4] Cover of *Sargassum muticum* to total cover of native species (extracted from their Figure 2).
[5] Cover of *Caulerpa racemosa* to total cover of native species (extracted from their text and Figure 1).
[6] Biomass of *Caulerpa racemosa* to total biomass of local taxa (extracted from their Figures 1 and 2).
[7] Biomass of *Caulerpa racemosa* to biomass of canopy, erect, turf and encrusting alga (extracted from their Figure 2 and Appendix Figure 1).
[8] Biomass of *Undaria pinnatifida* to total abundance of canopy, erect, turf and encrusting alga (extracted from their Figure 3).

Our analysis revealed that marine plant invaders have significant negative impacts on both local plant abundance (d = −1.78, 95% CL from −2.53 to −1.14, Figure 10.2a) and plant community richness (d = −0.67, 95% CL from −1.18 to −0.26, Figure 10.2b). These negative effects suggest that competition dominates interactions between invaders and natives within a trophic level. Importantly, the negative effects were cancelled out when the contribution of the invader itself was included in the estimates of total standing stock (d = 0.28, 95% CL from −0.67 to 1.63) and total community richness (d = −0.03, 95% CL from −0.44 to 0.39). Thus, at least for abundance and richness, marine plant invaders appear on average across studies, species and invaded systems, to substitute, rather than increase or decrease, total biodiversity and standing stocks. However, clearly more impact studies are needed, especially on functional community responses, such as total system productivity, respiration, decomposition rates and nutrient uptake to test if ecosystem functioning is typically unaffected, enhanced or reduced following invasions. For a few rare examples, see Altieri *et al.* (2009), Cacabelos *et al.* (2012) and Green *et al.* (2012, 2013).

10.5 A simple framework to discuss BEF and marine invasions

It is essential to understand how biodiversity affects ecosystem functioning (Chapters 1, 5) because ecosystem functions relate to provision of ecosystem services for humans (Hooper *et al.*, 2005; Barbier *et al.*, 2011; Hooper *et al.*, 2012; Chapters 1, 2). The first BEF studies originated in terrestrial grasslands, but in the last decade marine systems have also been studied in detail (reviewed in Chapter 5). Currently, the functioning of marine ecosystems is being strongly affected by impacts on patterns of biodiversity by human stressors, in particular climate change (Petchey *et al.*, 1999; Schmitz *et al.*, 2003; Wrona *et al.*, 2006; Wernberg *et al.*, 2012a), fisheries (Chapter 6), habitat alterations (Chapter 7), pollution (Chapters 8, 9) and species invasions (this chapter).

Recent reviews highlight that surprisingly little is known about how invaders influence ecosystem functioning (Vilà *et al.*, 2010; Simberloff, 2011; Strayer, 2012). Still, two aspects of invasions and BEF have received scrutiny. First, many studies have tested how biodiversity modifies community 'invasibility' (typically considered an ecosystem function) within the 'biotic resistance hypothesis' framework (Elton, 1958; Levine *et al.*, 2004; Parker and Hay, 2005). This theory suggests that high taxonomic richness correlates with functional diversity and efficient resource utilisation and less opportunity for new species to become established. Many marine studies have tested this theory, although often with opposing results (Stachowicz et al., 1999; Byers, 2002; Stachowicz et al., 2002a; Stachowicz and Byrnes, 2006; Ruesink, 2007). Today, it is acknowledged that there is no simple relationship between biodiversity and how easy it is for invasive species to invade a community, because numerous co-varying factors (e.g., trophic positions of species, potential presence of facilitator-species, spatio-temporal heterogeneity and scale) influence establishment (Fridley et al., 2007; Stachowicz et al., 2007; Catford et al., 2009). Our focus here is primarily on how invasive species modify ecosystem functioning via changes to biodiversity (i.e. we treat the invader as the independent variable), not how biodiversity and ecosystem processes modify invasions (i.e. where the invader is the dependent variable). For details on the latter see the literature listed above.

Second, there is an extensive literature documenting impacts of invasive 'ecologically important species' (e.g. ecosystem engineers, habitat-formers, foundation species, keystone consumers) on various ecosystem functions (Crooks, 1998, 2002a, 2009; Pedersen et al., 2005; Wallentinus and Nyberg, 2007; Thomsen, 2010; Thomsen et al., 2010a; Green et al., 2013). Typically, however, these studies do not frame invasion impacts within the context of BEF relationships. A first impression suggests that these and many other invasion impact studies are at odds with traditional BEF studies. Many BEF relationships follow saturation curves where the addition of species to communities with low diversity often results in large positive effects on ecosystem functioning as new 'functions' are added by keystone species and ecosystem engineers. However, addition of species to a high diversity community typically result only in *small positive effects*, as niches are increasingly 'filled out' by species that can substitute for each other (insurance/redundancy type hypotheses). In contrast, most invasion studies emphasise that addition of a single (invasive) species, also in a highly diverse community, has *dramatic, typically negative, effects*. However, several mechanisms may explain this difference (Ruesink et al., 2006). For example, invasion studies selectively target 'high profile' invaders, with 'easy-to-identify' high impacts, sometimes adding or destroying entire functional groups. By contrast, BEF studies typically document that addition of a 'randomly selected' species, within a functional group, has a low net effect (Scherer-Lorenzen, 2005). Furthermore, trophic position sometimes differs between invasion impact and BEF studies; large documented invasion impacts are often associated

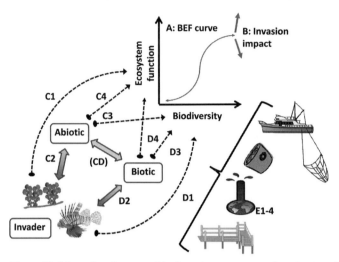

Figure 10.3 Interface between biodiversity–ecosystem functioning (A) and invasion impact (B). Marine invasive species (e.g. *Caulerpa* seaweeds or the Indo-Pacific lionfish, *Pterois volitans*) impact ecosystem function and local biodiversity (arrows) directly by its mere presence and metabolic activity (C1, D1) and indirectly by modifying the local abiotic and biotic environment (C2, D2). These interaction pathways are modified by human stressors such as fisheries, pollution, climate changes and habitat alterations (E1–4). Note that (a) invasion impact differ from the other human stressors by directly modifying BEF curves (C1, D1) and (b) local ecosystem functions and patterns of biodiversity also modify how successful the invader is (reversed arrows). See Section 10.5 for detail. Images used in the figure courtesy of the Integration and Application Network, University of Maryland Center for Environmental Science (ian.umces.edu/symbols/).

with consumer effects across trophic and/or functional levels, whereas many BEF studies that document saturation effects focus on plant-plant interactions within trophic and/or functional levels. Finally, it is of course also possible that a lack of co-evolution between invaders and native species creates fundamentally different interactions and thereby also different BEF effects, compared to relationships only involving co-evolved native species. Most importantly, none of the speculations above have been tested systematically or in any detail.

As well as affecting ecosystem functioning via changes in the abiotic environment (Section 10.3), species interactions (Section 10.2) and biodiversity (Section 10.4), marine NIS may also affect the relationship between biodiversity and functioning, further complicating prediction of their effects. Given that little is known about how invasive species modify BEF relationships, we here provide a simple schematic overview (Figure 10.3; capital letters in brackets correspond to the linkages on the figure). There are many studies that have quantified different BEF curve-shapes depending on specific operating mechanisms within an experimental context (Schulze, 1994; Hooper et al., 2005, 2012; Cardinale et al., 2006, 2011). A typical scenario involves a rapid initial increase in EF as keystone/dominant species are added from a random species pool, but approaching a saturating EF level as niches are slowly filled and resources used more efficiently (A). Superimposed on this pattern, and as discussed in the previous sections, invasion impacts are often documented as large negative or large positive effects even in rich and well-established communities, following the addition (or removal) of a single NIS (B). The invader itself (in Figure 10.3 exemplified by two high profile invaders, the green seaweed *Caulerpa racemosa* and the Indo-Pacific lionfish *Pterois volitans*) has direct impacts on ecosystem functioning (C1) and local biodiversity (D1) (arrows) by its mere presence and metabolic activity, for example by adding one species to the total species richness and adding its nutrient uptake to the system-wide nutrient uptake rates. Examples of how the addition of the invaders themselves modify biodiversity and ecosystem function were examined in Section 10.4 where we tested if invasion impact on plant richness and total standing stock changed depending on whether the invader itself was added to the community dataset. More importantly, invasive species can also have numerous impacts on the local abiotic (C2) and biotic (D2) environment, causing indirect and cascading impacts on the BEF curves (C3–4, D3–4). For example, *Pterois volitans* is a voracious consumer of herbivores on coral reefs (Albins and Hixon, 2008), which potentially could lead to extinctions of local populations and reduced biodiversity (D3) as well as expansive seaweed growth, followed by coral dieback and reduced calcification rates (D4). The processes whereby invaders affect local biota (D2) were reviewed in detail in Section 10.3, because only by understanding the breadth and scope of these processes are we able to understand the more complex cascading effects on BEF (and because these processes are much better studied than invasion vs. BEF curves). Invasive species also modify the local abiotic environment. For example, invasive *Caulerpa* seaweeds alter elemental fluxes in the sediment (C2), typically decreasing oxygen levels and redox potential (Wright et al., 2007; Gribben et al., 2009a). Reduced oxygen levels can then have a cascading impact on biodiversity, for example by killing off infauna that are susceptible to anoxia (C3) or by altering iron-precipitation rates (C4). Effects of invaders on abiotic conditions and biogeochemical fluxes (C2) were discussed in Section 10.4. More complex indirect interactions can also be envisioned between the invader and the invaded abiotic environment and biotic communities (CD), but are beyond the scope of this chapter.

Our emphasis has been on how marine NIS impact abiotic and biotic conditions and the invasive species was therefore representing an 'independent test variable'. The 'arrow' on Figure 10.3 therefore points towards the BEF curve (= 'invasion impact studies'). However, we emphasise that the invader itself is controlled by abiotic and biotic 'filters' (Richardson et al., 2000); that is, the invasive species is often considered the dependent response variable (= 'invasion success' studies). To fully understand the relationship between invasive species and BEF curves, reciprocal interactions between the invader, and the invaded abiotic and biotic environment also need to be considered (see Chapter 5).

10.6 Do invasion impacts co-occur with other human stressors?

Importantly, invasion impact does not operate in isolation from other human-induced stress and disturbances (for brevity here 'stressors', Figure 10.3, E1–4). Indeed, all pathways shown in Figure 10.3 are constantly modified by climate change (a global stressor), fisheries, pollution and habitat alterations (local stressors). In this section, we provide examples of possible interactions between invasive impacts, and other human stressors. We note that invasion impacts differ from the other human stressors by being of 'biological/taxonomic nature' and therefore (1) have a direct effect on BEF relationships (cf. arrow D1; irrelevant for abiotic stressors) and (2) experience strong reciprocal effects from the abiotic and biotic environment (by contrast, biodiversity has little reciprocal impact on how humans contribute to climate change or habitat alteration). If, where and when these stressors facilitate the survival, establishment and population expansions of marine NIS, impacts will also increase because effect sizes co-vary with invader abundance (Thomsen et al., 2011b). There are three fundamental ways these stressors can facilitate marine invaders: by (1) improving abiotic conditions and resource levels; (2) 'favouring friends' of the invader (Simberloff and Von Holle, 1999; Thomsen et al., 2010a); or (3) 'disfavouring enemies' of the invader. For example, eutrophication can increase resource availability for invasive plants (Davis et al., 2000), pollution with heavy metals can kill native competitors, and fisheries can reduce top-down control mechanism of native top predators. In these examples, space and other resources become more available and the biotic resistance of the system decreases, after which NIS can more readily establish new populations (Byers, 2002). A few studies have reviewed interactions between multiple human stressors (Crain et al., 2008; Darling and Cote, 2008), but with virtually no emphasis on marine NIS (only one marine invasion study was listed by Crain et al., 2008). Below we describe a few examples that suggest marine NIS impacts can increase when co-occurring with other human stressors.

Overfishing is a pressing conservation problem (Lotze et al., 2006) as food webs are fished down around the world, first reducing populations of apex predators and subsequently lower trophic levels. Large-scale reduction of predatory fish is likely to open up opportunities for marine invaders that typically are top-down controlled – a process that probably is more relevant for invasive animals than plants. For example, large groupers have been fished down in tropical systems potentially facilitating establishment of lionfish (Maljković et al., 2008) which then reduce abundances and diversity of native species (Albins and Hixon, 2008). Similarly, cod populations in the Baltic Sea have been severely reduced by fishing, thereby removing enemies for invasive round goby, that also can modify biodiversity (Almqvist et al., 2010; Kornis et al., 2012). Finally, it is possible that overfishing of California flounder, *Paralichthys californicus* and other fish in California has led to reduced predation on invasive bivalves (Crooks, 2002b), which also can have wide-ranging cascading impacts on biodiversity (Crooks, 1998; Crooks and Khim, 1999).

Pollution, in particular with excessive nutrients, also facilitates some marine invaders. Plant invaders with high nutrient uptake capacity and an ability to transform nutrients into rapid growth are likely to be facilitated by eutrophication. These types of invaders are typically ephemeral sheet-forming and filamentous seaweeds (often small epiphytic weedy species) that, when nutrients are abundant, gain a competitive advantage over slow-growing, large canopy-forming (native) seaweeds (Sand-Jensen and Borum, 1991). These fast-growing ephemerals are typically hull-fouling species that can be transported between ports and harbours on commercial and recreational vessels. We therefore expect a positive correlation between nutrient pollution, boat traffic and invasions, potentially leading to reduced local plant diversity if slow-growing native canopy species are outcompeted. Interestingly, decreased plant biodiversity and degradation of seagrass beds are often attributed to nutrient fuelled blooms of 'native' ephemeral seaweeds. However, many of these 'native' species could indeed be NIS, as they often are difficult to identify (simply being assumed to be native) and because these species have been translocated on ships between regions for centuries (i.e. they may have been introduced to new regions prior to scientific investigations). Thus, perhaps many impacts associated with 'weedy seaweeds' ultimately are driven by a combination of invasions and eutrophication, although detailed taxonomic, biogeographical and molecular studies are needed to test this hypothesis (Carlton, 1996; Thomsen et al., 2006a; Blakeslee et al., 2008). Other types of pollution, such as with metals and hydrocarbons, might also indirectly favour stress-resistant invasive sessile species, such as tunicates, bryozoa and hydrozoa. For example, many sessile fouling invaders are highly resistant to the toxins found in marine harbours, thereby indirectly facilitating them over less stress-resistant native species (Piola and Johnston, 2006, 2008; Dafforn et al., 2009).

Destruction and alteration of habitats modify invasion impacts; coastal forests have been destroyed around the world as they are converted to agricultural landscapes and urban centres. This process accelerates nutrient (see above) and sediment loadings (Syvitski et al., 2005), often with detrimental impacts on marine organisms (Airoldi, 2003; Fabricius, 2005). Heavy sediment load is likely to favour opportunistic invaders. Several invaders appear to be resistant to sediment stress, including certain bivalves (*Theora lubrica*), tunicates (*Styela clava, Ciona intestinalis, Ascediella adspersa*), bryozoa (*Watersipora subtorquata*) and polychaetes (*Sabella spallanzanii*). These species are typically associated with hull fouling and ballast water transport and are (as for weedy seaweeds) difficult to identify and may therefore have been mistaken for native species (Kremer and Rocha, 2011). Thus, as for nutrient pollution, habitat alteration may have large but potentially underestimated impacts on community structure and biodiversity by disadvantaging slower-growing native species. A more direct mechanism whereby habitat alteration modifies invasion impacts is by creation of hard substrates from wind farms, oil rigs, jetties and seawalls. These structures provide space for sessile species, and are often found in areas of intensive boat traffic and nutrient and sediment pollution, further favouring the establishment of invasive opportunistic species (Connell, 2001; Bulleri and Chapman, 2010; Chapter 7). Hard structures typically support a diverse flora and fauna, occasionally dominated by invaders, and can function as stepping stones for newly arrived NIS, facilitating a rapid secondary spread to nearby locations.

Finally, *global climate change* can modify invasion impacts, because climate change can reduce the resilience of communities to invasions and simultaneously increase the potency of invaders (Stachowicz et al., 2002b; Thuiller et al., 2007; Walther et al., 2009; Raitsos et al., 2010; Sorte et al.,

2012). Climate change modifies marine ecosystems through altered wave regimes (in some areas storm frequencies have increased), rising sea levels, and increasing acidification and temperature. Increased temperature will likely lower resistance to invasions as some native species become thermally stressed and, at the same time, facilitate survival and establishment of warm-tolerant invaders. For example, invasive lionfish are expected to 'perform' better and extend their invaded range pole-ward in the Atlantic with increasing temperatures (Cheung et al., 2009; Cote and Green, 2012). Invasion impacts might also become more severe under increased temperatures. For example, *Gracilaria vermiculophylla* had little impact on native seagrass under cold temperatures (under both low and high invasion densities), but had a strong negative impact under high temperatures in high densities, suggesting increased negative impacts in a future warmer world (Hoeffle et al., 2011). This strong context-dependent impact happened in part because the seagrass became less resistant to stress near its upper temperature tolerance level, in part because the effects of the invader became more severe as respiratory processes (leading to localised toxic anoxia) become increasingly important at high densities and high temperatures. An important large-scale, long-term effect of warming on species distributions is equatorial range-contraction and pole-ward range expansion (Raitsos et al., 2010; Wernberg et al., 2011; Wernberg et al., 2012b). However, in some cases species may not move fast enough and pole-ward movements can be interrupted by natural or human-created barriers. The interface between NIS invasion and climate change stress creates several conundrums for conservation biologists. First, biologists face a definition issue; are species that spread pole-ward by natural dispersal mechanisms, but only establish reproducing populations due to warming, actually NIS that need to be managed and perhaps even eradicated, or are they simply 'new natives'? Furthermore, as many native species become increasingly heat stressed the local communities are likely to become less resistant to invasions (Stachowicz et al., 2002b; Wernberg et al., 2010). From a practical perspective it can be difficult to distinguish if a new species has arrived by its own dispersal mechanisms and now survives due to warming or if it has arrived with human transport vectors and needs to be managed. Finally, if local species become heat stressed but are prohibited from moving pole-ward because of dispersal limitation, it has been argued that these species should be intentionally introduced to lower latitudes to save them from extinction. However, such 'assisted translocation' is resisted by invasion ecologists because adverse side-effects associated with intentional introductions can be expected (Ricciardi and

Simberloff, 2009; Sandler, 2010; Seddon, 2010; Seddon et al., 2011; Chauvenet et al., 2012).

Finally, we highlight that invasions typically co-occur not only with one but many human stressors, and more studies should aim to address invasion impacts in a multi-stressor framework. For example, it is likely that the infamous anchovy collapse in the Black Sea resulted as a combination of overfishing, eutrophication and climate-induced facilitation of the invasive comb jelly *Mnemiopsis leidyi* (Oguz et al., 2008). The similarly infamous invasion in the Mediterranean Sea of *Caulerpa* species – with potentially dramatic negative impacts on native seagrasses (Meinesz, 1999) – co-occurred with climate changes and increased temperature stress (Díaz-Almela et al., 2007), eutrophication and pollution with metals and hydrocarbons (Danovaro, 2003), excessive sediment loadings (Airoldi, 2003) and overfishing (Lejeusne et al., 2010). Importantly, these co-occurring processes all stress local seagrasses and likely caused them to be susceptible to *Caulerpa* invasions. The above examples of interactions between multiple human stressors and invasions also exemplify the difficulty of distinguishing whether the invader is a main cause (driver) of ecosystem change, mainly following another stressor (passenger) that causes ecosystem change, or interacting with the other stressor to cause ecosystem change (back-seat driver) (MacDougall and Turkington, 2005; Bulleri et al., 2010; Bauer, 2012). This distinction is vital from a conservation perspective because if the invader drives changes, the invaded ecosystem can be restored by controlling the invader, whereas if the invader is a passenger or back-seat driver, other management schemes (e.g. nutrient reduction) need to be implemented as well.

10.7 Invisible invasions: invasion impact is so much more than leaves and legs

Our review has overwhelmingly focused on invasion impacts associated with macroscopic plant and animal NIS – i.e. NIS with 'leaves and legs' (Cowan et al., 2013). We have done this because this is our area of expertise, but also because the vast majority of research focus has been on macroscopic NIS. However, there are potentially vast 'invisible' invasions associated with microbes (unicellular algae, protists, fungi, bacteria, viruses; Litchman, 2010) and parasites (Torchin et al., 2002) that should not be forgotten in future NIS research programs. Invasive terrestrial (e.g. chestnut blight, caused by the fungus *Cryphonectria parasitica*) and freshwater (e.g. crayfish plague caused by the water mold *Aphanomyces astaci*) microbial pathogens have long been recognised (Elton, 1958). However, even though one of the first documented examples of a marine invasion involved microbes (Ostenfeld, 1908), only recently have microbial marine NIS been appreciated for their ubiquity and importance (Cowan et al., 2013). A lack of appreciation of microbial invasion impacts may partly stem from the outdated 'everything is everywhere – but the environment selects' paradigm (De Wit and Bouvier, 2006). However, today it is well established that marine microbes also have specific biogeographical distributions with endemic genotypes, and therefore may be susceptible to invasions (Litchman, 2010). The most important transport vector for marine microbes is via shipping activities, and an increasing number of studies have quantified microbes in ballast water, ballast sediments, ship cavities and biofilms, suggesting that ballast water is the single most important vector (Ruiz et al., 2000; Drake et al., 2002, 2005, 2007). It has been suggested that up to 12 billion tons of ballast water is moved across the oceans annually (Ibrahim and El-Naggar, 2012), providing a potentially enormous microbial transport mechanism. Indeed, microbes associated with shipping may be particularly successful (and have dramatic impacts) due to microbial traits of high abundances, multiple modes of reproduction, high growth rates, ability to form resting stages, efficient resource utilisation, and potential toxicity (Hallegraeff, 1993; Ruiz et al., 2000; Drake et al., 2002, 2005, 2007; Litchman, 2010). For example, bacterial densities in ballast water can vary form 10^7–10^{10} cells per litre (Ruiz et al., 2000; Drake et al., 2002, 2005) and detailed sampling in the Chesapeake Bay region suggest that up 10^{20} bacteria and viruses are discharged annually from ballast waters (Ruiz et al., 2000; Drake et al., 2007), of which more than half survive the discharge into the new habitat. Of these, massive amounts of released microbes, pathogens and microbes that can cause 'harmful algal blooms' are of obvious concern (Hallegraeff, 1993, 2010; Landsberg, 2002; Butrón et al., 2011). For example, faecal streptococci, *Escherichia coli*, *Salmonella* spp., *Clostridium perfringens* and *Vibrio cholera* have all been detected in ballast water (McCarthy and Khambaty, 1994; Altug et al., 2012; Emami et al., 2012; Morris, 2013) and Ruiz et al. (2000) detected pathogenic strains of *V. cholerae* – the agents of human cholera – from >90% of the ships entering Chesapeake Bay. Taken in concert, these studies provide strong indications that invisible NIS may have dramatic impacts on marine biodiversity and ecosystem functioning. However, it will be up to future researchers to provide strong and novel data that link detected microbial and parasitic NIS associated with human vectors to actual effects in the invaded environment.

10.8 Discussion

Our review provides examples of the many ways that invaders modify the biotic and abiotic environment, and at first look, impacts appear to be idiosyncratic and almost impossible to predict. Still, many recent studies aim to see through this context dependency to establish first-order rules that can, at least partially, predict impact magnitude and direction. For example, it has been suggested that the 'history' of invasion impact can be a useful predictor of future impact (Kulhanek et al., 2011), but this rule has no inherent mechanisms associated with impacts and cannot predict impacts for species with no published impact data. A most obvious rule is that impacts increase with invaded range and invader density (Parker et al., 1999). Density-dependent impacts have subsequently been verified in meta-analyses across invasive species and invaded systems, although these also showed that impacts associated with species identity (also between ecologically similar invaders) are equally important (Thomsen et al., 2011a; Thomsen et al., 2011b). The density rule can be extended to include other attributes of the invader that relate to its 'fitness', such as that impacts increase with invader size and 'fitness'. Another commonly cited rule suggests that impacts are high when invaders are functionally different from native taxa (Ricciardi and Atkinson, 2004), although it appears that certain types of impacts instead may decrease with increasing differences (e.g. hybridisation effects and some types of interspecific competition). The functional dissimilarity rule has been confirmed by many studies that document strong impacts associated with large structural sessile NIS (plants and sessile invertebrates) invading unstructured environments (e.g. sedimentary ecosystems). We suggest that impact rules should be clear about whether they are 'universal' or apply within a specific context, such as between trophic groups (Thomsen et al., 2009c) or within or between function groups embedded within a trophic level (Thomsen et al., 2014). For example, within a trophic level, *Sargassum muticum* typically has negative impacts on functionally similar canopy-forming seaweeds that require hard substratum (Staehr et al., 2000), but positive impacts on functionally different epiphytic seaweed that use the invader as habitat (Thomsen et al., 2006c).

A key take-home message of our review is that, from a BEF perspective, invasion impact studies are few and theoretical linkages underexplored. Interestingly, we are not aware of field experiments that have documented 'massive' negative impacts on local patterns of diversity, such as through habitat destruction mechanisms (by contrast, it is relatively easy to demonstrate large positive impact through habitat formation) (Thomsen et al., 2010a). We suggest that abundant NIS that are known to consume habitat-formers from areas where habitat-formers already are few (e.g. in estuaries) would be good candidates on which to conduct field studies searching for strong negative impacts on diversity. Clearly, many more studies are needed to test how invaders modify BEF relationships. Importantly, none of the studies listed in our review compare NIS impacts of the same species from both its invaded and native region. Therefore, we simply do not know if NIS do something 'unusual' or just do what would be predicted from studies from its native region (Hierro et al., 2005). More studies are also needed that clearly address marine invasion impacts on a whole array of ecosystem functions, such as testing for multivariate complementarity effects (Stachowicz and Byrnes, 2006). It is vital that these studies always report the abundance of the invaders to allow for meaningful interpretations of impacts, a prerequisite that unfortunately sometimes is 'forgotten' (Thomsen et al., 2009c). Our review focused on invasion impacts on macrobiota because these organisms are typically targeted for study. However, it is likely that there is a world of hidden biotic impacts associated with how invaders modify microscopic bacterial (Green et al., 2012) and parasitic communities (Thieltges et al., 2009), probably with complex indirect impacts on the macrobiotic and abiotic environment. Clearly, more studies should address impacts on these microscopic communities and the linkages to better described macroscopic communities. It is also important that new studies report non-significant effects as well as effects from non-pooled treatments (e.g. in online appendixes and with associated data variability), because failing to do this limits synthetic advancement by biasing meta-analyses towards high impact invaders only, thereby making it difficult to identify NIS and environmental contexts that result in true weak impacts. In short, our ability to extrapolate impact assessments across space, time and taxa will increase significantly if above research gaps are targeted.

Our review has highlighted a few take-home messages for managers. For example, some invader are 'drivers' of ecological changes (e.g. *Gracilaria vermiculophylla*; Thomsen et al., 2010a), whereas others more often are 'passengers' (Jaubert et al., 2003) or 'back-seat drivers' (Bulleri et al., 2010). This distinction is important because if the invader drives changes, the invaded ecosystem can be restored by controlling the invader, whereas if the invader is a passenger or back-seat driver, other management schemes need to be implemented as well to control the invader (e.g. nutrient reductions). Second, it is clear that invasions result in both 'winners' and 'losers' – negative impacts on one group of native species often lead to positive impacts (quantified or not) on another group of native species. Note that these scientific insights do not equate to whether invasions are 'good' or 'bad', simply that managers and politicians need clear conservation goals. Such goals could be that invaders need to be eradicated irrespective of costs if they have well-documented negative impacts on a highly valued habitat, or that invaders may be preserved if they facilitate a threatened native species (Schlaepfer et al., 2011; Thomsen et al., 2012a). Managers also need to be aware that ecosystem functions may decrease if successful eradication occurs. For example, if *Gracilaria vermiculophylla* or *Spartina alterniflora* are removed from invaded mudflats, nutrient filtering would be reduced. Managers also need to acknowledge that impacts reported in the scientific literature typically focus on mean effects on aggregated responses, cancelling out potentially dramatic negative and positive effects reported on sub-aggregate entities. Perhaps more scientific studies need to analyse and report maximum (or quartile) impacts, so that managers can address invasions more conservatively. Managers should also note whether impacts have been evaluated from mensurative or manipulative experiments (see Box 10.1). For example, strong negative impacts observed in small experimental plots do not necessarily translate into regional scales if larger-scale refuge habitats exist. Similarly, positive effects observed on regional scales from mensurative data are no safeguard against neighbourhood-scale impacts by invaders, because positive impacts reported on large scales typically reflect a complex array of co-varying factors, such as increased habitat complexity, area and sampling effect (Fridley et al., 2007; Powell et al., 2011).

Managers have several options to control invaders. For example, where eradication is impossible or not merited, impacts may be reduced by (1) eliminating other stressors to strengthen the 'biotic resistance' of local communities; (2) controlling vectors (i.e. not allowing establishment in the first place or minimising the chance for multiple entries); (3) establishing continuous control actions, such as constant removals (as done for marine nuisance weeds); or (4) facilitating enemies, such as grazers or predators of the invaders (e.g. sea urchins are being trialled to control *Undaria pinnatifida* in Fjordland, New Zealand (http://www.biosecurity.govt.nz/media/08-07-11/working-together-in-fiordland). Finally, there is, in rare cases, the option of eradication. Many eradication attempts on marine invaders have failed, particularly where invaders have been well established or are abundant in nearby locations. Failed eradication on abundant NIS have been documented for *Sargassum muticum* (Critchley et al., 1986), *Undaria pinnatifida* (Hewitt *et al.*, 2005; Hunt, 2009) and *Pterois volitans* (Barbour *et al.*, 2011). In addition, intensive eradication campaigns on a very rare NIS that did not have nearby populations have also failed (Read *et al.*, 2011). Still, a growing number of studies suggest that eradication is possible in marine systems, in particular if the NIS exists in low population sizes, with no nearby populations, and near its environmental tolerance limits. Such successful (so far) eradications include NIS like *Ascophyllum nodosum* (Miller *et al.*, 2004), *Terebrasabella heterouncinata* (Culver and Kuris, 2000), *Caulerpa taxifolia* (Neverauskas, 2005), *Perna perna* (Hopkins *et al.*, 2011) and *Undaria pinnatifida* (Wotton *et al.*, 2004).

We finally note that a conservative 'warning' regarding marine NIS impact is merited. We have previously stated that we are not familiar with global extinctions associated with marine invasion (Briggs, 2010) and we have listed many examples of how local species can be facilitated by NIS. Politicians, managers and the public are therefore likely to ask 'why worry?' However, there are several reasons to support a precautionary conservation principle to marine invasion impacts:

1. Any invasion can cause negative impacts on some local species, simply because all invasions have both winners and losers.
2. With each new invasion, global homogenisation increases, as new species are added locally from the global species pool, i.e. the local biota slowly become more similar across biogeographical realms.
3. Scientists know little about possible local extinctions following invasions because distributions of rare species from open subtidal marine ecosystems are notoriously difficult to monitor.
4. Many studies report mean impacts on aggregated responses, implying that stronger negative impacts almost always go unreported as they are averaged out from positive effects.
5. Even if invaders, in isolation, do not cause extinctions, invasions co-occur with other human stressors and the combined multiple-stress effects may be the 'nail in the coffin' for local species.
6. In contrast to many other stress effects, marine invaders typically cause irreversible effects, i.e. whereas managers may be able to clean up pollution and rebuild destroyed habitats, marine NIS are difficult to eradicate and even if successfully removed there is no guarantee that local species will bounce back (Klein and Verlaque, 2011).
7. Ecological interactions are not in equilibrium, causing effects to have time lags, and invasion-driven extinction 'debts' could occur (i.e. small invasion effects occurring today may accumulate to dramatic effects over long time scales) (Tilman *et al.*, 1994; Kuussaari *et al.*, 2009).

10.9 Key findings and recommendations

- More than 1800 marine NIS exist, being particularly common in the Mediterranean Sea, Australia and New Zealand, the South Pacific, the north-east Pacific and the north-east Atlantic, and in estuaries centred around metropolitan areas.
- Marine NIS can impact local biota negatively directly through competition, habitat destruction, disturbance and consumption, and indirectly through keystone competition, keystone habitat destruction, keystone consumption and cascading consumption.
- Marine NIS can impact local biota positively directly through habitat formation, habitat modification and mutualism, and indirectly through cascading habitat formation, cascading consumption, cascading competition, keystone consumption and keystone competition.
- Marine NIS also modify abiotic conditions, resource levels and elemental cycling, directly by their metabolic and behavioural activities and indirectly by interacting with local species, e.g. by altering pools and fluxes of macro nutrients, light and oxygen levels, the organic matter, texture and erosion rates of sediments, and temperature and moisture levels in the intertidal zone.
- Invasion impacts co-occur with other human stressors, such as pollution, eutrophication, habitat destruction and climate changes. These stressors can increase NIS abundances and, at the same time, stress local communities, thereby making the local biota more susceptible to invasion effects.
- Several reviews of marine invasion impacts exist but none has emphasised impacts on biodiversity. In a meta-analysis of field experiments, we documented that marine plant invaders have negative effects on total plant abundance and plant richness but that these effects were nullified when the invaders' own abundance and taxonomic identity was added to the community data. This suggests that marine invasive plants generally substitute standing stocks and richness.
- Two aspects of invasions and ecosystem functioning have received some scrutiny: (1) how biodiversity modifies community invasibility, often finding opposing results, and (2) how invasive ecologically important species (e.g. ecosystem engineers) impact different ecosystem functions – although these studies have rarely been explicitly framed in a BEF context. Thus, BEF-related invasion impact studies are very few and theoretical linkages under-explored.
- More studies are needed to address how invaders modify BEF relationships, and, if possible, compare impacts from the same species in the

invaded versus native region, impacts on a whole array of ecosystem functions, as well as impacts on microscopic communities and linkages to better described macroscopic communities.
- It is important to know if invaders are drivers or passengers of ecological change; in the first case, ecosystems can be restored by controlling the invader, whereas in the latter additional management schemes also need to be implemented.
- Every invasion results in winners and losers in local communities, but this does not equate to whether invasions are 'good' or 'bad'. Given the complex ways that invaders impact communities, managers need clearly defined goals.
- Managers have several options to control NIS. For example, (1) eliminating other stressors to strengthen the resistance of local communities, (2) controlling vectors, (3) establishing continuous control actions, such as constant removals, (4) facilitating enemies, such as consumers of the invaders, (5) attempting eradication, in particular if the NIS exists in low population sizes with no nearby populations and near its environmental tolerance limits.
- We are not aware of any global extinctions driven by marine NIS but the precautionary conservation principle remains important because (1) any invasion causes negative impacts on some species, (2) global homogenisation increases, (3) negative effects on inconspicuous/rare species may go unreported (including local extinctions), (4) invasions co-occur with other human stressors and the combined effects may be dramatic and difficult to predict, (5) impacts are typically irreversible, and (6) effects may follow time lags, i.e. present-day small local effects may accumulate to dramatic effects over long and large spatio-temporal scales.

Acknowledgements

DRS gratefully acknowledges the continued support by the New Zealand Ministry of Science and Innovation and the National Institute of Water and Atmospheric Research (contract C01×0501). TW was supported by a Future Fellows grant from the Australian Research Council. MST was supported by the Marsden Fund of New Zealand.

References

Agosta, S. J. and Klemens, J. A. (2008). Ecological fitting by phenotypically flexible genotypes: implications for species associations, community assembly and evolution. *Ecology Letters*, 11, 1123–1134.

Airoldi, L. (2003). The effects of sedimentation on rocky coast assemblages. *Oceanography and Marine Biology: An Annual Review*, 41, 161–236.

Albins, M. A. (2013). Effects of invasive Pacific red lionfish *Pterois volitans* versus a native predator on Bahamian coral-reef fish communities. *Biological Invasions*, 15, 29–43.

Albins, M. A. and Hixon, M. A. (2008). Invasive Indo-Pacific lionfish *Pterois volitans* reduce recruitment of Atlantic coral-reef fishes. *Marine Ecology Progress Series*, 367, 233–238.

Alcoverro, T. and Mariani, S. (2004). Patterns of fish and sea urchin grazing on tropical Indo-Pacific seagrass beds. *Ecography*, 27, 361–365.

Almqvist, G., Strandmark, A. and Appelberg, M. (2010). Has the invasive round goby caused new links in Baltic food webs? *Environmental Biology of Fishes*, 89, 79–93.

Altieri, A., Trussell, G., Ewanchuck, P. and Bernatchez, G. (2009). Consumers control diversity and functioning of a natural marine ecosystem. *PLoS Biology*, 4, e5291.

Altieri, A. H., Silliman, B. and Bertness, M. D. (2007). Hierarchical organization via a facilitation cascade in intertidal cordgrass bed communities. *American Naturalist*, 169, 195–206.

Altug, G., Gurun, S., Cardak, M., Ciftci, P. S. and Kalkan, S. (2012). The occurrence of pathogenic bacteria in some ships' ballast water incoming from various marine regions to the Sea of Marmara, Turkey. *Marine Environmental Research*, 81, 35–42.

Baldwin, J. R. and Lovvorn, J. R. (1994). Expansion of seagrass habitat by the exotic *Zostera japonica*, and its use by dabbling ducks and brant in Boundary Bay, British Colombia. *Marine Ecology Progress Series*, 103, 119–127.

Barbier, E. B., Hacker, S. D., Kennedy, C. *et al.* (2011). The value of estuarine and coastal ecosystem services. *Ecological Monographs*, 81(2), 169–193.

Barbour, A. B., Allen, M. S., Frazer, T. K. and Sherman, K. D. (2011). Evaluating the potential efficacy of invasive lionfish (*Pterois volitans*) removals. *PLoS ONE*, 6, e19666.

Bauer, J. (2012). Invasive species: 'back-seat drivers' of ecosystem change? *Biological Invasions*, 14, 1295–1304.

Bell, S. S. (1991). Amphipods as insect equivalents? An alternative view. *Ecology*, 72, 350–354.

Berkenbusch, K. and Rowden, A. A. (2007). An examination of the spatial and temporal generality of the influence of ecosystem engineers on the composition of associated assemblages. *Aquatic Ecology*, 41, 129–147.

Berkenbusch, K., Rowden, A. A. and Myers, T. E. (2007). Interactions between seagrasses and burrowing ghost shrimps and their influence on infaunal assemblages. *Journal of Experimental Marine Biology and Ecology*, 341, 70–84.

Bilio, M. and Niermann, U. (2004). Is the comb jelly really to blame for it all? *Mnemiopsis leidyi* and the ecological concerns about the Caspian Sea. *Marine Ecology Progress Series*, 269, 173–183.

Blakeslee, A. M. H., Byers, J., Lesser, M. P. (2008). Solving cryptogenic histories using host and parasite molecular genetics: the resolution of *Littorina littorea*'s North American origin. *Molecular Ecology*, 17, 3684–3696.

Bollens, S., Cordell, J., Avent, S. and Hooff, R. (2002). Zooplankton invasions: a brief review, plus two case studies from the northeast Pacific Ocean. *Hydrobiologia*, 480, 87–110.

Borenstein, M., Hedges, L. V., Higgins, J. P. T. and Rothstein, H. R. (2009). *Introduction to Meta-analysis*. West Sussex: John Wiley and Sons Ltd.

Boudouresque, C. F., Lemée, R., Mari, X. and Meinesz, A. (1996). The invasive alga *Caulerpa taxifolia* is not a suitable diet for the sea urchin *Paracentrotus lividus*. *Aquatic Botany*, 53, 245–250.

Briggs, J. C. (2010). Marine biology: the role of accommodation in shaping marine biodiversity. *Marine Biology*, 157, 2117–2126.

Britton-Simmons, K. H. (2004). Direct and indirect effects of the introduced alga *Sargassum muticum* on benthic, subtidal communities of Washington State, USA. *Marine Ecology Progress Series*, 277, 61–78.

Bulleri, F., Airoldi, L., Branca, G. M. and Abbiati, M. (2006). Positive effects of the introduced green alga, *Codium fragile* ssp. *tomentosoides*, on recruitment and survival of mussels. *Marine Biology*, 148, 1213–1220.

Bulleri, F., Balata, D., Bertocci, I., Tamburello, L. and Benedetti-Cecchi, L. (2010). The seaweed *Caulerpa racemosa* on Mediterranean rocky reefs: from passenger to driver of ecological change. *Ecology*, 91, 2205–2212.

Bulleri, F. and Chapman, M. G. (2010). The introduction of coastal infrastructure as a driver of change in marine environments. *Journal of Applied Ecology*, 47, 26–35.

Buschbaum, C., Dittmann, S., Hong, J. S. *et al.* (2009). Mytilid mussels: global habitat engineers in coastal sediments. *Helgoland Marine Research*, 63, 47–58.

Butrón, A., Orive, E. and Madariaga, I. (2011). Potential risk of harmful algae transport by ballast waters: the case of Bilbao Harbour. *Marine Pollution Bulletin*, 62, 747–757.

Byers, J. (2000). Competition between two estuarine snails: implications for invasions of exotic species. *Ecology*, 81, 1225–1239.

Byers, J. E. (2002). Physical habitat attribute mediates biotic resistance to non-indigenous species invasions. *Oecologia*, 130, 146–156.

Byers, J. E. (2009). Competition in marine invasions. In *Biological Invasions in Marine Ecosystems: Ecological, Management, and Geographic Perspectives*, ed. G. Rilov and J. A. Crooks. Heidelberg, Germany: Springer-Verlag, pp. 245–260.

Byers, J. E., Reichard, S., Randall, J. M. *et al.* (2002). Directing research to reduce the impact of nonindigenous species. *Conservation Biology*, 16, 630–640.

Byers, J. E. Wright, J. T. and Gribben, P. E. (2010). Variable direct and indirect effects of a habitat-modifying invasive species on mortality of native fauna. *Ecology*, 91, 1787–1798.

Byers, J. E., Gribben, P., Yeager, C. and Sotka, E. (2012). Impacts of an abundant introduced ecosystem engineer within mudflats of the southeastern US coast. *Biological Invasions*, 14, 2587–2600.

Cacabelos, E., Engelen, A. H., Mejia, A. and Arenas, F. (2012). Comparison of the assemblage functioning of estuary systems dominated by the seagrass *Nanozostera noltii* versus the invasive drift seaweed *Gracilaria vermiculophylla*. *Journal of Sea Research*, 72, 99–105.

Cardinale, B. J., Matulich, K. L., Hooper, D. U. et al. (2011). The functional role of producer diversity in ecosystems. *American Journal of Botany*, 98, 572–592.

Cardinale, B. J., Srivastava, D. S., Duffy, J. E. et al. (2006). Effects of biodiversity on the functioning of trophic groups and ecosystems. *Nature*, 443, 989–992.

Carlton, J. T. (1979). History, biogeography, and ecology of the introduced marine and estuarine invertebrates of the Pacific coast of North America. PhD dissertation, University of California, Davis.

Carlton, J. T. (1996). Biological invasions and cryptogenic species. *Ecology*, 77, 1653–1655.

Casas, G., Scrosati, R. and Piriz, M. L. (2004). The invasive kelp *Undaria pinnatifida* (Phaeophyceae, Laminariales) reduces native seaweed diversity in Nuevo Gulf (Patagonia, Argentina). *Biological Invasions*, 6, 411–416.

Casini, M., Lövgren, J., Hjelm, J. et al. (2008). Multi-level trophic cascades in a heavily exploited open marine ecosystem. *Proceedings of the Royal Society B: Biological Sciences*, 275, 1793–1801.

Castilla, J. C., Lagos, N. A. and Cerda, M. (2004). Marine ecosystem engineering by the alien ascidian *Pyura praeputialis* on a mid-intertidal rocky shore. *Marine Ecology Progress Series*, 268, 119–130.

Catford, J. A., Jansson, R. and Nilsson, C. (2009). Reducing redundancy in invasion ecology by integrating hypothesis into a single theoretical framework. *Diversity and Distributions*, 15, 22–40.

Cebrian, E., Ballesteros, E., Linares, C. and Tomas, F. (2011). Do native herbivores provide resistance to Mediterranean marine bioinvasions? A seaweed example. *Biological Invasions*, 13, 1397–1408.

Cebrian, E., Linares, C., Marschal, C. and Garrabou, J. (2012). Exploring the effects of invasive algae on the persistence of gorgonian populations. *Biological Invasions*, 14, 2647–2656.

Ceccherelli, G. and Cinelli, F. (1997). Short-term effects of nutrient enrichment of the sediment and interactions between the seagrass *Cymodocea nodosa* and the introduced green alga *Caulerpa taxifolia* in a Mediterranean bay. *Journal of Experimental Marine Biology and Ecology*, 217, 165–177.

Ceccherelli, G. and Sechi, N. (2002). Nutrient availability in the sediment and the reciprocal effects between the native seagrass *Cymodocea nodosa* and the introduced green alga *Caulerpa taxifolia* in a Mediterranean bay. *Hydrobiologia*, 474, 57–66.

Chauvenet, A. L. M., Ewen, J. G., Armstrong, D. P., Blackburn, T. M. and Pettorelli, N. (2012). Maximizing the success of assisted colonizations. *Animal Conservation*, 16, 161–169.

Cheung, W. W. L., Lam, V. W. Y., Sarmiento, J. L. et al. (2009). Projecting global marine biodiversity impacts under climate change scenarios. *Fish and Fisheries*, 10. 235–251.

Connell, S. D. (2001). Urban structures as marine habitats: an experimental comparison of the composition and abundance of subtidal epibiota among pilings, pontoons and rocky reefs. *Marine Environmental Research*, 52, 115–125.

Cote, I. and Green, S. J. (2012). Potential effects of climate change on a marine invasion: the importance of current context. *Current Zoology*, 58, 1–8.

Cowan, D. A., Rybicki, E. P., Tuffin, M. I., Valverde, A. and Wingfield, M. J. (2013). Biodiversity: so much more than legs and leaves. *South African Journal of Science*, 109, 1–9.
Crain, C. M., Kroeker, K. and Halpern, B. S. (2008). Interactive and cumulative effects of multiple human stressors in marine systems. *Ecology Letters*, 11, 1304–1315.
Critchley, A. T., De Visscher, P. R. M. and Nienhuis, P. H. (1990). Canopy characteristics of the brown alga *Sargassum muticum* (Fucales, Phaeophyta) in Lake Greveling, southwest Netherlands. *Hydrobiologia*, 204/205, 211–217.
Critchley, A. T., Farnham, W. F. and Morrell, S. L. (1986). An account of the attempted control of an introduced marine alga *Sargassum muticum* in southern England UK. *Biological Conservation*, 35, 313–332.
Crooks, J. A. (1998). Habitat alteration and community-level effects of an exotic mussel, *Musculista senhousia*. *Marine Ecology Progress Series*, 162, 137–152.
Crooks, J. A. (2002a). Characterizing ecosystem-level consequences of biological invasions: the role of ecosystem engineers. *Oikos*, 97, 153–166.
Crooks, J. A. (2002b). Predators of the invasive mussel *Musculista senhousia* (Mollusca: Mytilidae). *Pacific Science*, 56, 49–56.
Crooks, J. A. (2009). The role of exotic marine ecosystem engineers. In *Biological Invasions in Marine Ecosystems: Ecological, Management, and Geographic Perspectives*, ed. G. Rilov and J. A. Crooks. Heidelberg, Germany: Springer-Verlag, pp. 287–304.
Crooks, J. A. and Khim, H. S. (1999). Architectural vs. biological effects of a habitat-altering, exotic mussel, *Musculista senhousia*. *Journal of Experimental Marine Biology and Ecology*, 240, 53–75.
Culver, C. S. and Kuris, M. A. (2000). The apparent eradication of a locally established introduced marine pest. *Biological Invasions*, 2, 245–253.
Dafforn, K. A., Glasby, T. M. and Johnston, E. L. (2009). Links between estuarine condition and spatial distributions of marine invaders. *Diversity and Distributions*, 15, 807–821.
Danovaro, R. (2003). Pollution threats in the Mediterranean Sea: an overview. *Chemistry and Ecology*, 19, 15–32.
Darling, E. S. and Cote, E. S. (2008). Quantifying the evidence for ecological synergies. *Ecology Letters*, 11, 1278–1286.
Daskalov, G. M., Grishin, A. N., Rodionov, S. and Mihneva, V. (2007). Trophic cascades triggered by overfishing reveal possible mechanisms of ecosystem regime shifts. *Proceedings of the National Academy of Sciences*, 104, 10518–10523.
Davidson, T., Shanks, A. and Rumrill, S. (2010). The composition and density of fauna utilizing burrow microhabitats created by a non-native burrowing crustacean (*Sphaeroma quoianum*). *Biological Invasions*, 12, 1403–1413.
Davidson, T. M. and de Rivera, C. E. (2010). Accelerated erosion of saltmarshes infested by the non-native burrowing crustacean *Sphaeroma quoianum*. *Marine Ecology Progress Series*, 419, 129–136.
Davies, B., Stuart, V. and De Villiers, M. (1989). The filtration activity of a serpulid polychaete population (*Ficopomatus enigmaticus* (Fauvel)) and its effects on water quality in a coastal marina. *Estuarine, Coastal and Shelf Science*, 29, 613–620.

Davis, M. A., Grime, J. P. and Thompsen, K. (2000). Fluctuating resources in plant communities: a general theory of invasibility. *Journal of Ecology*, 88, 528–534.

Davis, R. C., Short, F. T. and Burdick, D. M. (1998). Quantifying the effects of green crab damage to eelgrass transplants. *Restoration Ecology*, 6, 297–302.

DeGraaf, J. D and Tyrrell, M. C. (2004). Comparison of the feeding rates of two introduced crab species, *Carcinus maenas* and *Hemigrapsus sanguineus*, on the blue mussel, *Mytilus edulis*. *Northeastern Naturalist*, 11, 163–166.

Demopoulos, A. W. J. and Smith, C. R. (2010) Invasive mangroves alter macrofaunal community structure and facilitate opportunistic exotics. *Marine Ecology Progress Series*, 404, 51–67.

DeRivera, C. E., Ruiz, G. M., Hines, A. H. and Jivoff, P. (2005). Biotic resistance to invasion: native predator limits abundance and distribution of an introduced crab. *Ecology*, 86, 3364–3376.

de Villele, X. and Verlaque, M. (1995). Changes and degradation in a Posidonia oceanica bed invaded by the introduced tropical alga *Caulerpa taxifolia* in the North Western Mediterranean. *Botanica Marina*, 38, 79–87.

De Wit, R. and Bouvier, T. (2006). 'Everything is everywhere, but, the environment selects'; what did Baas, Becking and Beijerinck really say? *Environmental Microbiology*, 8, 755–758.

Díaz-Almela, E., Marba, N. and Duarte, C. M. (2007). Consequences of Mediterranean warming events in seagrass (*Posidonia oceanica*) flowering records. *Global Change Biology*, 13, 224–235.

Drake, L. A., Doblin, M. A and Dobbs, F. C. (2007). Potential microbial bioinvasions via ships' ballast water, sediment, and biofilm. *Marine Pollution Bulletin*, 55, 333–341.

Drake, L. A., Meyer, A. E., Forsberg, R. L. *et al.* (2005). Potential invasion of microorganisms and pathogens via 'interior hull fouling': biofilms inside ballast water tanks. *Biological Invasions*, 7, 969–982.

Drake, L. A., Ruiz, G. M., Galil, B. S. *et al.* (2002). Microbial ecology of ballast water during a transoceanic voyage and the effects of open-ocean exchange. *Marine Ecology Progress Series*, 233, 13–20.

Eastwood, M. M., Donahue, M. J. and Fowler, A. E. (2007). Reconstructing past biological invasions: niche shifts in response to invasive predators and competitors. *Biological Invasions*, 9, 397–407.

Ekloef, J. S., Henriksson, R. and Kautsky, N. (2006). Effects of tropical open-water seaweed farming on seagrass ecosystem structure and function. *Marine Ecology Progress Series*, 325, 73–84.

Elton, C. S. (1958). *The Ecology of Invasions by Animals and Plants*. London: Mathuess.

Emami, K., Askari, V., Ullrich, M. *et al.* (2012). Characterization of bacteria in ballast water using MALDI-TOF mass spectrometry. *PLoS ONE*, 7:e38515.

Engelen, A., Henriques, N., Monteiro, C. and Santos, R. (2011). Mesograzers prefer mostly native seaweeds over the invasive brown seaweed *Sargassum muticum*. *Hydrobiologia*, 669, 157–165.

Estes, J. A. and Palmisano, J. F. (1974). Sea otters: their role in structuring nearshore communities. *Science*, 185, 1058–1060.

Fabricius, K. E. (2005). Effects of terrestrial runoff on the ecology of corals and coral reefs: review and synthesis. *Marine Pollution Bulletin*, 50, 125–146.

Fridley, J. D., Stachowicz, J. J., Sax, D. F. et al. (2007). The invasion paradox: reconciling pattern and process in species invasions. *Ecology*, 88, 3–17.

Gennaro, P. and Piazzi, L. (2011). Synergism between two anthropic impacts: *Caulerpa racemosa* var. cylindracea invasion and seawater nutrient enrichment. *Marine Ecology Progress Series*, 427, 59–70.

Gollan, J. R. and Wright, J. T. (2006). Limited grazing pressure by native herbivores on the invasive seaweed *Caulerpa taxifolia* in a temperate Australian estuary. *Marine and Freshwater Research*, 57, 685–694.

Green, D. S., Boots, B. and Crowe, T. P. (2012). Effects of non-indigenous oysters on microbial diversity and ecosystem functioning. *PLoS ONE*, 7, e48410.

Green, D., Rocha, C. and Crowe, T. (2013). Effects of non-indigenous oysters on ecosystem processes vary with abundance and context. *Ecosystems*, 16, 881–893.

Gribben, P. E., Byers, J., Clements, M. et al. (2009a). Behavioural interactions between ecosystem engineers control community species richness. *Ecology Letters*, 12, 1127–1136.

Gribben, P. E. and Wright, J. T. (2006). Sublethal effects on reproduction in native fauna: are females more vulnerable to biological invasion? *Oecologia*, 149, 352–361.

Gribben, P. E., Wright, J. T., O'Connor, W. A. et al. (2009b). Reduced performance of native infauna following recruitment to a habitat-forming invasive marine alga. *Oecologia*, 158, 733–745.

Grosholz, E. D. and Ruiz, G. (2009). Multitrophic effects of invasion in marine and estuarine systems. In *Biological Invasions in Marine Ecosystems: Ecological, Management, and Geographic Perspectives*, ed. G. Rilov and J. A. Crooks. Heidelberg, Germany: Springer-Verlag, pp. 305–324.

Grosholz, E. D. and Ruiz, G. M. (1996). Predicting the impact of introduced marine species: lessons from the multiple invasions of the European green crab *Carcinus maenas*. *Biological Conservation*, 78, 59–66.

Guerra-García, J. M., Ros, M., Izquierdo, D. and Soler-Hurtado, M. M. (2012). The invasive *Asparagopsis armata* versus the native *Corallina elongata*: differences in associated peracarid assemblages. *Journal of Experimental Marine Biology and Ecology*, 416/417, 121–128.

Hallegraeff, G. M. (1993). A review of harmful algal blooms and their apparent global increase. *Phycologia*, 32, 79–99.

Hallegraeff, G. M. (2010). Ocean climate change, phytoplankton community responses and harmful algal blooms: a formidable predictive challenge. *Journal of Phycology*, 46, 220–235.

Hairston, N. G., Smith, F. E. and Slobodkin, L. S. (1960). Community structure, population control, and competition. *American Naturalist*, 94, 421–425.

Harris, L. G. and Jones, A. C. (2005). Temperature, herbivory and epibiont acquisition as factors controlling the distribution and ecological role of an invasive seaweed. *Biological Invasions*, 7, 913–924.

Hewitt, C. L. and Campbell, M. (2010). The relative contribution of vectors to the introduction and translocation of marine invasive species. Australian Department of Agriculture, Fisheries and Forestry, Canberra, 56.

Hewitt, C. L., Campbell, M. L., McEnnulty, F. *et al.* (2005). Efficacy of physical removal of a marine pest: the introduced kelp *Undaria pinnatifida* in a Tasmanian Marine Reserve. *Biological Invasions*, 7, 251–263.

Hewitt, C. L., Campbell, M. L., Thresher, R.E. *et al.* (2004). Introduced and cryptogenic species in Port Phillip Bay, Victoria, Australia. *Marine Biology*, 144, 183–202.

Hierro, J. L., Maron, J. L. and Callaway, R. M. (2005). A biogeographical approach to plant invasions: the importance of studying exotics in their introduced and native range. *Journal of Ecology*, 93, 5–15.

Hoeffle, H., Thomsen, M. S. and Holmer, M. (2011). High mortality of *Zostera marina* under high temperature regimes but minor effects of the invasive macroalgae *Gracilaria vermiculophylla*. *Estuarine Coastal Shelf Science*, 92, 35–46.

Hoeffle, H., Wernberg, T., Thomsen, M. S. and Holmer, M. (2012). Drift algae, an invasive snail and elevated temperature reduce the ecological performance of a warm-temperate seagrass via additive effects. *Marine Ecology Progress Series*, 450, 67–80.

Hooper, D. U., Adair, E. C., Cardinale, B. J. *et al.* (2012). A global synthesis reveals biodiversity loss as a major driver of ecosystem change. *Nature*, 486, 105–108.

Hooper, D. U., Chapin, F. S., Ewel, J. J. *et al.* (2005). Effects of biodiversity on ecosystem functioning: a consensus of current knowledge. *Ecological Monographs*, 75, 3–35.

Hopkins, G. A., Forrest, B. M., Jiang, W., Gardner, J. P. A. (2011). Successful eradication of a non-indigenous marine bivalve from a subtidal soft-sediment environment. *Journal of Applied Ecology*, 48, 424–431.

Hunt, L. (2009). Results of an attempt to control and eradicate *Undaria pinnatifida* in Southland, New Zealand, April 1997-November 2004. New Zealand. Dept. of Conservation, Department of Conservation, 48.

Huston, M. A. (1994). *Biological Diversity: The Coexistence of Species on Changing Landscapes*. Cambridge: Cambridge University Press.

Ibrahim, A. M. and El-Naggar, M. M. (2012). Ballast water review: impacts, treatments and management. *Middle-East Journal of Scientific Research*, 12, 976–984.

Janiak, D. S. and Whitlatch, R. B. (2012). Epifaunal and algal assemblages associated with the native *Chondrus crispus* (Stackhouse) and the non-native *Grateloupia turuturu* (Yamada) in eastern Long Island Sound. *Journal of Experimental Marine Biology and Ecology*, 413, 38–44.

Jaubert, J. M., Chisholm, J. R. M., Ducrot, D. *et al.* (1999). No deleterious alterations in *Posidonia* beds in the Bay of Menton (France) eight years after *Caulerpa taxifolia* colonization. *Journal of Phycology*, 35, 1113–1119.

Jaubert, J. M., Chisholm, J. R. M., Minghelli-Roman, A. *et al.* (2003). Re-evaluation of the extent of *Caulerpa taxifolia* development in the northern Mediterranean using airborne spectrographic sensing. *Marine Ecology Progress Series*, 263, 75–82.

Kideys, A. E. (2002). The comb jelly *Mnemiopsis leidyi* in the Black Sea. In *Invasive Aquatic Species of Europe. Distribution, Impact and Management*, ed. E. Leppakoski S. Gollasch and S. Olenin. Dordrecht, The Netherlands: Kluwer Academic Publisher, pp. 56–61.

Klein, J. C. and Verlaque, M. (2011). Experimental removal of the invasive *Caulerpa racemosa* triggers partial assemblage recovery. *Journal of the Marine Biological Association of the UK*, 91, 117–125.

Kochmann, J., Buschbaum, C., Volkenborn, N. and Reise, K. (2008). Shift from native mussels to alien oysters: differential effects of ecosystem engineers. *Journal of Experimental Marine Biology and Ecology*, 364, 1–10.

Kornis, M. S., Mercado-Silva, N. and Vander Zanden, M. J. (2012). Twenty years of invasion: a review of round goby *Neogobius melanostomus* biology, spread and ecological implications. *Journal of Fish Biology*, 80, 235–285.

Kremer, L. P. and Rocha, R. M. (2011). The role of *Didemnum perlucidum* F. Monniot, 1983 (Tunicata, Ascidiacea) in a marine fouling community. *Aquatic Invasions*, 6, 441–449.

Kristensen, E., Hansen, T., Delefosse, M., Banta, G. T. and Quintana, C. O. (2011). Contrasting effects of the polychaetes *Marenzelleria viridis* and *Nereis diversicolor* on benthic metabolism and solute transport in sandy coastal sediment. *Marine Ecology Progress Series*, 425, 125–139.

Kulhanek, S. A., Ricciardi, A. and Leung, B. (2011). Is invasion history a useful tool for predicting the impacts of the world's worst aquatic invasive species? *Ecological Applications*, 21. 189–202.

Kuussaari, M., Bommarco R. K. R, Heikkinen R. K. et al. (2009). Extinction debt: a challenge for biodiversity conservation. *Trends in Ecology and Evolution*, 24, 564–571.

Landsberg, J. H. (2002). The effects of harmful algal blooms on aquatic organisms. *Reviews in Fisheries Science*, 10, 113–390.

Lang, A. C. and Buschbaum, C. (2010). Facilitative effects of introduced Pacific oysters on native macroalgae are limited by a secondary invader, the seaweed *Sargassum muticum*. *Journal of Sea Research*, 63, 119–128.

Lejeusne, C., Chevaldonné, P., Pergent-Martini, C., Boudouresque, C. F. and Pérez, T. (2010). Climate change effects on a miniature ocean: the highly diverse, highly impacted Mediterranean Sea. *Trends in Ecology and Evolution*, 25, 250–260.

Lesser, M. and Slattery, M. (2011). Phase shift to algal dominated communities at mesophotic depths associated with lionfish (*Pterois volitans*) invasion on a Bahamian coral reef. *Biological Invasions*, 13, 1855–1868.

Levin, P. S., Coyer, J. A., Petrik, R. and Good, T. P. (2002). Community-wide effects of nonindigenous species on temperate rocky reefs. *Ecology*, 83, 3182–3193.

Levine, J. (1999). Indirect facilitation: evidence and predictions from a riparian community. *Ecology*, 80, 1762–1769.

Levine, J. M., Adler, P. B. and Yelenik, S. G. (2004). A meta-analysis of biotic resistance to exotic plant invasions. *Ecology Letters*, 7, 975–989.

Levine, J. M., D'Antonio, C. M., Dukes, J. S. et al. (2003). Mechanisms underlying the impacts of exotic plant invasions. *Proceedings of the Royal Society of London Series B: Biological Sciences*, 270, 775–781.

Linares, C., Cebrian, E. and Coma, R. (2012). Effects of turf algae on recruitment and juvenile survival of gorgonian corals. *Marine Ecology Progress Series*, 452, 81–88.

Litchman, E. (2010). Invisible invaders: non-pathogenic invasive microbes in aquatic and terrestrial ecosystems. *Ecology Letters*, 13, 1560–1572.

Lotze, H. K., Lenihan, H. S., Bourque, B. J. et al. (2006). Depletion, degradation, and recovery potential of estuaries and coastal seas. *Science*, 312, 1806–1809.

Lubchenco, J. (1978). Plant species diversity in a marine intertidal community: importance of herbivore food preference and algal competitive abilities. *The American Naturalist*, 112, 23–39.

Lubchenco, J. (1983). *Littorina* and *Fucus*: effects of herbivores, substratum heterogeneity, and plant escapes during succession. *Ecology*, 64, 1116–1123.

Lubchenco, J. and Menge, B. A. (1978). Community development and persistence in a low rocky intertidal zone. *Ecological Monographs*, 59, 67–94.

MacDonald, J. A., Roudez, R., Glover, T. and Weis, J. S. (2007). The invasive green crab and Japanese shore crab: behavioral interactions with a native crab species, the blue crab. *Biological Invasions*, 9, 837–848.

MacDougall, A. S. and Turkington, R. (2005). Are invasive species the drivers or passengers of change in degraded ecosystems? *Ecology*, 86. 42–55.

Maljković, A., Leeuwen, T. E. and Cove, S. N. (2008). Predation on the invasive red lionfish, *Pterois volitans* (Pisces: Scorpaenidae), by native groupers in the Bahamas. *Coral Reefs*, 27, 501.

Markert, A., Wehrmann, A. and Kroncke, I. (2010). Recently established *Crassostrea*-reefs versus native *Mytilus*-beds: differences in ecosystem engineering affects the macrofaunal communities (Wadden Sea of Lower Saxony, southern German Bight). *Biological Invasions*, 12, 15–32.

McCarthy, S. A. and Khambaty, F. M. (1994). International dissemination of epidemic *Vibrio cholerae* by cargo ship ballast and other nonpotable waters. *Applied and Environmental Microbiology*, 60, 2597–2601.

McClanahan, T. R., Kamukuru, A. T., Muthiga, N. A., Yebio, M. and Obura, D. (1996). Effect of sea urchin reductions on algae, coral, and fish populations. *Conservation Biology*, 10, 136–154.

McKinnon, J. G., Gribben, P. E., Davis, A. R., Jolley, D. F. and Wright, J. T. (2009). Differences in soft-sediment macrobenthic assemblages invaded by *Caulerpa taxifolia* compared to uninvaded habitats. *Marine Ecology Progress Series*, 380, 59–71.

Meinesz, A. (1999). *Killer algae: the true tale of a biological invasion*. Chicago, IL: University of Chicago Press.

Miller, A. W., Chang, A. L., Cosentino-Manning, N. and Ruiz, G. M. (2004). New record and eradication of the northern Atlantic alga *Ascophyllum nodosum* (Phaeophyceae) from San Francisco Bay, California, USA. *Journal of Phycology*, 40, 1028–1031.

Monteiro, C., Engelen, A. H. and Santos, R. O. (2009). Macro- and mesoherbivores prefer native seaweeds over the invasive brown seaweed *Sargassum muticum*: a potential regulating role on invasions. *Marine Biology*, 156, 2505–2515.

Morris, T. L. (2013). Evaluation of ships' ballast water as a vector for transfer of pathogenic bacteria to marine protected areas in the Gulf of Mexico. Master's thesis, Texas A&M University.

Neira, C., Grozholz, E. D., Levin, L. A. and Blake, R. (2006). Mechanisms generating modification of benthos following tidal flat invasion by a Spartina hybrid. *Ecological Applications*, 16, 1391–1404.

Neira, C., Levin, L. A., Grosholz, E. D. and Mendoza, G. (2007). Influence of invasive *Spartina* growth stages on associated macrofaunal communities. *Biological Invasions*, 9, 975–993.

Nejrup, L., Pedersen, M. and Vinzent, J. (2012). Grazer avoidance may explain the invasiveness of the red alga *Gracilaria vermiculophylla* in Scandinavian waters. *Marine Biology*, 159, 1703–1712.

Nejrup, L. B. and Pedersen, M. F. (2010). Growth and biomass development of the introduced red alga *Gracilaria vermiculophylla* is unaffected by nutrient limitation and grazing. *Aquatic Biology*, 10, 249–259.

Neverauskas, V. (2005). Eradication of Caulerpa taxifolia from West Lakes, South Australia, using urban stormwater. *Abstracts, Conference Programme, 4th International Conference on Marine Bioinvasion*, Wellington, New Zealand 22–26 August, 151.

Norkko, J., Reed, D. C., Timmermann, K. et al. (2012). A welcome can of worms? Hypoxia mitigation by an invasive species. *Global Change Biology*, 18, 422–434.

Nyberg, C. D., Thomsen, M. S. and Wallentinus, I. (2009). Flora and fauna associated with the introduced red alga *Gracilaria vermiculophylla*. *European Journal of Phycology*, 44, 395–403.

Nylund, G. M., Weinberger, F., Rempt, M. and Pohnert, G. (2011). Metabolomic assessment of induced and activated chemical defence in the invasive red alga *Gracilaria vermiculophylla*. *PLoS ONE*, 6, e29359.

O'Leary, J. K. and McClanahan, T. R. (2010). Trophic cascades result in large-scale coralline algae loss through differential grazer effects. *Ecology*, 91, 3584–3597.

Oguz, T., Fach, B. and Salihoglu, B. (2008). Invasion dynamics of the alien ctenophore *Mnemiopsis leidyi* and its impact on anchovy collapse in the Black Sea. *Journal of Plankton Research*, 30, 1385–1397.

Olabarria, C., Rodil, I. F., Incera, M. and Troncoso, J. S. (2009). Limited impact of *Sargassum muticum* on native algal assemblages from rocky intertidal shores. *Marine Environmental Research*, 67, 153–158.

Olden, J. D. and Rooney. T. P. (2006). On defining and quantifying biotic homogenization. *Global Ecology and Biogeography*, 15, 113–120.

Ostenfeld, C. H. (1908). On the immigration of *Biddulphia sinensis* Grev. and its occurrence in the North Sea during 1903–1907 and on its use for the study of the direction and rate of flow of the currents. *Meddelelser fra Kommissionen for Danmarks Fiskeri- og Havundersøgelser: Serie Plankton*, 6, 1–44.

Pacciardi, L., de Biasi, N. M. and Piazzi, L. (2011). Effects of *Caulerpa racemosa* invasion on soft-bottom assemblages in the Western Mediterranean Sea. *Biological Invasions*, 13, 2677–2690.

Padilla, D. K. (2010). Context-dependent impacts of non-native ecosystem engineers, the Pacific oyster *Crassostrea gigas*. *Integrative and Comparative Biology*, 50, 213–225.

Paine, R. T. (1966). Food web complexity and species diversity. *American Naturalist*, 100, 65–75.
Parker, I. M., Simberloff, D., Lonsdale, W. M. *et al.* (1999). Impact: toward a framework for understanding the ecological effects of invaders. *Biological Invasions*, 1, 3–19.
Parker, J. D. and Hay, M. E. (2005). Biotic resistance to plant invasions? Native herbivores prefer non-native plants. *Ecological Letters*, 8, 959–967.
Parks, J. R. (2006). Shorebird use of smooth cordgrass (*Spartina alterniflora*) meadows in Willapa Bay, Washington. MSc thesis, Environmental Studies, The Evergreen State College.
Pedersen, M. F., Stæhr, P. A., Wernberg, T. and Thomsen, M. (2005). Biomass dynamics of exotic *Sargassum muticum* and native *Halidrys siliquosa* in Limfjorden, Denmark: implications of species replacements on turnover rates. *Aquatic Botany*, 83, 31–47.
Petchey, O. L., McPhearson, P. T., Casey, T. M. and Morin, P. J. (1999). Environmental warming alters food-web structure and ecosystem function. *Nature*, 402, 69–72.
Petersen, J. K., Hansen, J. W., Laursen, M. B. *et al.* (2008). Regime shift in a coastal marine ecosystem. *Ecological Applications*, 18, 497–510.
Piazzi, L., Balata, D., Ceccherelli, G. and Cinellia, F. (2005). Interactive effect of sedimentation and *Caulerpa racemosa* var. *cylindracea* invasion on macroalgal assemblages in the Mediterranean Sea. *Estuarine, Coastal and Shelf Science*, 64, 467–474.
Piazzi, L. and Ceccherelli, G. (2006). Persistence of biological invasion effects: recovery of macroalgal assemblages after removal of *Caulerpa racemosa* var. *cylindracea*. *Estuarine Coastal and Shelf Science*, 68, 455–461.
Piola, R. F. and Johnston, E. L. (2006). Differential resistance to extended copper exposure in four introduced bryozoans. *Marine Ecology Progress Series*, 311, 103–114.
Piola, R. F. and Johnston, E. L. (2008). Pollution reduces native diversity and increases invader dominance in marine hard-substrate communities. *Diversity and Distributions*, 14, 329–342.
Posey, M. H. (1988). Community changes associated with the spread of an introduced seagrass, *Zostera japonica*. *Ecology*, 69, 974–983.
Powell, K. I., Chase, J. M. and Knight, T. M. (2011). A synthesis of plant invasion effects on biodiversity across spatial scales. *American Journal of Botany*, 98, 539–548.
Pyšek, P., Jarošík, V., Hulme, P. J. *et al.* (2012). A global assessment of invasive plant impacts on resident species, communities and ecosystems: the interaction of impact measures, invading species' traits and environment. *Global Change Biology*, 18, 1725–1737.
Raitsos, D. E., Beaugrand, G., Georgopoulos, D. *et al.* (2010). Global climate change amplifies the entry of tropical species into the eastern Mediterranean Sea. *Limnology and Oceanography*, 55, 1478–1484.
Read, G. B., Inglis, G., Stratford, P. and Ahyong, S. T. (2011). Arrival of the alien fanworm *Sabella spallanzanii* (Gmelin, 1791) (Polychaeta: Sabellidae) in two New Zealand harbours. *Aquaculture*, 6, 273–279.

Reusch, T. B. H. and Williams, S. L. (1998). Variable response of native eelgrass *Zostera marina* to a non-indigenous bivalve *Musculista senhousia*. *Oecologia*, 113, 428–441.
Reynolds, L. K., Carr, L. A. and Boyer, K. E. (2012). A non-native amphipod consumes eelgrass inflorescences in San Francisco Bay. *Marine Ecology Progress Series*, 451, 107–118.
Ricciardi, A. and Atkinson, S. K. (2004). Distinctiveness magnifies the impact of biological invaders in aquatic ecosystems. *Ecology Letters*, 7, 781–784.
Ricciardi, A. and Simberloff, D. (2009). Assisted colonization is not a viable conservation strategy. *Trends in Ecology and Evolution*, 24, 248–253.
Richardson, D. M., Pysek, P., Rejmánek, M. et al. (2000). Naturalization and invasion of plants: concepts and definitions. *Diversity and Distributions*, 6, 93–107.
Rilov, G. (2009). Predator-prey interactions of marine invaders. In *Biological Invasions in Marine Ecosystems: Ecological, Management, and Geographic Perspectives*, ed. G. Rilov and J. A. Crooks. Heidelberg, Germany: Springer-Verlag, pp. 261–285.
Rodriguez, L. F. (2006). Can invasive species facilitate native species? Evidence of how, when, and why these impacts occur. *Biological Invasions*, 8, 927–939.
Roscher, C., Temperton, V. M., Scherer-Lorenzen, M. et al. (2005) Overyielding in experimental grassland communities – irrespective of species pool or spatial scale. *Ecology Letters*, 8, 419–429.
Rosenberg, M. S., Adams, D. C. and Gurevitch, J. (2000). *Metawin: Statistical Software for Meta-analysis*. Sunderland, MA: Sinauer Associates.
Rossong, M. A., Williams, P. J., Comeau, M., Mitchell, S. C, and Apaloo, J. (2006). Agonistic interactions between the invasive green crab, *Carcinus maenas* (Linnaeus) and juvenile American lobster, *Homarus americanus* (Milne Edwards). *Journal of Experimental Marine Biology and Ecology*, 329, 281–288.
Rudnick, D. A., Chan, V. and Resh, V. H. (2005). Morphology and impacts of the burrows of the Chinese mitten crab, *Eriocheir sinensis* H. Milne Edwards (Decapoda, Grapsoidea), in south San Francisco Bay, California, USA. *Crustaceana*, 78, 787–807.
Ruesink, J., Hong, J.-S., Wisehart, L. et al. (2010). Congener comparison of native (*Zostera marina*) and introduced (*Z. japonica*) eelgrass at multiple scales within a Pacific Northwest estuary. *Biological Invasions*, 12, 1773–1789.
Ruesink, J. L. (2007). Biotic resistance and facilitation of a non-native oyster on rocky shores. *Marine Ecology Progress Series*, 331, 1–9.
Ruesink, J. L., Feist, B. E., Harvey, C. J. et al. (2006). Changes in productivity associated with four introduced species: ecosystem transformation of a 'pristine' estuary. *Marine Ecology Progress Series*, 311, 203–215.
Ruiz, G. M., Rawlings, T. K., Dobbs, F. C. et al. (2000). Global spread of microorganisms by ships. *Nature*, 408, 49–50.
Ruiz, G. M., Fofonoff, P. W., Steves, B., Foss, S. F. and Shiba, S. N. (2011). Marine invasion history and vector analysis of California: a hotspot for western North America. *Diversity and Distributions*, 17, 362–373.
Sánchez, I. and Fernández, C. (2005). Impact of the invasive seaweed *Sargassum muticum* (Phaeophyta) on an intertidal macroalgal assemblage. *Journal of Phycology*, 41, 923–930.

Sand-Jensen, K. and Borum, J. (1991). Interactions among phytoplankton, periphyton, and macrophytes in temperate freshwaters and estuaries. *Aquatic Botany*, 41, 137–175.

Sandler, R. (2010). The value of species and the ethical foundations of assisted colonization. *Conservation Biology*, 24, 424–431.

Saunders, M. and Metaxas, A. (2008). High recruitment of the introduced bryozoan *Membranipora membranacea* is associated with kelp bed defoliation in Nova Scotia, Canada. *Marine Ecology Progress Series*, 369, 139–151.

Schaffelke, B. and Hewitt, C. L. (2007). Impacts of introduced seaweeds. *Botanica Marina*, 50. 397–417.

Scheibling, R. E. and Anthony, S. X. (2001). Feeding, growth and reproduction of sea urchins (*Strongylocentrotus droebachiensis*) on single and mixed diets of kelp (*Laminaria* spp.) and the invasive alga *Codium fragile* spp. *tometosoides*. *Marine Biology*, 139, 139–146.

Scheibling, R. E., Lyons, D. A. and Sumi, C. B. T. (2008). Grazing of the invasive alga *Codium fragile* ssp. *tomentosoides* by the common periwinkle *Littorina littorea*: effects of thallus size, age and condition. *Journal of Experimental Marine Biology and Ecology*, 355, 103–113.

Scherer-Lorenzen, M. (2005). Biodiversity and ecosystem functioning: basic principles. Biodiversity: structure and function. *Encyclopedia of Life Support Systems (EOLSS)*, developed under the Auspices of the UNESCO. Oxford: Eolss Publishers. Available at: http://www.eolss.net.

Schlaepfer, M. A., Sax, D. F. and Olden, J. D. (2011). The potential conservation value of non-native species. *Conservation Biology*, 25, 428–437.

Schmidt, A. L. and Scheibling, R. E. (2006). A comparison of epifauna and epiphytes on native kelps (*Laminaria species*) and an invasive alga (*Codium fragile* ssp. *tomentosoides*) in Nova Scotia, Canada *Botanica Marina*, 49, 315–330.

Schmitz, O. J., Post, E., Burns, C. E. and Johnston, K. M. (2003). Ecosystem responses to global climate change: moving beyond color mapping. *Bioscience*, 53, 1199–1205.

Schulze, E. D. (1994). *Biodiversity and Ecosystem Function*. Heidelberg, Germany: Springer.

Seddon, P. J. (2010). From reintroduction to assisted colonization: moving along the conservation translocation spectrum. *Restoration Ecology*, 18, 796–802.

Seddon, P. J., Price, M. S., Launay, F. et al. (2011). Frankenstein ecosystems and 21st century conservation agendas: reply to Oliveira-Santos and Fernandez. *Conservation Biology*, 25, 212–212.

Shurin, J. B., Borer, E. T., Seabloom, E. W. et al. (2005). A cross-ecosystem comparison of the strength of trophic cascades. *Ecology Letters*, 5, 785–791.

Simberloff, D. (2011). How common are invasion-induced ecosystem impacts? *Biological Invasions*, 13, 1255–1268.

Simberloff, D. and Von Holle, B. (1999). Positive interactions of nonindigenous species: invasional meltdown? *Biological Invasions*, 1, 21–32.

Sjotun, K., Eggereide, S. F. and Hoisaeter, T. (2007). Grazer-controlled recruitment of the introduced *Sargassum muticum* (Phaeophycae, Fucales) in northern Europe. *Marine Ecology Progress Series*, 342, 127–138.

Sorte, C. J. B., Ibáñez, I., Blumenthal, D. M. et al. (2012). Poised to prosper? A cross-system comparison of climate change effects on native and non-native species performance. *Ecology Letters*, 16, 261–270.

Sousa, R., Gutierrez, J. L. and Aldridge, D. C. (2009). Non-indigenous invasive bivalves as ecosystem engineers. *Biological Invasions*, 11, 2367–2385.

Stachowicz, J. J. and Byrnes, J. E. (2006). Species diversity, invasion success, and ecosystem functioning: disentangling the influence of resource competition, facilitation, and extrinsic factors. *Marine Ecology Progress Series*, 311, 251–262.

Stachowicz, J. J., Whitlatch, R. B. and Osman, R. W. (1999). Species diversity and invasion resistance in a marine ecosystem. *Science*, 286, 1577–1579.

Stachowicz, J. J., Fried, H., Osman, R. W. and Whitlatch, R. B. (2002a). Biodiversity, invasion resistance, and marine ecosystem function: reconciling pattern and process. *Ecology*, 83, 2575–2590.

Stachowicz, J. J., Terwin, J. R., Whitlatch, R. B. and Osman, R. W. (2002b). Linking climate change and biological invasions: ocean warming facilitates nonindigenous species invasions. *PNAS (USA)*, 99, 15497–15500.

Stachowicz, J. J., Bruno, J. F. and Duffy, J. E. (2007). Understanding the effects of marine biodiversity on communities and ecosystems. *Annual Review of Ecology, Evolution, and Systematics*, 38, 739–766.

Staehr, P. A., Pedersen, M. F., Thomsen, M. S., Wernberg, T. and Krause-Jensen, D. (2000). Invasion of *Sargassum muticum* in Limfjorden (Denmark) and its possible impact on the indigenous macroalgal community. *Marine Ecology Progress Series*, 207, 79–88.

Steffani, C. N. and Branch, G. M. (2005). Mechanisms and consequences of competition between an alien mussel, *Mytilus galloprovincialis*, and an indigenous limpet, *Scutellastra argenvillei*. *Journal of Experimental Marine Biology and Ecology*, 317, 127–142.

Strayer, D. L. (2012). Eight questions about invasions and ecosystem functioning. *Ecology Letters*, 15, 1199–1210.

Strayer, D. L., Eviner, V. T., Jeschke, J. M. and Pace, M. L. (2006). Understanding the long-term effects of species invasions. *Trends in Ecology and Evolution*, 21, 645–651.

Strong, J. A., Dring, M. J. and Maggs, C. A. (2006). Colonisation and modification of soft substratum habitats by the invasive macroalga *Sargassum muticum*. *Marine Ecology Progress Series*, 321, 87–97.

Syvitski, J. P. M., Vörösmarty, C. J., Kettner, A. J. and Green, P. (2005). Impact of humans on the flux of terrestrial sediment to the global coastal ocean. *Science*, 308, 376–380.

Thieltges, D. W., Reise, K., Prinz, K. and Jensen, K. T. (2009). Invaders interfere with native parasite–host interactions. *Biological Invasions*, 11, 1421–1429.

Thomsen, M. S. (2010). Experimental evidence for positive effects of invasive seaweed on native invertebrates via habitat-formation in a seagrass bed. *Aquatic Invasions*, 5, 341–346.

Thomsen, M. S. and McGlathery, K. J. (2006). Effects of accumulations of sediments and drift algae on recruitment of sessile organisms associated with oyster reefs. *Journal of Experimental Marine Biology and Ecology*, 328, 22–34.

Thomsen, M. S. and McGlathery, K. J. (2007). Stress tolerance of the invasive macroalgae *Codium fragile* and *Gracilaria vermiculophylla* in a soft-bottom turbid lagoon. *Biological Invasions*, 9, 499–513.

Thomsen, M. S., Gurgel, C. F. D., Fredericq, S. and McGlathery, K. J. (2006a). *Gracilaria vermiculophylla* (Rhodophyta, Gracilariales) in Hog Island Bay, Virginia: a cryptic alien and invasive macroalga and taxonomic correction. *Journal of Phycology*, 42, 139–141.

Thomsen, M. S., McGlathery, K. J. and Tyler, A. C. (2006b). Macroalgal distribution patterns in a shallow, soft-bottom lagoon, with emphasis on the nonnative *Gracilaria vermiculophylla* and *Codium fragile*. *Estuaries and Coasts*, 29, 470–478.

Thomsen, M. S., Wernberg, T., Stæhr, P. A. and Pedersen, M. F. (2006c). Spatiotemporal distribution patterns of the invasive macroalga *Sargassum muticum* within a Danish Sargassum-bed. *Helgoland Marine Research*, 60, 50–58.

Thomsen, M. S., Stæhr, P., Nyberg, C. D. *et al.* (2007). *Gracilaria vermiculophylla* in northern Europe, with focus on Denmark, and what to expect in the future. *Aquatic Invasions*, 2, 83–94.

Thomsen, M. S., Adam, P. and Silliman, B. (2009a). Anthropogenic threats to Australasian coastal salt marshes. In *Anthropogenic Modification of North American Salt Marshes*, ed. B. R. Silliman, M. D. Bertness, D. Strong. Oakland, CA: University of California Press, pp. 361–390.

Thomsen, M. S., McGlathery, K. J., Schwarzschild, A. and Silliman, B. R. (2009b). Distribution and ecological role of the non-native macroalga *Gracilaria vermiculophylla* in Virginia salt marshes. *Biological Invasions*, 11, 2303–2316.

Thomsen, M. S., Wernberg, T., Tuya, F. and Silliman, B. R. (2009c). Evidence for impacts of non-indigenous macroalgae: a meta-analysis of experimental field studies. *Journal of Phycology*, 45, 812–819.

Thomsen, M. S., Wernberg, T., Altieri, A. *et al.* (2010a). Habitat cascades: the conceptual context and global relevance of facilitation cascades via habitat formation and modification. *Integrative and Comparative Biology*, 50, 158–175.

Thomsen, M. S., Wernberg, T., Tuya, F. and Silliman, B. R. (2010b). Ecological performance and possible origin of a ubiquitous but under-studied gastropod. *Estuarine Coastal and Shelf Science*, 87, 501–509.

Thomsen, M. S., Olden, J. D., Wernberg, T., Griffin, J. N. and Silliman, B. R. (2011a). A broad framework to organize and compare ecological invasion impacts. *Environmental Research*, 111, 899–908.

Thomsen, M. S., Wernberg, T., Olden, J. D., Griffin, J. N. and Silliman, B. R. (2011b). A framework to study the context-dependent impacts of marine invasions. *Journal of Experimental Marine Biology and Ecology*, 400, 322–327.

Thomsen, M. S., de Bettignies, T., Wernberg, T., Holmer, M. and Debeuf, B. (2012a). Harmful algae are not harmful to everyone. *Harmful Algae*, 16, 74–80.

Thomsen, M. S., Wernberg, T., Engelen, A. H. *et al.* (2012b). A meta-analysis of seaweed impacts on seagrasses: generalities and knowledge gaps. *PLoS ONE*, 7, e28595.

Thomsen, M. S., Staehr, P. A., Nejrup, L. B. and Schiel, D. R. (2013). Effects of the invasive macroalgae *Gracilaria vermiculophylla* on two co-occurring foundation species and associated invertebrates. *Aquatic Invasions*, 8, 1–13.

Thomsen, M. S., Byers, J. E., Schiel, D. R. et al. (2014). Impacts of marine invaders on biodiversity depend on trophic position and functional similarity. *Marine Ecology Progress Series*, 495, 39–47.

Thornber, C. S., Kinlan, B. P., Graham, M. H. and Stachowicz, J. J. (2004). Population ecology of the invasive kelp *Undaria pinnatifida* in California: environmental and biological controls on demography. *Marine Ecology Progress Series*, 268, 69–80.

Thuiller, W., Richardson, D. M. and Midgley, G. F. (2007). Will climate change promote alien plant invasions? *Biological Invasions, Ecological Studies*, 193, 197–211.

Thyrring, J., Thomsen, M. S. and Wernberg, T. (2013). Large-scale facilitation of a sessile community by an invasive habitat-forming snail. *Helgoland Marine Research*, 67, 789–794.

Tilman, D., May, R. M., Lehman, C. L. and Nowak, M. A. (1994). Habitat destruction and the extinction debt. *Nature*, 371, 65–66.

Tomas, F., Cebrian, E. and Ballesteros, E. (2011). Differential herbivory of invasive algae by native fish in the Mediterranean Sea. *Estuarine, Coastal and Shelf Science*, 92, 27–34.

Torchin, M. E., Lafferty, K. D. and Kuris, A. M. (2002). Parasites and marine invasions. *Parasitology*, 124, 137–151.

Tronstad, M., Hall, R. O., Koel, T. M. and Gerow, K. G. (2010). Introduced lake trout produced a four-level trophic cascade in Yellowstone Lake. *Transactions of the American Fisheries Society*, 139, 1536–1550.

Trowbridge, C. D. (2002). Local elimination of *Codium fragile* ssp. *tomentosoides*: indirect evidence of sacoglossan herbivory. *Journal of the Marine Biological Association of the UK*, 82, 1029–1030.

Trowbridge, C. D. and Todd, C. D. (2001). Host-plant change in marine specialist herbivores: sacoglossan sea slugs on introduced macroalgae. *Ecological Monographs*, 71, 219–243.

Trussell, G., Ewanchuck, P. and Bertness, M. D. (2002). Field evidence of trait-mediated indirect interactions in a rocky intertidal food web. *Ecology Letters*, 5. 241–245.

Trussell, G., Ewanchuk, P., Bertness, M. D., Silliman, B. R. (2004). Trophic cascades in rocky shore tide pools: distinguishing lethal and non-lethal effects. *Oecologia*, 139, 427–432.

Tsai, C., Yang, S., Trimble, A. C. and Ruesink, J. L. (2010). Interactions between two introduced species: *Zostera japonica* (dwarf eelgrass) facilitates itself and reduces condition of *Ruditapes philippinarum* (Manila clam) on intertidal mudflats. *Marine Biology*, 157, 1929–1936.

Tyler, A. C. and McGlathery, K. J. (2006). Uptake and release of nitrogen by the macroalgae *Gracilaria vermiculophylla* (Rhodophyta). *Journal of Phycology*, 42, 515–525.

Tyler, A. C., McGlathery, K. J. and Macko, S. A. (2005). Uptake of urea and amino acids by the macroalgae *Ulva lactuca* (Chlorophyta) and *Gracilaria vermiculophylla* (Rhodophyta). *Marine Ecology Progress Series*, 294, 161–172.

Valentine, J. P. and Johnson, C. R. (2005). Persistence of the exotic kelp *Undaria pinnatifida* does not depend on sea urchin grazing. *Marine Ecology Progress Series*, 285, 43–55.

Valentine, J. P., Magierowski, R. H. and Johnson, C. R. (2007). Mechanisms of invasion: establishment, spread and persistence of introduced seaweed populations. *Botanica Marina*, 50, 351–360.

Viejo, R. M. (1997). The effects of colonization by Sargassum muticum on tidepool macroalgal assemblages. *Journal of the Marine Biological Association of the UK*, 77, 325–340.

Viejo, R. M. (1999). Mobile epifauna inhabiting the invasive *Sargassum muticum* and two local seaweeds in northern Spain. *Aquatic Botany*, 64, 131–149.

Vilà, M., Basnou, C., Pysek, P. *et al.* (2010). How well do we understand the impacts of alien species on ecosystem services? A pan-European, cross-taxa assessment. *Frontiers in Ecology and the Environment*, 8, 135–144.

Vilà, M., Espinar, J. L., Hejda, M., *et al.* (2011). Ecological impacts of invasive alien plants: a meta-analysis of their effects on species, communities and ecosystems. *Ecology Letters*, 14, 702–708.

Wallentinus, I. and Nyberg, C. D. (2007). Introduced marine organisms as habitat modifiers. *Marine Pollution Bulletin*, 55, 323–332.

Walther, G. R., Roques, A., Hulme, P. E. *et al.* (2009). Alien species in a warmer world: risks and opportunities. *Trends in Ecology and Evolution*, 24, 686–693.

Wernberg, T., Russell, B., Thomsen, M. S. *et al.* (2011). Seaweeds in retreat from ocean warming. *Current Biology*, 21, 1–5.

Wernberg, T., Russell, B. D., Thomsen, M. S., Connell, S. D. (2012a). Marine biodiversity and climate change. In *Global Environmental Change*, ed. B. Freedman. Heidelberg, Germany: Springer, pp. 181–187.

Wernberg, T., Smale, D. A., Tuya, F. *et al.* (2012b). An extreme climatic event alters marine ecosystem structure in a global biodiversity hotspot. Nature Climate Change advance online publication.

Wernberg, T., Thomsen, M. S., Stæhr, P. A. and Pedersen, M. F. (2001). Comparative phenology of *Sargassum muticum* and *Halidrys siliquosa* (Phaeophyceae: Fucales) in Limfjorden, Denmark. *Botanica Marina*, 44, 31–39.

Wernberg, T., Thomsen, M. S., Stæhr, P. A. and Pedersen, M. F. (2004). Epibiota communities of the introduced and indigenous macroalgal relatives *Sargassum muticum* and *Halidrys siliquosa* in Limfjorden (Denmark). *Helgoland Marine Research*, 58, 154–161.

Wernberg, T., Thomsen, M. S., Tuya, F. *et al.* (2010). The resilience of Australasian kelp beds decrease along a latitudinal gradient in ocean temperature. *Ecology Letters*, 13, 685–694.

White, E. M., Wilson, J. C. and Clarke, A. R. (2006). Biotic indirect effects: a neglected concept in invasion biology. *Diversity and Distributions*, 12, 443–455.

White, L. F. and Shurin, J. B. (2011). Density dependent effects of an exotic marine macroalga on native community structure. *Journal of Experimental Marine Biology and Ecology*, 405, 111–119.

Whitfield, P. E., Gardner, T., Vives, S. P. *et al.* (2002). Biological invasion of the Indo-Pacific lionfish *Pterois volitans* along the Atlantic coast of North America. *Marine Ecology Progress Series*, 235, 289–297.

Willette, D. A. and Ambrose, R. F. (2009). The distribution and expansion of the invasive seagrass *Halophila stipulacea* in Dominica, West Indies, with a preliminary report from St. Lucia. *Aquatic Botany*, 91, 137–142.

Willette, D. A. and Ambrose, R. F. (2012). Effects of the invasive seagrass *Halophila stipulacea* on the native seagrass, *Syringodium filiforme*, and associated fish and epibiota communities in the Eastern Caribbean. *Aquatic Botany*, 103, 74–82.

Williams, P. J., Floyd, T. A. and Rossong, M. A. (2006). Agonistic interactions between invasive green crabs, *Carcinus maenas* (Linnaeus), and sub-adult American lobsters, *Homarus americanus* (Milne Edwards). *Journal of Experimental Marine Biology and Ecology*, 329, 66–74.

Williams, S. L. and Smith, J. E. (2007). A global review of the distribution, taxonomy, and impacts of introduced seaweeds. *Annual Review of Ecology, Evolution, and Systematics*, 38, 327–359.

Wilson E. E. (2011). The facilitative role of an introduced bryozoan (*Watersipora* spp.): structuring fouling community assemblages within Humboldt Bay. MSc Thesis The Faculty of Humboldt State University.

Wonham, M. J., O'Connor, M. and Harley, C. D. G. (2005). Positive effects of a dominant invader on introduced and native mudflat species. *Marine Ecology Progress Series*, 289, 109–116.

Wootton, J. T. (1994). The nature and consequences of indirect effects in ecological communities. *Annual Review of Ecology and Systematics*, 25, 443–466.

Wotton, D. M., O'Brien, C. and Stuart, M. D. (2004). Eradication success down under: heat treatment of a sunken trawler to kill the invasive seaweed *Undaria pinnatifida*. *Marine Pollution Bulletin*, 49, 844–849.

Wright, J. T., McKenzie, L. A. and Gribben, P. E. (2007). A decline in the abundance and condition of a native bivalve associated with *Caulerpa taxifolia* invasion. *Marine and Freshwater Research*, 58, 263–272.

Wright, J. T., Byers, J. E., Koukoumaftsis, L. P., Ralph, P. J. and Gribben, P. E. (2010). Native species behaviour mitigates the impact of habitat-forming invasive seaweed. *Oecologia*, 163, 527–534.

Wrona, F. J., Prowse, T. D., Reist, J. D. *et al.* (2006). Climate change effects on aquatic biota, ecosystem structure and function. *AMBIO: A Journal of the Human Environment*, 35, 359–369.

Wu, Y. T., Wang, C. H., Zhang, X. D. *et al.* (2009). Effects of saltmarsh invasion by *Spartina alterniflora* on arthropod community structure and diets. *Biological Invasions*, 11, 635–649.

York, P. H., Booth, D. J., Glasby, T. M. and Pease, B. C. (2006) Fish assemblages in habitats dominated by *Caulerpa taxifolia* and native seagrasses in south-eastern Australia. *Marine Ecology Progress Series*, 312, 223–234.

Part III
Synthesis and conclusions

11 · Human activities and ecosystem service use: impacts and trade-offs

MELANIE AUSTEN, CAROLINE HATTAM
AND SAMANTHA GARRARD

11.1 Introduction

Ecosystem services provided by the marine environment are fundamental to human health and well-being. Despite this, many marine systems are being degraded to an extent that may reduce their capacity to provide these ecosystem services. The ecosystem approach is a strategy for the integrated management of land, water and living resources that promotes conservation and sustainable use in an equitable way (UN Convention on Biological Diversity, 2000). Its application to marine management and spatial planning has been proposed as a means of maintaining the economic and social value of the oceans, not only in the present but for generations to come. Characterising the susceptibility of services (and combinations of services) to particular human activities based on knowledge of impacts on biodiversity and ecosystem functioning (as described in preceding chapters) is a challenge for future management of the oceans.

In this chapter, we highlight the existing, but limited knowledge of how ecosystem services may be impacted by different human activities. We discuss how impacts on one service can impact multiple services and explore how the impacts on services can vary both spatially and temporally and according to context. We focus particularly on the effects on ecosystem services of activities whose impacts on biodiversity and ecosystem functioning have already been considered in previous chapters. Some of these activities are associated with poor management of ecosystem benefits, for example, from provisioning services (aquaculture

Marine Ecosystems: Human Impacts on Biodiversity, Functioning and Services, eds T. P. Crowe and C. L. J. Frid. Published by Cambridge University Press. © Cambridge University Press 2015.

and fisheries), or with excessive input of wastes, fertilisers and contaminants into the system overburdening the waste treatment and assimilation services. Other impacts are associated with the construction of structures or use of space designed to generate benefits from environmental services such as the presence of water as a carrier for shipping, or sources of wind, wave and tidal power.

We discuss the trade-offs that are made, consciously or otherwise, between different ecosystem services, which arise from human activities to optimise or manage specific ecosystem services. We go on to look at the implications of trade-offs among ecosystem services for a multi-sector approach to management as implemented through marine planning. Finally we discuss the role and limitations of ecosystem benefit valuation in informing understanding of the trade-offs.

11.2 Impacts on the relationships between biodiversity, ecosystem functioning and ecosystem services

Ecosystem functioning is the sum of ecosystem processes in a system where the processes involve fluxes of energy and matter between trophic levels and the environment (Chapters 1, 5). Ecosystem services are the direct and indirect contributions of ecosystems to human well-being (TEEB, 2010) and are described in Chapter 2. Ecosystem benefits are the outcomes of ecosystem services that make contributions to human welfare. It is these benefits that can be valued through monetary or non-monetary approaches.

Anthropogenic and climate change pressures can act directly on any part of the biodiversity–ecosystem functioning–ecosystem service–ecosystem benefit continuum (see Chapter 2), either separately or simultaneously. The impacts on biodiversity and on ecosystem functioning are probably the most well understood within this continuum. However, as ecosystem services are exploited and goods extracted to generate ecosystem benefits there are likely to be feedbacks that impact upon biodiversity and ecosystem functioning. These feedbacks often go unrecognised, but may affect not just the service being exploited, but potentially other services as well.

Initial research on the relationship between biodiversity and ecosystem service provision suggests that ecosystem service provision decreases with a decrease in biodiversity (Balvanera *et al.*, 2006; Worm *et al.*, 2006; Micheli *et al.*, 2014). The relationship between biodiversity, ecosystem functioning and ecosystem services probably depends on a dynamic network of interactions between species. This network will vary according to the context of the ecosystem being investigated, the biological structure of the community present and other environmental variables. The impact of loss of one or more species due to human activities will therefore be dependent upon a number of different factors including the role of the species within the ecosystem, their positive or negative interactions with other species and compensation dynamics. In some cases, perturbations caused by human activities (singly or multiply) can result in changes that cross a threshold and lead to a phase shift to an alternative steady state. The ensuing biodiversity and processes are different from the previous steady state, producing a different (and usually more limited) type of ecosystem functioning and potentially a change in service provision.

An example of such a phase shift resulted from the drastic decline in sea otter populations observed along the north-west coast of the United States as an indirect consequence of human exploitation of marine resources (Springer et al., 2003). Sea urchins were released from predation by sea otters and thus kelp forests faced extreme herbivory by the increasing sea urchin populations. This prevented the re-growth of the foliose macroalgae, leading to a transformation of the coastal areas from a lush forest to urchin barrens (Stewart and Konar, 2012). The loss of kelp resulted in a reduction in primary productivity and simplification of the food web (Graham, 2004), leading to a decrease in ecosystem functioning. More recently, Wilmers et al. (2012) found this has an enormous impact on a coastal area's carbon dynamics and thus its potential to regulate climate. In areas where abundant kelp forests were present, net primary productivity (NPP) was 313–900 g C m^{-2} y^{-1}, whilst in areas where the sea otters were absent and the kelp forest had been transformed into an urchin barren NPP was 25–70 g C m^{-2} y^{-1}, a more than 90% decrease. Similarly, the biomass of carbon stored in kelp was 101–180 g C m^{-2} when abundant kelp forests were present but decreased to 8–14 g C m^{-2} in the urchin barrens. Furthermore, Harrold et al. (1998) calculated that there is approximately 16.5 g C m^{-2} y^{-1} long-term removal of carbon from the system (i.e. carbon sequestration) as it is transported to the deep sea from kelp beds, although this value is likely to vary with species, currents and proximity to deep waters. These results show that the climate regulation service provided by kelp beds has been severely depleted with the loss of sea otters and the transformation of kelp forests to urchin barrens. Climate regulation, however, is just one of many services provided by kelp forests. Others include the provision of food, biotic raw materials, migratory and nursery habitat, and gene pool protection, the regulation of waste treatment and assimilation, as well as its contribution to tourism and recreation. Deforestation will lead to reduction or loss of a number of these services. This example highlights that whilst diversity may play a role in provision of ecosystem services, loss of key species can have severe negative impacts on ecosystem functioning and service provision.

For other ecosystem services, and in different habitats, the role of biodiversity in service delivery may be more pronounced. As with terrestrial studies that indicate that the more diverse a plant community, the more aesthetically pleasing it is to the public (Lindemann-Matthies et al., 2010), preliminary studies indicate that enhanced biodiversity in marine aquaria increases positive feelings and emotions (Cracknell et al., submitted). Similarly, increasing species diversity is expected to increase cultural services such as tourism and recreation or aesthetic experiences. For example, coral and fish diversity have been shown to be important to the holiday destination choice for tourists visiting the Bonaire in the Caribbean (Uyarra et al., 2005). In other cases, the role of biodiversity may be a contribution to the resilience of ecosystems to environmental and human perturbation. In turn, ecosystem resilience may govern the fundamental ecosystem dynamics which provide ecosystem services (Folke et al., 2004). Coral and herbivore diversity, alongside the number of resistant coral species and herbivore biomass have been implicated as key indicators of coral reef resilience (McClanahan et al., 2012). In the face of perturbation, for example due to climate change or ocean acidification, as reef resilience deteriorates a phase shift to a macroalgal-dominated system can occur (Cheal et al., 2010) and the ecosystem services that the area provide inevitably decrease and/or change (Moberg and Folke, 1999).

The relationship between marine ecosystem services, ecosystem functions and biodiversity can be difficult to elucidate. There are numerous linkages between the different components of an ecosystem that lead to the services and the benefits that those services provide. Whale watching, a benefit of the cultural service 'provision of leisure, recreation and tourism', provides a useful example. It is a growing industry and was estimated to be worth £2.1 billion globally in 2008 (O'Connor et al., 2009). Assuming that the benefits of whale watching depend on the presence and abundance of populations of whales, these benefits are likely to be affected by any direct anthropogenic effects on whale populations themselves. For example, until the 1970s whaling severely depleted many whale populations to less than 10% of their original numbers, although there is evidence that some stocks are beginning to recover (Best, 1993). Pollution (Béland et al., 1993) and underwater noise, for example from seismic surveys (Richardson et al., 1999), may have negative impacts on this group of taxa. Whale populations are also likely to be affected by changes to other elements of biodiversity and ecosystem functions that hence have a role in providing this benefit through the various trophic links and food webs that support the whales. These include the availability of pelagic prey such as euphausiids (Stevick et al., 2008), copepods (Davies et al., 2013), squid (Abend and Smith, 1995) and fish (Ford and Ellis, 2006; Ainley and Pauly, 2014) that whales feed on. In turn these food sources will all be affected by phytoplankton primary production, secondary production and trophic dynamics, including nutrient cycling. All of these links are in turn vulnerable to anthropogenic disturbance as presented in previous chapters in this book.

The links between biodiversity, ecosystem functions and ecosystem services are rarely simple. A complex mix of environmental and human perturbations must be taken into consideration when considering these linkages (Chapters 3, 4, 5). The reality is that our knowledge of the impacts of different stressors on the different parts of the interlinked biodiversity–ecosystem functioning–ecosystem services–ecosystem benefits system is highly fragmented and piecemeal. It is a long way from the integrated understanding that could optimise management of human use of marine ecosystems.

11.3 Impacts of human activities on ecosystem services

In this section we review our understanding of the ecosystem impacts of different human activities, especially those considered in previous chapters, from the perspective of their implications for provision of each category of ecosystem services (see summary in Table 11.1).

11.3.1 Provisioning services

Impacts of fishing and aquaculture
Capture fisheries and aquaculture are human activities that are undertaken to exploit the globally important provisioning services of food and non-food biotic raw materials. The ecosystem service can be considered as the provision and maintenance of populations (stocks) of different species of fish and shellfish so that they are available for direct consumption or to support aquaculture. The FAO (2012) estimated that in 2010 capture

Table 11.1 *Summary of likely impacts on ecosystem service provision of stressors reviewed in this volume.*

Ecosystem service	Fishing and aquaculture	Physical structures	Eutrophication	Contamination	Invasive species
PROVISIONING SERVICES					
Food provision	—	++	— +	—	+ —
Biotic raw materials (non-food and not supporting food provision)	?	?	?	?	?
REGULATING SERVICES					
Air purification	—	?	—	—	+ —
Climate regulation	—	—	—	—	+ —
Disturbance prevention or moderation	?	—	—	—	+ —
Regulation of water flows	?	—	?	—	+ —
Waste treatment and assimilation	—	—	—	—	+ —
Coastal erosion prevention	?	+ —	—	—	++ —
Biological control	—	?	—	—	—
HABITAT SERVICES					
Migratory and nursery habitat	—	++ —	— +	—	— +
Gene pool protection	—	?	?	?	?
CULTURAL SERVICES					
Leisure, recreation and tourism	— +	—	—	—	— +
Aesthetic experience	?	— +	—	—	—
Inspiration for culture, art and design	+	—	—	—	—
Cultural heritage	?	—	—	—	—
Cultural diversity	+	+	?	?	?
Spiritual experience	+ —	?	?	?	?

Table 11.1 (cont.)

Ecosystem service	Fishing and aquaculture	Physical structures	Eutrophication	Contamination	Invasive species
Information for Cognitive Development	++	+	+	+	+

Fishing (but not aquaculture) impacts are likely to be widespread given that this activity is widespread. Impacts of physical structures and of contamination, and sometimes of eutrophication, are likely to be spatially localised. Invasive species impacts may be localised or widespread depending on the mobility and rate of spread of the species. Shading denotes relative extent of available data to support our understanding of impacts with darker shades indicating more data is available and no shading indicating very little or no data available. In the case of white shading, impact directions are of low confidence being based on little data and/or hypothesised effects. Negative and positive impacts are denoted by '−' and '+' symbols respectively with the number of symbols providing an indication of the estimated severity of the impact. Where impacts are not known and cannot be readily hypothesised a '?' symbol is used. Where there is evidence of either positive or negative impacts in different contexts both '−' and '+' symbols are shown.

fisheries and aquaculture combined were worth US$217.5 billion. They were also thought to contribute to the livelihoods of 660–820 million people worldwide through direct and ancillary occupations (about 10–12% of the world's population) and provide 16.5% of global animal protein intake (6.5% of all protein intake). Despite this importance, human activities impact on the provision of the very services on which these benefits are dependent. Poor management of both capture fisheries and aquaculture can lead to significant environmental impacts that may negatively affect biodiversity, ecosystem function and ecosystem services including the provisioning services themselves (Chapter 6).

Capture fisheries impact directly on provisioning services through removal of organisms which are either targeted fish or as bycatch. The removal of commercial fish species is extensive and whilst fishing pressure has not led to any known global extinctions, many populations of fish, sharks and rays have been declared locally extinct (Dulvy et al., 2003). Overfishing can reduce stock size and sustainability and over 85% of the world's fish stocks have either collapsed, are overexploited or are fully exploited (Froese et al., 2012). The biomass of large predatory fish is thought to be approximately 10% that of pre-industrial levels (Myers and Worm, 2003) and as exploitation continues and long-lived high trophic level fish become depleted, fisheries are targeting shorter-lived, lower trophic level species to maintain fisheries landings (Pauly et al., 1998). This exploitation also indirectly decreases the potential food provision of the oceans through changes in secondary production, trophic dynamics and genetic diversity of the ecosystem. Landings of fish and other seafood into the UK have more than halved since 1948, through reduced stock sizes but also fisheries management efforts to sustain and restore the stocks, although the human population and food requirements have grown in this time (Austen et al., 2011).

It is not only active fishing that impacts fish populations. Marine debris from fisheries, including fishing gear that has not been retrieved and other debris that is purposefully discarded at sea may continue to catch fish. So-called 'ghost fishing' is a major issue (Laist, 1987). Research suggests that deep-water ghost nets (>500 m depth) can continue to catch organisms more than 8 years after they are discarded (Brown et al., 2005). Ghost fishing will have similar implications for commercial fishing without the benefit to humans, although some scavenging species may benefit from increased food supply.

Different fishing gears also differ in their impact on marine ecosystems. Soft-bottom communities are vulnerable to the physical consequences of bottom fishing, where gear is dragged across the seafloor. This can significantly alter seafloor complexity through the removal of both sedimentary and biogenic structures (Auster et al., 1996), altering the productivity of seabed communities and the settlement and growth of sessile organisms (Jennings and Kaiser, 1998). These communities play an important role in ecosystem processes such as nutrient and carbon cycling that support food webs and the fish populations that underpin fisheries. Benthic species can also be an important component of the diet of fish (Link and Garrison, 2002), and a decrease in their abundance can decrease fish larval survival and recruitment success (Hussy et al., 1997).

Declining fish stocks and the growing demand for protein for human consumption has led to a sharp increase in aquaculture in the past 20 years (Naylor et al., 2000). Many countries have benefited in terms of food security as a result, and especially from the culture of low trophic level species (Rice and Garcia, 2011; FAO, 2012; Frid and Paramor, 2012). However, aquaculture operations can have negative ecosystem impacts as well. For example, many cultured species have large requirements for fishmeal and fish oils supplied by capture fisheries (Chapter 6). Aquaculture installations are also responsible for the release of organic waste, antimicrobials and pathogens, habitat destruction, and the escape of cultivated species that may mix with native stocks (affecting gene pools) or invade nearby communities. Escape of cultured fish is a common issue, particularly documented for farmed salmon. Escaped fish can displace wild populations and cause a decrease in their fitness and fecundity through inter-breeding (Jonsson and Jonsson, 2006). Diseases of cultured finfish can also be transferred to conspecifics, further decreasing fitness of wild populations. Escapees from aquaculture may therefore decrease the potential for recovery of already overexploited stocks. The extent to which these pressures from aquaculture impact on biodiversity, ecosystem function and ecosystem services will depend on the species farmed (Naylor et al., 2000), the farming methods and spatial differences in environmental conditions.

Maintaining food security in an ecologically and socially sustainable manner is a challenge for the coming decades (Chapter 1), in which wild capture fisheries and aquaculture will play a major role (Godfray et al., 2010). Whilst impacts on provisioning and other ecosystem processes and services are unavoidable, improving management to reduce overexploitation, destructive fishing practices, intensive farming methods and poor animal husbandry would minimise these impacts, enabling the maximisation of ecosystem service provision of the oceans.

Impacts of coastal development: physical structures, eutrophication, contamination and invasive species

Understanding of impacts of built structures, eutrophication, contaminants and invasive species on provisioning services is somewhat piecemeal. There is very little evidence of the impacts of built structures (Chapter 7). Species richness and abundance of fish can be higher on estuarine breakwaters than adjacent natural rocky reefs (Folpp et al., 2013; Fowler and Booth, 2013) but, in Taiwan, fish species richness and abundance did not differ between breakwaters and adjacent rocky reefs, although a decrease in reef specialists and an increase in generalist species was observed (Wen et al., 2010). Armouring of cables with rocks around offshore wind arrays enhances crab and possibly also lobster populations potentially increasing stocks for food (Hooper and Austen, 2014). Other semi-permanent structures, such as oil platforms are known to act as fish aggregating devices suggesting that wind turbine bases will do the same (De Troch et al., 2013; Krone et al., 2013; Reubens et al., 2013) and that they could be designed in such a way as to encourage fish aggregation.

The extent of eutrophication influences its impacts on fish populations, and hence their use as a source of food (Caddy, 2000). Mild eutrophication can enhance production of demersal and pelagic species as a response to an increased availability of food (Chapter 8). Increasing eutrophication, however, coupled with hypoxic events, causes declines of demersal fish but continued increases in pelagic fish species. At very high levels of eutrophication there is a general decline in total fish production under conditions of broadly deteriorating water and habitat quality and reduced prey availability. Where hypoxia and anoxia (reduced and zero dissolved oxygen respectively) are experienced as a consequence of eutrophication (sometimes in combination with climate change), mobile species such as fish and some of the invertebrates they will feed on tend to move away (Craig, 2012) or in extreme cases suffer from mortality. In the Baltic Sea, oxygen depletion has been shown to lead to increased catches of crustaceans and fish while these animals attempt to escape low oxygen conditions (HELCOM, 2009). Again in the Baltic, Österblom et al. (2007) proposed that the exploitation of fish combined with eutrophication has caused a series of impacts, such as population crashes of seals (due to a loss of prey species), increased primary production and overfishing of cod. However, the complex feedback loops between different components of the ecosystem mean the overall picture was not so simple since this increased primary production during 1950–70 in turn enhanced the abundance of fish in the Baltic.

The presence and abundance of many types of algae increases during the early stages of eutrophication with varying impacts. For example, loose-lying macroalgae can clog gillnets (especially in estuaries and shallow seas) reducing fish catch and increasing maintenance efforts for these nets. Eutrophication can also potentially increase blooms of harmful toxin-producing algae (harmful algal blooms; HABs) (Anderson et al., 2010), although the evidence supporting this is somewhat contradictory (e.g. Gowen et al., 2012). The toxins produced can accumulate in filter-feeding shellfish which, when consumed by humans, can affect the human health benefits derived from marine food provision. If poisoned shellfish are consumed, either because of a screening failure or unregulated harvesting, the human consequences can be severe, ranging from diarrhoea, to memory loss, paralysis and death. Similarly, phytoplankton blooms that arise due to eutrophication often comprise dinoflagellates and cyanobacteria, many of which are toxic to marine mammals, fish and invertebrates. Fish poisoning can result from consuming contaminated algae either directly or indirectly by eating prey that have consumed contaminated algae. Some harmful algal blooms can indirectly lead to fish and invertebrate mortality by clogging their gills or blooming to such an extent that they remove oxygen from the water column, causing hypoxia (Hallegraeff, 1993). The result can be reduced wild and farmed fisheries production in the case of direct kills (e.g. fish and shellfish) or through closure of both wild and aquaculture shellfisheries when accumulated toxins have rendered the harvested shellfish unfit for human consumption (Hallegraeff, 1993).

There is little research on the effects of contaminants on food provision services although meta-analysis by McKinley and Johnston (2010) showed that contaminants negatively affect the abundances of fish but do not have a strong effect on their species richness. One of the clearest effects of a contaminant on a marine fishery is the case of tributyltin (TBT). This was introduced as an active, biocidal, ingredient in a range of antifouling paints in the late 1970s and is known to have caused reproductive failure and shell thickening abnormalities in cultured oysters in France and the UK. The impact in France was so severe that emergency legislation was introduced in 1982 banning the use of TBT on static structures and small craft (i.e. those that spent a large proportion of their time on moorings in bays close to the oyster beds) (Alzieu, 2000). This legislation was adopted EU-wide in 1986 and subsequently the International Maritime Organisation (IMO) introduced a global ban on TBT, initially on small vessels and subsequently on all vessels. TBT interferes with molluscs' endocrine systems and disrupts natural hormone systems. A range of other contaminants have now been identified as having similar effects and they are collectively known as endocrine disruptors (EDs). They include plasticisers, industrial cleaners, paints and pharmaceutical residues including human reproductive hormones released from sewage treatment works. EDs have been shown to impact on estuarine fish populations (Lye et al., 1997) and on freshwater fish, including economically important populations of trout, although, to date, effects on wild marine populations have not been identified.

The effects of contaminants on invertebrate communities (Chapter 9) may have indirect effects on fish populations, although these impacts can be highly variable between different contaminants. For example the herbicide triazine leads to a decrease in phytoplankton productivity (Bester et al., 1995) and hence reduced energy available for organisms further up the food chain. Copper contamination, on the other hand, does not affect phytoplankton productivity (Coale, 1991), but leads to negative impacts on the feeding rate of zooplankton (Reeve et al., 1977). This may similarly limit energy transport up the food chain to larger organisms and may lead to a greater abundance of phytoplankton as they are released from predation. As an indirect effect of pollution, a decrease in energy transfer will lead to reduced food availability for fish and potentially decrease fisheries production.

A greater issue is the effect of consumption of contaminated food on human health, which reduces the benefits obtained from the food provisioning service. Much of the marine food chain has become contaminated directly and indirectly as a result of anthropogenic activities. Bioaccumulation through the food chain of pollutants such as dioxins, heavy metals (e.g. methyl mercury) and organic compounds (e.g. polycyclic aromatic hydrocarbons (PAHs), brominated flame retardants), and of natural toxins produced by HAB organisms can impact on human health when fish and shellfish are consumed (Fleming et al., 2006; Neisheim et al., 2006; Kite-Powell et al., 2008; Bushkin-Bedient and Carpenter, 2010; Rossini and Hess, 2010; Twiner et al., 2008; Angeletti et al., 2014). For example, contaminants such as methyl mercury have been known to accumulate in the flesh of long-lived predatory fish such as sharks, marlin and swordfish at levels above recommended guidelines (Park et al., 2011). Exposure to high levels of methyl mercury can lead to constriction of the visual fields, hearing loss, ataxia and paresthesia in adults and in severe cases cerebral palsy in prenatal infants (Clarkson, 1993). The Joint FAO/WHO Expert Committee on Food Additives, as well as many national and international organisations, such as the European Food Safety Authority, the US Environment Protection Agency and the UK's Food Standards Agency, all provide guidance on tolerable daily or weekly intakes of seafood in the light of bioaccumulated contaminants.

Very little evidence is available on the impacts of invasive species on provisioning services. The jellyfish *Mnemiopsis leidyi* has, via its predation on fish larvae, been implicated in the collapse of the anchovy fishery in the Black Sea (Kideys, 2002) although other factors are likely to have contributed to the decline of this fishery (Bilio and Niermann, 2004). Elsewhere, sea squirts are reported to be threatening shellfish aquaculture in New Zealand. Not all invasions are necessarily reported as being harmful. The king crab, a species of crab normally found in the North Pacific was introduced into the Barents Sea in the 1960 and has since spread into Norwegian waters. Although the king crab is reported to affect other fish species and clams, in some locations it is considered a lucrative source of income (Falk-Peterson et al., 2011). However, the ecological impacts of the species are not fully known and the lack of apparent impact does not necessarily mean no impact is present.

11.3.2 Regulating services

Regulating services, such as climate regulation and waste treatment and assimilation, are dependent on organism-mediated particulate filtering, bioturbation and biogeochemical cycling within the water column and at the seabed. These ecosystem functions enable processing and sequestration of climatically important carbon and constituents of other climate gases, and the transformation of contaminants and wastes to less toxic forms, or their burial and sequestration. Another important function for regulatory services such as disturbance prevention or moderation, regulation of water flows and coastal erosion prevention is the construction and maintenance of biogenic habitat. Such habitats act to reduce hydrodynamic forces and absorb excess flood water and reduce erosion. Impacts of human activities and non-indigenous species on these functions or the elements of biodiversity that undertake them are likely to impact on delivery of regulating services.

Impacts of fishing and aquaculture

Fisheries-induced changes in trophic dynamics, community biomass, and both pelagic and benthic diversity (Chapter 6) will not only impact on provisioning ecosystem services. They also affect regulating services such as waste treatment and assimilation and climate regulation. Capture fisheries using demersal (seabed) fishing gear can disturb the structure of the seafloor, affecting benthic communities whose functioning contributes to regulation services (Jennings and Kaiser, 1998). Soft-bottom benthic communities occur where the seabed is comprised of sediment such as sand or mud, or a mixture of the two. These benthic communities play an important role in ecosystem processes such as nutrient and carbon cycling, seafloor stability, contaminant sequestration, water column turbidity and primary and secondary production (Thrush and Dayton, 2002). Demersal fishing can impact on the important bioturbation role played by many soft-bottom invertebrates, which influences biogeochemical processes such as nutrient and carbon cycling that underpin many regulating services (Thrush and Dayton, 2002; Mermillod-Blondin and Rosenberg, 2006; Austen et al., 2011; Mermillod-Blondin, 2011). Contaminant sequestration and reduction in water column turbidity are aspects of waste treatment and assimilation which are also mediated by bioturbation and filter feeding of benthic organisms. More directly, bottom fishing can overburden these services through disturbance of the seafloor, resuspending sediment and releasing carbon and waste stored in seafloor sediments into the water column. Furthermore benthic organisms can help to maintain seafloor stability, which can contribute to disturbance prevention or moderation and coastal erosion prevention services (Kaiser et al., 2002). Ghost fishing by lost or abandoned fishing gear and debris can locally overburden waste treatment and assimilation services as trapped fish die and decay in situ.

Similarly, whilst the impact of aquaculture activity on many ecosystem services has not been investigated, it is likely that changes to biodegradation of waste and climate regulation may occur. The most widely documented impact of coastal finfish farming cages is the accumulation of waste products on the seafloor beneath the cages, leading to degradation of the seabed (Chapter 6). This can result in a reduction in benthic diversity and an increase in the abundance of opportunistic species (Karakassis et al., 1999), which may alter biogeochemical cycling on the seafloor. Anaerobic metabolism may increase mineralisation leading to eutrophication of surrounding waters (Holmer and Kristensen, 1992). Nutrient and carbon cycling, and waste treatment through bioremediation may further be altered by the use of antimicrobials, which may alter microbial community structure in the local area. However, not all fish farms appear to have large environmental consequences and diversity has been shown to remain high under some cages (Apostolaki et al., 2007).

Impacts of coastal development: physical structures, eutrophication, contamination and invasive species
A diffuse effect of eutrophication (Chapter 8) is the greater abundance of plankton that arises as a result of mild eutrophication which increases suspended material and reduces light penetration. Filter-feeding organisms in hard-bottom habitats will remediate against these effects (thus performing the waste treatment and assimilation service), but in cases of more extensive eutrophication the service capability will be overburdened. In this case excess particles are deposited in the seabed affecting nutrient cycling, and by extension, the other regulating services that depend upon this ecosystem function. Similarly, extreme eutrophication causing oxygen deficiency or eutrophication-enhanced harmful algal blooms can cause mortality and thus alter benthic and pelagic communities, again with associated impacts on biogeochemical cycling and bioturbation that can in turn affect regulatory services (Cloern, 2001).

Coastal aquatic vegetative habitats such as seagrass beds, salt marshes, kelp forests and mangroves are important carbon sinks, through the fixation of CO_2 by photosynthesis and storage of carbon in living biomass, and its subsequent transfer into sediments or the deep sea, essentially removing it from the system (Laffoley and Grimsditch, 2009). Storage of carbon by marine (and terrestrial) ecosystems can lead to a reduction in the atmospheric concentrations of CO_2, regulating our climate and reducing the impact of climate change (McLeod et al., 2011). Coastal systems provide other important regulating ecosystem services such as reduction of excessive coastal nutrient concentrations, and protection of the shoreline from erosion and storm surges (Laffoley and Grimsditch, 2009). Coastal eutrophication can lead to significant negative effects on these coastal carbon sinks. Seagrass density is reduced in areas affected by eutrophication (Cardoso et al., 2004), as opportunistic free-floating and epiphytic macroalgae and phytoplankton decrease light penetration to the seagrass (Burkholder et al., 2007). Eutrophication is thought to significantly contribute to seagrass loss around the globe (Waycott et al., 2009). Opportunistic macroalgae and phytoplankton will similarly outcompete benthic, perennial macroalgae such as kelps and fucoids (Duarte, 1995). In saltmarshes and mangroves, eutrophication works on the macrophytes directly by altering energetic distribution and has been shown to reduce the root:shoot biomass (Darby and Turner, 2008; Lovelock et al., 2009; Deegan et al., 2012). In salt marshes, this leads to a reduction in geomorphologic stability and loss of saltmarsh (Deegan et al., 2012), whilst in mangroves this leads to increased vulnerability to environmental perturbations and increased mortality (Lovelock et al., 2009). Loss or decline of these habitats will lead to negative impacts or loss of associated ecosystem services.

Built structures that alter hydrodynamics or replace soft-sediment habitat with hard substrate inevitably change the ecosystem (Chapter 7) and its constituent communities. Such changes will affect the functioning of communities especially with respect to the flows and stocks of carbon, nutrients and contaminants affecting regulating services of climate regulation and waste treatment and assimilation. For example, in coastal situations, reductions in flows can allow outbreaks of opportunistic macroalgae (Chapter 7) with similar impacts on regulating services to those described above.

Contaminants have been associated with reductions in biodiversity within every marine habitat (Johnston and Roberts, 2009) and with negative impacts on many ecosystem functions (Chapter 9). However, as discussed in Chapter 9, the impacts of contaminants on biodiversity and ecosystem functioning are highly context-dependant. They are influenced by factors such as the composition of species, the type of toxicant and abiotic factors (Hooper *et al.*, 2005; McMahon *et al.*, 2012). As such it appears to be difficult to consider generic examples of contaminant impacts on regulating ecosystem services at this stage. It is important to note, however, that absence of proof does not necessarily indicate absence of impact.

Depending on the invasive species considered, the regulatory service of coastal erosion protection can be enhanced or degraded. Invasive *Spartina* species can increase the extent of salt marsh thereby improving the levels of coastal erosion protection compared to unvegetated mud flats. In contrast, through bioturbation and burrowing some mobile invaders can change sediment stability and reduce coastal erosion protection. Two species invasive to Europe and N. America, the Chinese mitten crab *Eriocheir sinensis* and the isopod *Sphaeroma quoyanum*, make burrows in estuarine banks, potentially causing bank erosion or converting marsh to mud (Rudnick *et al.*, 2005; Davidson and de Rivera, 2010).

Invasive species appear to inevitably affect the stocks of carbon and other nutrients in different compartments of ecosystems as well as fluxes of nutrients within and in some cases exported from ecosystems (Chapter 10). They do this often by nature of their habitat modification where structural species are involved, but sometimes by nature of their greater uptake of nutrients, growth and decomposition compared to the native species which they replace (see Chapter 10 for detail). Given that climate regulation and waste treatment and assimilation services are underpinned by biologically mediated stocks and flows of nutrients (Austen *et al.*, 2011), it can therefore be assumed these services are likely to be affected by invasive species.

11.3.3 Habitat services

Habitat services are the provision by ecosystems of living space for resident and migratory species (thus maintaining the gene pool and nursery service). Migratory and nursery habitat services are provided, for example, by biogenic reefs, seagrass beds, macroalgae stands, and estuaries and their salt marshes for commercially valuable species that are harvested elsewhere. Most biogenic marine habitats help to maintain viable gene pools through natural selection and evolutionary processes and this enhances the adaptability of species to environmental changes, as well as the resilience of the ecosystem and these underpin all marine ecosystem services. More directly, marine habitats contribute to the maintenance of inter- and intraspecific genetic diversity of species with potential benefits and/or commercial use. Any activities that disturb or degrade habitats that have strong biogenic components therefore reduce the extent of provision of this service.

Impacts of fishing and aquaculture
Capture fisheries utilising bottom-fishing gear can disrupt seafloor structure directly as well as negatively impacting the benthic organisms that maintain this structure through their functioning. Where biogenic habitats such as seagrass beds, maerl beds or deep-water corals are present, habitat destruction can be significant (Hall-Spencer and Moore, 2000; Hall-Spencer *et al.*, 2002; Kaiser *et al.*, 2002) and recovery can take a considerable time (Kaiser *et al.*, 2006). Complex seabed habitat provides niches and crevices that act as shelter from predators and reduce hydrodynamic flows. Many types of bottom trawling reduce habitat complexity, by removal of biogenic habitats and smoothing of surface components such as burrows. This leads to a decrease in nursery habitat for juvenile fish and invertebrates (Kaiser *et al.*, 2002) potentially compounding the impacts on fish stocks of overexploitation and other stressors such as climate change. Overexploitation of fish stocks has resulted in an observed decrease in genetic diversity of red snapper, orange roughy and flounder populations suggesting that the habitat service to maintain viable gene pools can be challenged (Hauser *et al.*, 2002; Hoarau *et al.*, 2005; Smith *et al.*, 1991).

Many aquaculture operations have led to a loss of important coastal nursery habitats. In Asia and Latin America the lack of regulation and appropriate management of aquaculture led to clearance of 1–1.5 million hectares of coastal lowlands (primarily mangroves and salt marshes) for construction of shrimp ponds (Páez-Osuna, 2001). Even when regulation is in place and habitat removal is not a direct action, studies have shown that fish farming leads to localised decline in nursery habitats such as seagrass and maerl beds (Delgado *et al.*, 1999; Cancemi *et al.*, 2003; Hall-Spencer *et al.*, 2006). For example, experimental studies have shown that where oysters are farmed by suspending them in bags they decrease the amount of light reaching the seabed which causes degradation of seagrass beds and reduces photosynthesis (Skinner *et al.*, 2013).

Impacts of coastal development: physical structures, eutrophication, contamination and invasive species
As discussed in Chapter 7, coastal structures can cause erosion or deposition of sediments on adjacent intertidal and subtidal habitats. This affects

natural habitats such as kelp beds that are known to provide valuable habitat services (Smale et al., 2013). Artificial structures, however, create novel habitat in their own right and in some cases this might be viewed as an ecosystem service. For example, offshore oil and gas structures and wind energy arrays restrict access to fishing vessels and provide refugia. The presence of hard structures such as wind turbines can create novel reefs with localised upwelling that can provide shelter and habitat for other species (De Troch et al., 2013; Krone et al., 2013; Reubens et al., 2013).

In contrast, eutrophication (see Chapter 8) changes the biodiversity of keystone species that create biogenic habitat. For example, in rocky shores perennial macroalgae can be replaced with fast-growing, opportunistic and filamentous annual species or with filter-feeding bivalve or barnacle reefs (Worm, 2000; Korpinen, 2007a, 2007b; Westerbom, 2006). This changes the nature of the habitat service made available. In the case of opportunistic macroalgae these are often associated with reduced dissolved oxygen due to decomposition as the algae smothers underlying organisms and also as it dies back seasonally due to low light levels or exhaustion of nutrients. Eutrophication-induced hypoxia and anoxia affect habitat conditions potentially leading to loss of suitable spawning grounds for commercially exploited fish species. Yet below the photic zone hard-bottom, filter-feeding, reef-building invertebrate communities benefit from additional nutrients under conditions of mild eutrophication, and the biogenic structure can be enhanced simultaneously providing more habitat service (Worm, 2000; Westebom, 2006). An example is the increase in the abundance of blue mussels and roach in the northern Baltic Sea as a consequence of eutrophication (Lappalainen et al., 2005).

The impacts of contaminants on biogenic habitat-formers (e.g. kelps, seagrasses and mussels) and their functioning are largely negative (Chapter 9, Figure 9.3) although among the different types of autotrophic habitat-formers there are differences in the susceptibility of their ecosystem functioning to contaminants. Generally it is therefore assumed that contaminants reduce habitat services. However, metacombinations of contaminants, such as metals and organic compounds, seem to have no effects on primary productivity, possibly due to the occurrence of antagonistic effects of different types of contaminants (Chapters 4 and 9). Thus impacts of contaminants on this service will vary and understanding of the specific effects is required on a contaminant by contaminant basis.

There are various mechanisms by which invasive species can impact on and alter biogenic habitats (Chapter 10 and references therein) and thus affect habitat services although the extent of the effect of many of these is highly context specific. For example, habitat-forming species can be outcompeted by invasive species, resulting in a change from one habitat to another and presumably one type of habitat service to another. *Caulerpa* is a macroalgae genus that can outcompete native seagrasses (de Villele and Verlaque, 1995; Ceccherelli and Cinelli, 1997). Since *Caulerpa* is structurally rather different to seagrass the communities which associate with it will change and hence the habitat service will change. Some invasive species can provide new biogenic habitat by converting unvegetated sediment to vegetated or reef habitat thus providing habitat services that were previously unavailable, or enhancing existing biogenic structure. Examples include non-indigenous oysters, seaweeds, seagrasses and saltmarshes (Chapter 10). Whilst this may seem to be beneficial often the unvegetated sediments provide important physical habitat in their own right, particularly in estuaries where they support burying fish like flounders and stargazers, as well as many birds that feed on mudflats, particularly during migrations (Parks, 2006). The context will determine whether changes from one habitat service to another are positive or negative.

Invasive plants may affect animals that prefer to forage around and inhabit native plants (York *et al.*, 2006; Schmidt and Scheibling, 2006; Willette and Ambrose, 2012), but these effects appear to be relatively minor, probably because marine animals are often generalists that can survive on different vegetation (Bell, 1991). Consumption by invasive species of biogenic habitat-forming species in estuarine soft-sediment systems can be much more problematic and have a large impact. Invasive crabs such as *Carcinus maenas* or *Hemigrapsus sanguineus* consume large quantities of sessile animals, in particular mussels (Grosholz and Ruiz, 1996; DeGraaf and Tyrrell, 2004) that are important biogenic habitat-formers supporting other sessile animals and plants (Altieri *et al.*, 2007; Lang and Buschbaum, 2010).

11.3.4 Cultural services

Cultural ecosystem services are provided by ecosystems as a whole, as well as by the species that live within them. They contribute to many cultural benefits such as leisure, recreation and tourism and aesthetic experiences. The generation of these benefits is enabled and enhanced by the presence of charismatic megafauna such as dolphins, seals, whales, basking sharks, turtles and seabirds. Human activities that reduce the populations of such organisms can therefore reduce the benefits derived from these wildlife related services.

Impacts of fishing and aquaculture
Capture fisheries impact directly on many cultural services through substantial removal of charismatic megafauna as non-target bycatch (Lewison et al., 2004). Around the world non-target bycatch goes largely unrecorded and is extremely difficult to evaluate. Nevertheless, estimates suggest that over 650 000 cetaceans and seals (pinnipeds) are unintentionally caught each year by gills nets (Read et al., 2006), whilst sharks, turtles and seabirds are frequently caught in longlines and trawls (Casale et al., 2007). Ghost fishing can also have a substantial impact on charismatic species; for example, one 1500 m long net which was estimated to have only been at sea for one month already contained 2 sharks, 99 seabirds and 75 salmon (EPA, 1992). Up to 100 000 marine mammal deaths a year were estimated to be caused by entanglement in discarded fisheries nets and lines (Wallace, 1985).

There are also many indirect impacts of capture fisheries on marine megafauna. For example, competition for target species between fisheries and marine megafauna may lead to declines in the latter. This is particularly pertinent for fisheries which target small fish such as anchovies, capelins and sandeels for feed and fertiliser. In the North Sea, fisheries' exploitation of sandeels led to a decline in black-legged kittiwake populations (Daunt et al., 2008), whilst exploitation of Peruvian anchovies led to a decrease in seabird and marine mammal populations (Pauly, 1987; Jahncke et al., 2004). Conversely, removal of predatory fish such as cod can lead to a substantial increase in these small fish (Bundy and Fanning, 2005), so the effect of fisheries on the prey abundance of marine megafauna can vary spatially and temporally, dependent on which fish are targeted by local fisheries. Where local declines of large megafauna such as whales and sharks have occurred, this can lead to a reduction in the benefits derived from these species, as evidenced by a reduction in local tourism (Hoyt and Hvenegaard, 2002; Topelko and Dearden, 2005).

It is not only megafauna that capture fisheries impact, but also many culturally valued ecosystems. For example, indiscriminate fishing techniques such as dynamiting coral reefs have resulted in substantial destruction of coral reef communities, transforming reef into rubble and reducing reef fish abundance and diversity (Riegl, 2001; Fox, 2004). Bottom trawling is known to impact fragile ecosystems such as seamounts (Clark and O'Driscoll, 2003), cold water corals and sponge fields (Roberts, 2002). As already mentioned in Section 11.2, overfishing can also contribute to changes in species diversity that result in phase shifts, for example, from coral reefs to systems dominated by macroalgae (McManus and Polsenber, 2004), and from kelp forests to urchin barrens (Ling et al., 2009). The degree of impact of these changes to cultural ecosystem services is unclear, but where individuals' livelihoods and local economies are dependent on such ecosystems, for example, for tourism, it is not unreasonable to assume that their loss will have social and economic consequences.

Impacts of coastal development: physical structures, eutrophication, contamination and invasive species
Demand for coastal tourism and recreation has grown substantially in recent decades and marine coastal tourism is one of the fastest growing sectors of contemporary tourism (Hall, 2001). This growing demand, together with a growing coastal population as a whole, is associated with increased coastal development, exerting pressure on coastal ecosystems. Where poorly planned, such developments can result in negative impacts. For example, rapid development of coastal Turkey outstripped the construction of the sewerage system leading to the disposal of sewage out at sea and a reduction in water quality (Burak et al., 2004). Offshore artificial structures such as wind turbines and oil and gas installations can have both positive and negative impacts on charismatic species such as seals, cetaceans and birds. These impacts vary during the life cycle of the structures. Seals and cetaceans may be adversely affected during the construction phase but then either unaffected during the operational phase (Tougaard et al., 2006a; Tougaard et al., 2006b) or else, in the case of seals, potentially benefit from this phase because of the artificial reef fish-aggregating effect which attracts them to feed at the installations (Russell et al., 2014; Wilhelmsson et al., 2006). Bird mortality through collision with turbines can occur, although the level of risk is species specific and risk can be reduced through careful selection of location (Garthe and Huppop, 2004).

A growing coastal population and associated development makes it easier for contaminants and excess nutrients to enter the marine environment. In areas where eutrophication is a problem, it can impact upon many cultural services. For example, in eutrophic areas where hypoxia and anoxia are experienced, mobile charismatic species such as turtles (as well as mobile fish and invertebrates) tend to move away (reviewed by Craig et al., 2001), potentially affecting nature watching opportunities in these areas. In the Gulf of Mexico, benthic-feeding sea turtles did not occur in hypoxic areas, even though they were abundant in nearby areas (Craig et al., 2001). Aesthetic experiences can also be impacted by eutrophication. During the early stages of eutrophication, increased loose-lying macroalgae and seagrass detritus can accumulate on beaches reducing beach value and increasing cleaning costs. In many coastal areas, eutrophication has caused unsightly algal foams to appear on beaches resulting in negative impacts on leisure and recreation services (OSPAR, 2010). A number of studies have attempted to estimate the value of the effects of such foams and eutrophication on recreation (see Taylor and Longo, 2010 for a review). For example, in Sweden, Sandström (1996) found that a 50% reduction in nutrient loading (and hence eutrophication) along the entire Swedish coast would generate seaside recreation benefits of between €29.6 million and €66.5 million. In Bulgaria, the residents of one community affected by foams would be willing to pay approximately €1.3 million for the elimination of the blooms (Taylor and Longo, 2010). In the Caribbean Sea eutrophication impacts (caused by untreated sewage waters) include overgrowth of coral reefs by macroalgae, fish mortalities, decline of seagrass meadows and red tides (UNEP CEP, 2013) all of which can impact on local opportunities for leisure, recreation and tourism and aesthetic experience. Here and elsewhere in tropical and subtropical zones, marine biodiversity and abundance play an important role in the underwater experience of both tourists and local populations (Principe et al., 2011).

Other forms of contamination, such as oil spills and marine debris also affect cultural ecosystem services. Oil spills may conspicuously impact on species that are important for cultural services causing the death of sea birds and marine mammals such as sea otters. For example, 31 000 oiled birds were found after the Exxon Valdez oil spill, but estimated mortality was ten times higher (Piatt et al., 1990). Marine debris, especially plastics, is an increasing problem for the marine environment. Between 50% and 90% of shoreline debris is made up of plastics (Barnes et al., 2009). Marine fish, invertebrates and megafauna can be negatively affected by ingestion of such debris (Laist, 1987; Daunt et al., 2008). Large pieces of plastic are known to kill and injure marine mammals and sea birds via ingestion or entanglement (Rochman and Browne, 2013). The implications of this for cultural ecosystem services is unknown, but when beaches are persistently plagued by plastics and litter, losses do occur to the tourism industry (Defeo et al., 2009).

Invasive species can also impact on cultural services. Invasive lionfish in the Caribbean exert strong top-down control on juvenile tropical reef fish (Albins and Hixon, 2008; Albins, 2013) potentially altering the diversity of reef fish communities that are attractive to tourists. Pacific oysters, slipper limpets and razor shells have invaded estuaries and shallow sandflats along the eastern edge of the North Sea. Particularly in the Wadden See they have displaced and disrupted large mussel beds that previously supported diverse seabird populations (Baird, 2012, although see also Isacch et al., 2004 for another example in Patagonia where this impact was not observed with the invasion of Pacific oysters). Areas such as the Wadden See are important conservation sites and are also attractive for birdwatchers and eco-tourists. These invasives are therefore threatening the cultural services in this area. Promoted partially as a response to address the problem in the Wadden See a new form of tourism has emerged: gathering of oysters for personal consumption ('oyster safari'). This suggests that invasive species can both detract from and enhance different cultural services; the net benefit from the invasion, however, is unclear.

11.4 Influences of interactions between services, spatial and temporal scales and context

Delivery of many marine ecosystem services is strongly interlinked and synergistic. The biological activity and ecosystem functions of the same, or very similar, organisms underpin several regulatory and food provision services (Austen et al., 2011; Chapter 2). Impacts on particular elements of biodiversity or functions can affect several of these services simultaneously. The impacts on these services can then have further effects on other services. For example, cultural services such as leisure and recreation are dependent on clean, functioning seas. Hence the functioning of organisms that deliver waste treatment and assimilation services also underpins cultural services. Thus some cultural services are indirectly vulnerable to the same impacts that directly affect waste treatment and assimilation services. Similarly, habitats such as salt marshes and biogenic intertidal reefs that prevent or moderate disturbance by mitigating the hazards of flooding and wave damage also provide supporting habitat for birds and juvenile fish which may be of importance for cultural services or for food provision, therefore any impacts on these habitats will impact multiple services. Locally caught fish and shellfish are a realised benefit of food provisioning services. Since in some places people like to eat locally caught food, or catch their own food, the local populations of fish and shellfish provide both food provision services and cultural ecosystem services of cultural heritage and/or as an element of leisure, recreation and tourism services. Thus all of these services are simultaneously vulnerable to impacts on targeted fish populations.

The interlinked and synergistic nature of the delivery of many marine ecosystem services is also evident spatially, as would be expected when considering ecosystem services in interconnected marine habitats from estuaries to coasts and through to the open ocean. Thus impacts on biodiversity and functions in one location, or diffusely across larger areas, may have effects on services elsewhere. For example, changes in land use may affect water quality at the coast, or changes in coral reefs and seagrass meadows may affect the delivery of habitat services and storm protection services of mangroves (Barbier, 2009). These effects may also arise at different points in time. Large, global-scale processes such as carbon sequestration may be slow to change and lag behind the initial shock that brought about the change. In contrast, relatively smaller processes, such as localised algal blooms, may occur quickly with little lag (MA, 2003).

To better manage human activities it is particularly important to improve our current knowledge and understanding of the extent to which impacts occurring at different spatial or temporal scales affect ecosystem services and benefits, and where these effects take place. Key research questions that can be raised across different marine ecosystems and habitats and the different ecosystems services they provide include: How much service does an area provide? Does provision of service vary temporally? Does this provision coincide with demand for the service and its benefits? What are the critical spatial scales at which services need to be delivered to enable benefits? What are the critical times to avoid impacting on ecosystems to ensure continued provision of services and benefits? How resistant and resilient are the ecosystems to the impacts? How do the impacts on the services affect the benefits they generate?

Context is also likely to be important in assessing and managing the impacts of human activities and other stressors on different ecosystem services. For example, there are geographic and social differences in the importance of fish populations for culture and for food. Compared to many developed countries, fish and other seafood contribute a higher proportion of dietary protein intake in some developing countries (especially small island developing states), and consequently play an important role in local food security (FAO, 2014). In some cases aquaculture makes a substantial contribution to fish protein, especially in the Asia-Pacific region (Chapter 6). Key habitats such as mangroves are often cleared to provide space for ponds to grow species. In China, large bays and areas of coastline are densely covered with suspended pens, nets and bags for the culture of fish, shellfish and seaweed. This prioritisation of food provision leads to trade-offs with other ecosystem services (see Section 11.6). Elsewhere, fish and other marine biodiversity may make a greater contribution to the cultural services of leisure, tourism and recreation. In many tropical or subtropical regions marine biodiversity related tourism can provide significant local income. Global coral reef tourism was estimated to generate US$9.6 billion in annual net benefits (Cesar et al., 2003). For the Great Barrier Reef alone, Carr and Mendelsohn (2003) estimate that the total consumer surplus (the amount people would be willing to pay over and above what they actually pay) associated with tourism was between US$710 million and US$1.6 billion.

11.5 Trade-offs in services, benefits and values

Trade-offs between ecosystem services occur when attempts are made to optimise or manage a single service, often leading to a reduction or loss in one or more of the other services provided by that ecosystem (Holling and Meffe, 1996; Rodríguez et al., 2006). Trade-offs can influence the type, magnitude and relative mix of ecosystem services, and although trade-offs are sometimes made consciously, their occurrence is often unrecognised (Rodríguez et al., 2006). This lack of recognition stems from knowledge deficits about the interdependent nature and nonlinear relationships between ecosystem services and our limited understanding of how management actions may influence these services.

Rodríguez et al. (2006) identify three key types of trade-off: in space, in time and in terms of reversibility (although in reality trade-offs are rarely so clear cut). Trade-offs in space occur when an area is managed for a particular service (e.g. food provision) at the expense of other services. Such a trade-off is demonstrated by the intensive use of terrestrial ecosystems for the production of food and the concomitant overburdening of the waste treatment and assimilation service provided downstream by marine ecosystems. Application of fertilisers and livestock manure on farmland promotes increased terrestrial food provision, but excess nutrients and nutrient-rich effluent from the storage of silage are conveyed, via freshwater runoff, into estuarine and coastal areas (Austen et al., 2011). These inputs drive much of the eutrophication in estuarine and coastal waters leading to impacts described in Section 11.3.

Another example results from the conflict between wildlife watching (a popular leisure activity and benefit provided by cultural services) and fisheries overexploitation. Commercial fisheries for small fish species, such as sandeels, may reduce food availability for seabirds (Frederiksen et al., 2004; Frederiksen et al., 2007; Wanless et al., 2005), marine mammals and predatory fishes (MacLeod et al., 2007). Poor breeding success at many seabird colonies in the North Sea was attributed to a lack of sandeel prey resources. A ban was imposed on sandeel fishing off eastern Scotland and north-east England. Here the trade-off was made in favour of species of cultural service importance rather than food provision. Reductions in breeding success and survival of seabirds are probably now being caused by increases in sea surface temperature as a result of climate change (Mitchell et al., 2010) but if the ban on sandeel fishing were reversed it would potentially exacerbate the decline in seabird populations.

Trade-offs in time result when current ecosystem service delivery is favoured at the expense of future services. Such trade-offs may occur between services, but also within a service. For example, poor management of wild capture fishing has resulted in unsustainable and excessive fish extraction. This impacts negatively not only on other ecosystem services, but also on the capacity of the food provision service to continue providing food into the future. The gradual loss of wetlands (e.g. from the Mississippi delta) and of mangroves (e.g. in Thailand) in favour of improving access to ports and aquaculture has meant that the ability of such habitats to protect against storms has dramatically decreased over time. The devastating effects of this loss were clearly demonstrated by Hurricane Katrina and the Asian tsunami of 2004 (Costanza and Farley, 2007).

Drawing on the fishing-seabird example above, temporal effects can also be seen. While fisheries have negatively affected some seabird and charismatic species populations, at the same time, they have been benefiting other seabirds by providing them with food as discharged offal and discarded undersize fish. Over time, this has supported populations of scavenging species (e.g. great skua, northern fulmar) above levels that natural food sources could sustain. The introduction of measures to conserve fish stocks has consequently reduced the amount of discards. This in turn is thought to have contributed to a population downturn of northern fulmars and other offshore surface-feeders since the mid-1990s (Mitchell et al., 2010). This situation is likely to become more widespread with the gradual introduction of a ban on fisheries' discards across the EU through the Common Fisheries Policy.

Construction of artificial structures in coastal areas to allow access to ships and boats (e.g. for recreation and fishing as well as transport), or protection of the shoreline from erosion creates further temporal effects and trade-offs among ecosystem services. Construction of jetties allowing access to coastal ecosystems for recreation has led to erosion of 85% of Florida's beaches (Finkl, 1996), impacting on cultural services such as aesthetic experience, tourism, leisure and recreation, and the regulating service of coastal erosion prevention (Barbier et al., 2011). In areas such as Florida, where beach-related activities bring in an extremely high income, beach nourishment (removal of sediment from elsewhere and redeposition on the beach) to maintain beaches is required (Finkl, 1996). The action of nourishing the beach, as well as being an extremely costly exercise, has its own ecosystem impacts and hence leads to further trade-offs in the provision of ecosystem services (e.g. increased erosion or sedimentation at the source site, associated mortality of benthic invertebrates, and removal of prey species for birds, Speybroeck et al., 2006). Unfortunately, temporal trade-offs will continue to be the norm while short-term needs drive management decisions.

The third type of trade-off focuses on the extent to which changes to ecosystem services are reversible. In ecology we rarely know where the tipping points to irreversibility are, which is an important issue for valuation. If the tipping points for ecosystem services are not known the value of ecosystem services will always be substantially underestimated. The collapse of the Grand Banks cod fishery provides an example of such a trade-off that was unknowingly made. The Grand Banks, off the coast of Newfoundland, were rich fishing grounds, especially for cod. At its peak in the 1960s, fishing supported whole communities and the industry employed 40 000 people. Overfishing and fishery management failures devastated the fishery and thousands of people lost their livelihoods. There has been a moratorium on cod fishing since 1992, but only very recently do cod stocks on the Grand Bank appear to be starting to recover, although the population that is recovering comprises much smaller specimens than previously (Frank et al., 2011).

Abandoned aquaculture ponds for shrimp also demonstrate irreversible impacts (if only because of the high cost of restoration). Shrimp ponds typically have a productive life of only 5 years, after which producers move to new areas. For mangroves to re-establish, the soils need to be detoxified and mangrove seedlings need to be protected for many years (Barbier, 2009). High restoration costs mean that countries like Thailand have lost 50–65% of their mangrove cover to shrimp farm conversion

(Aksornkoae and Tokrisna, 2004). Such irreversibility is often due to the presence of thresholds or limits to the current state of the ecosystem. Much effort has been placed in trying to understand and identify signs that indicate these thresholds (Dobson et al., 2006; Thrush et al., 2009), but currently the only sure way of identifying a threshold is to actually cross it (Carpenter et al., 2005).

Many of the trade-offs between different ecosystem services go barely recognised and there are few formal attempts to address them. To rectify this situation marine planning has been introduced in various parts of the world. This is accompanied with integrated legislation and policies to address use and exploitation of marine resources, both internationally (e.g. UN Convention on Biological Diversity, EU Integrated Maritime Policy, EU Marine Strategy Framework Directive) and nationally (e.g. Canada Oceans Act 1996, UK Marine and Coastal Access Act 2009, US National Ocean Policy 2013). Awareness has also been growing of the role of ecosystems in providing services and benefits (e.g. MA (2003), TEEB (2010) and the establishment of the Intergovernmental Panel on Biodiversity and Ecosystem Services (IPBES) in 2012). It is anticipated that the trade-offs among different ecosystem services and the human activities that impact upon them will become more widely recognised and managed.

An example of how marine planning might assist with management of future trade-offs comes with the rapid development of offshore marine renewable energy. Increasing emplacement of large hard structures in the sea to enable extraction of energy supports both energy security and low carbon emissions goals. At the same time it reduces access to fishing grounds. Current policy simply aims to minimise the placement of devices in favoured fishing grounds. Future policy might seek to actively utilise such areas to promote marine protection, or to enhance shellfish populations and allow fixed gear fishing only, or to promote such areas for recreational angling, or to use the structures to suspend aquaculture lines for bivalve production for sea pens for fish culture, or some combination of all of these. Marine planning should enable the different trade-offs among these activities to be recognised. The most favourable balance of different ecosystem outcomes should be achieved according to current policy needs, while at the same time bearing in mind future needs and potentially irreversible changes.

Monetary (and non-monetary) valuation of the benefits from ecosystem services is seen as a mechanism for enabling the trade-offs between different management objectives to become more easily recognised. By allowing the gains and losses from ecosystem exploitation to be quantified the implications of social and economic uses of the marine environment can be more explicitly identified. Effective valuation requires adequate knowledge of key ecological and economic relationships. As can be seen in this chapter, much of the detailed understanding of how ecosystems and especially their services are affected by different impacts is lacking. This has been highlighted as one of the main challenges for the incorporation of ecosystem service assessments and valuation into marine planning (Börger et al., 2014). Using valuation in decision-making is also complicated by different sectors of society, and even individuals, placing different values on different services and benefits. This raises questions over whose values are then considered the most legitimate and how they are used in the decision-making process (O'Neill and Spash, 2000).

In the context of trade-offs, Barbier (2009) identifies a number of additional challenges. For example there is considerable uncertainty over future ecosystem benefits, meaning ecosystems services today may remain undervalued and too much development or exploitation may occur. If ecosystem services are too heavily impacted today, it may not be possible to restore them to their previous state. Estimating the value of the option to avoid such future losses is problematic, not least because of uncertainty about the state of future ecosystems, but also because of uncertainty over future preferences of individuals. Another challenge that remains is how to measure the value of ecosystem resilience. The resilience of an ecosystem should enter into ecosystem management decisions, but requires knowledge of the probability of crossing a threshold and the implications of breaching this threshold in ecosystem terms. Significant further research is required to elucidate many of the unknowns that have been raised here.

11.6 Key Points and recommendations

- Understanding how different stressors affect biodiversity and how that links to changes in ecosystem function is highly complex. Although some stressors may impact several species (causing biodiversity loss), these may not always include species that are important for ecosystem functions and hence service provision. Conversely, a stressor may impact a single species which plays an important role in ecosystem functions and services (such as seagrass and kelp forests as providers of habitat services) leading to significant negative impacts on service provision.

- In some cases, the key service providers and their sensitivities are known, although in many other cases there is currently only piecemeal and largely qualitative information on how different stressors impact on the different ecosystem functions and services.
- Extraction of fish through capture fisheries and aquaculture to realise food provision services has documented impacts on all other services.
- Impacts of other stressors on ecosystem services are less well documented, particularly impacts of built structures, contaminants and invasive species.
- Contaminants particularly impact on food provision services by affecting the safety, or perception of safety of marine food for human consumption.
- Substitutability of biodiversity and functions to maintain delivery of services is poorly understood. This is evident especially in relation to the impact of invasive species where one type of biogenic habitat may be replaced with another, but it is uncertain how or whether the habitat service is continued.
- There are many interactions between the different services. The biological activity and ecosystem functions of the same, or very similar, organisms underpin several services and several ecosystem services are interdependent. Hence impacts on one service will likely impact others either directly or indirectly.
- Delivery of many ecosystem services is also linked spatially and temporally, this will affect the relationship between stressors and ecosystem services. However the nature and importance of interconnectedness at different spatial and temporal scales for the delivery of ecosystem services is also currently poorly researched and requires elucidation.
- In assessing and managing the impacts of human activities and other stressors on different ecosystem services, geographical and social context, as well as the environment, need to be considered.
- Many trade-offs between ecosystem services occur, most frequently when attempts are made to optimise or manage a single service, often leading to a reduction or loss in one or more of the other services provided by that ecosystem. Trade-offs can influence the type, magnitude and relative mix of ecosystem services, and although trade-offs are sometimes made consciously, their occurrence is often unrecognised.
- Trade-offs between ecosystem services occur in space, time and in consideration of their irreversibility. They are particularly evident when considering utilisation of food provisioning services by fisheries and aquaculture but also terrestrial food provisioning which adversely

impacts on other ecosystem services. Marine planning will facilitate more systematic consideration of these trade-offs.
- Monetary (and non-monetary) valuation of the benefits from ecosystem services is seen as a mechanism for enabling the trade-offs between different management objectives to become more easily recognised. Effective valuation requires adequate knowledge of key ecological and economic relationships but currently much of the detailed understanding of how ecosystems and especially their services are affected by different impacts is lacking and is a major challenge for the incorporation of ecosystem service assessments and valuation into marine planning.
- Better modelling and predictive tools are required: that link biodiversity to function, provision of ecosystem services and value of benefits; that can incorporate temporal and spatial scale issues; that identify the interactions between ecosystem services; and that can model the impact of stressors according to different scenarios. To support development of such models further primary data is required at different spatial and temporal scales, on impacts of stressors on ecosystems and their functioning, and of values of marine ecosystem benefits.

References

Abend, A. G. and Smith, T. D. (1995). Differences in ratios of stable isotopes of nitrogen in long-finned pilot whales (*Globicephala melas*) in the Western and Eastern North Atlantic. *ICES Journal of Marine Science*, 52(5), 837–841.

Ainley, D. G. and Pauly, D. (2014). Fishing down the food web of the Antarctic continental shelf and slope. *Polar Record*, 50, 92–107.

Aksornkoae, S. and Tokrisna, R. (2004). Overview of shrimp farming and mangrove loss in Thailand. In *Shrimp Farming and Mangrove Loss in Thailand,* ed. E. B. Barbier and S. Sathirathai. London: Edward Elgar, pp. 37–51.

Albins, M. A. (2013). Effects of invasive Pacific red lionfish *Pterois volitans* versus a native predator on Bahamian coral-reef fish communities. *Biological Invasions*, 15(1), 29–43.

Albins, M. A. and Hixon, M. A. (2008). Invasive Indo-Pacific lionfish *Pterois volitans* reduce recruitment of Atlantic coral-reef fishes. *Marine Ecology Progress Series*, 367, 233–238.

Altieri, A. H., Silliman, B. R. and Bertness, M. D. (2007). Hierarchical organization via a facilitation cascade in intertidal cordgrass bed communities. *American Naturalist*, 169(2), 195–206.

Alzieu, C. (2000). Environmental impact of TBT: the French experience. *Science of the Total Environment*, 258, 99–102.

Anderson, D. M., Pitcher, G. C. and Enevoldsen, H. O. (2010). The IOC International Harmful Algal Bloom Program: history and science impacts. *Oceanography*, 23(3), 72–85.

Angeletti, R., Binato, G., Guidotti, M. et al. (2014). Cadmium bioaccumulation in Mediterranean spider crab (*Maya squinado*): human consumption and health implications for exposure in Italian population. *Chemosphere*, 100, 83–88.

Apostolaki, E. T., Tsagaraki, T., Tsapaki, M. and Karakassis, I. (2007). Fish farming impact on sediments and macrofauna associated with seagrass meadows in the Mediterranean. *Estuarine Coastal and Shelf Science*, 75(3), 408–416.

Austen, M. C., Malcolm, S. J., Frost, M. et al. (2011). Marine. In *The UK National Ecosystem Assessment Technical Report*. UNEP-WCMC, pp. 459–499.

Auster, P. J., Malatesta, R. J., Langton, R. W. et al. (1996). The impacts of mobile fishing gear on seafloor habitats in the Gulf of Maine (Northwest Atlantic): Implications for conservation of fish populations. *Reviews in Fisheries Science*, 4(2), 185–202.

Baird, D. (2012). Assessment of observed and perceived changes in ecosystems over time, with special reference to the Sylt-Romo Bight, German Wadden Sea. *Estuarine Coastal and Shelf Science*, 108, 144–154.

Balvanera, P., Pfisterer, A. B., Buchmann, N. et al. (2006). Quantifying the evidence for biodiversity effects on ecosystem functioning and services. *Ecology Letters*, 9(10), 1146–1156.

Barbier, E. B. (2009). Ecosystem service trade-offs. In *Ecosystem-Based Management for the Oceans*, ed. K. L. McLeod and H. M. Leslie. Washington DC: Island Press, pp. 129–144.

Barbier, E. B., Hacker, S. D., Kennedy, C. et al. (2011). The value of estuarine and coastal ecosystem services. *Ecological Monographs*, 81(2), 169–193.

Barnes, D. K. A., Galgani, F., Thompson, R. C. and Barlaz, M. (2009). Accumulation and fragmentation of plastic debris in global environments. *Philosophical Transactions of The Royal Society B*, 364(1526), 1985–1998.

Béland, P., DeGuise, S., Girard, C. et al. (1993). Toxic compounds and health and reproductive effects in St. Lawrence beluga whales. *Journal of Great Lakes Research*, 19(4), 766–775.

Bell, S. S. (1991). Amphipods as insect equivalents? An alternative view. *Ecology*, 72, 350–354.

Best, P. B. (1993). Increase rates in severely depleted stocks of baleen whales. *ICES Journal of Marine Science*, 50(2), 169–186.

Bester, K., Hühnerfuss, H., Brockmann, U. and Rick, H. J. (1995). Biological effects of triazine herbicide contamination on marine phytoplankton. *Archives of Environmental Contamination and Toxicology*, 29(3), 277–283.

Bilio, M. and Niermann, U. (2004). Is the comb jelly really to blame for it all? *Mnemiopsis leidyi* and the ecological concerns about the Caspian Sea. *Marine Ecology Progress Series*, 269, 173–183.

Börger, T., Beaumont, N. J., Pendleton, L. et al. (2014). Incorporating ecosystem services in marine planning: the role of valuation, *Marine Policy*, 46, 161–170.

Brown, J., Macfadyen, G., Huntington, T., Magnus, J. and Tumilty, J. (2005). Ghost fishing by lost fishing gear. Joint Report, Institute for European Environmental Policy/Poseidon Aquatic Resource Management Ltd.

Bundy, A. and Fanning, L. P. (2005). Can Atlantic cod (*Gadus morhua*) recover? Exploring trophic explanations for the non-recovery of the cod stock on the

eastern Scotian Shelf, Canada. *Canadian Journal of Fisheries and Aquatic Sciences*, 62(7), 1474–1489.
Burak, S., Dogan, E. and Gazioglu, C. (2004). Impact of urbanization and tourism on coastal environment. *Ocean and Coastal Management*, 47(9–10), 515–527.
Burkholder, J. M., Tomasko, D. A. and Touchette, B. W. (2007). Seagrasses and eutrophication. *Journal of Experimental Marine Biology and Ecology*, 350(1–2), 46–72.
Bushkin-Bedient, S. and Carpenter, D. O. (2010). Benefits versus risks associated with consumption of fish and other seafood. *Reviews on Environmental Health*, 25, 161–191.
Caddy, J. F. (2000). Marine catchment basin effects versus impacts of fisheries on semi-enclosed seas. *ICES Journal of Marine Science*, 57(3), 628–640.
Cancemi, G., De Falco, G. and Pergent, G. (2003). Effects of organic matter input from a fish farming facility on a *Posidonia oceanica* meadow. *Estuarine Coastal and Shelf Science*, 56(5–6), 961–968.
Cardoso, P. G., Pardal, M. A., Lillebø, A. I. *et al.* (2004). Dynamic changes in seagrass assemblages under eutrophication and implications for recovery. *Journal of Experimental Marine Biology and Ecology*, 302(2), 233–248.
Carpenter, S. R., Westley, F. and Turner, M. G. (2005). Surrogates for resilience of social–ecological systems. *Ecosystems*, 8(8), 814–944.
Carr, L. and Mendelsohn, R. (2003). Valuing coral reefs: a travel cost analysis of the Great Barrier Reef. *Ambio*, 32(5), 353–357.
Casale, P., Cattarino, L., Freggi, D., Rocco, M. and Argano, R. (2007). Incidental catch of marine turtles by Italian trawlers and longliners in the central Mediterranean. *Aquatic Conservation: Marine and Freshwater Ecosystems*, 17(7), 686–701.
Ceccherelli, G. and Cinelli, F. (1997). Short-term effects of nutrient enrichment of the sediment and interactions between the seagrass *Cymodocea nodosa* and the introduced green alga *Caulerpa taxifolia* in a Mediterranean bay. *Journal of Experimental Marine Biology and Ecology*, 217(2), 165–177.
Cesar, H. S. J., Burke, L. and Pet-Soede, L. (2003). *The Economics of Worldwide Coral Reef Degradation*. Arnhem, The Netherlands: WWF and International Coral Reef Action Network.
Cheal, A. J., MacNeil, M. A., Cripps, E. *et al.* (2010). Coral-macroalgal phase shifts or reef resilience: links with diversity and functional roles of herbivorous fishes on the Great Barrier Reef. *Coral Reefs*, 29(4), 1005–1015.
Clark, M. and O'Driscoll, R. (2003). Deepwater fisheries and aspects of their impact on seamount habitat in New Zealand. *Journal of Northwest Atlantic Fisheries Science*, 31, 441–458.
Clarkson, T. W. (1993). Mercury: major issues in environmental health. *Environmental Health Perspectives*, 100, 31–38.
Cloern, J. E. (2001). Our evolving conceptual model of the coastal eutrophication problem. *Marine Ecology Progress Series*, 210, 223–253.
Coale, K. H. (1991). Effects of iron, manganese, copper, and zinc enrichments on productivity and biomass in the sub-arctic Pacific. *Limnology and Oceanography*, 36(8), 1851–1864.

Costanza, R. and Farley, J. (2007). Ecological economics of coastal disasters: Introduction to the special issue. *Ecological Economics*, 63(2–3), 249–253.

Cracknell, D., White, M. P., Pahl, S., Nichols, W. J., and Depledge, M. H. (submitted). Species diversity and psychological well-being: a preliminary examination of dose-response effects in an aquarium setting. *Journal of Environmental Psychology*.

Craig, J. K. (2012). Aggregation on the edge: effects of hypoxia avoidance on the spatial distribution of brown shrimp and demersal fishes in the northern Gulf of Mexico. *Marine Ecology Progress Series*, 445, 75–95.

Craig, J. K., Crowder, L. B., Gray, C. D. *et al.* (2001). Ecological Effects of Hypoxia on Fish, Sea Turtles, and Marine Mammals in the Northwestern Gulf of Mexico. In *Coastal Hypoxia: Consequences for Living Resources and Ecosystems*, ed. N. N. Rabalais and R. E. Turner. Washington DC: American Geophysical Union, pp. 269–292.

Darby, F. A. and Turner, R. E. (2008). Effects of eutrophication on salt marsh root and rhizome biomass accumulation. *Marine Ecology Progress Series*, 363, 63–70.

Daunt, F., Wanless, S., Greenstreet, S. P. R. *et al.* (2008). The impact of the sandeel fishery closure on seabird food consumption, distribution, and productivity in the northwestern North Sea. *Canadian Journal of Fisheries and Aquatic Sciences*, 65(3), 362–381.

Davidson, T. M. and de Rivera, C. E. (2010). Accelerated erosion of saltmarshes infested by the non-native burrowing crustacean *Sphaeroma quoianum*. *Marine Ecology Progress Series*, 419, 129–136.

Davies, K. T. A., Ross, T. and Taggart, C. T. (2013). Tidal and subtidal currents affect deep aggregations of right whale prey, *Calanus* spp., along a shelf-basin margin. *Marine Ecology Progress Series*, 479, 263–282.

Deegan, L. A., Johnson, D. S., Warren, R. S. *et al.* (2012). Coastal eutrophication as a driver of salt marsh loss. *Nature*, 490(7420), 388–392.

Defeo, O., McLachlan, A., Schoeman, D. S. *et al.* (2009). Threats to sandy beach ecosystems: a review. *Estuarine, Coastal and Shelf Science*, 81, 1–12.

DeGraaf, J. D. and Tyrrell, M. C. (2004). Comparison of the feeding rates of two introduced crab species, *Carcinus maenas* and *Hemigrapsus sanguineus*, on the blue mussel, *Mytilus edulis*. *Northeastern Naturalist*, 11(2), 163–166.

Delgado, O., Ruiz, J., Perez, M., Romero, J. and Ballesteros, E. (1999). Effects of fish farming on seagrass (*Posidonia oceanica*) in a Mediterranean bay: seagrass decline after organic loading cessation. *Oceanologica Acta*, 22(1), 109–117.

De Troch, M., Reubens, J. T., Heirman, E., Degraer, S. and Vincx, M. (2013). Energy profiling of demersal fish: A case-study in wind farm artificial reefs. *Marine Environmental Research*, 92, 224–233.

de Villele, X. and Verlaque, M. (1995). Changes and degradation in a *Posidonia oceanica* bed invaded by the introduced tropical alga *Caulerpa taxifolia* in the North Western Mediterranean. *Botanica Marina*, 38(1), 79–87.

Dobson, A., Lodge, D., Alder, J. *et al.* (2006). Habitat loss, trophic collapse, and the decline of ecosystem services. *Ecology*, 87(8), 1915–1924.

Duarte, C. M. (1995). Submerged aquatic vegetation in relation to different nutrient regimes. *Ophelia*, 41, 87–112.

Dulvy, N. K., Sadovy, Y. and Reynolds, J. D. (2003). Extinction vulnerability in marine populations. *Fish and Fisheries*, 4(1), 25–64.
EPA (1992). *Turning the Tide on Trash. A Learning Guide on Marine Debris*. Washington DC: United States Environmental Protection Agency. Available at: http://water.epa.gov/type/oceb/marinedebris/marine_contents.cfm, accessed 12 August 2014.
Falk-Petersen, J., Renaud, P. and Anisimova, N. (2011). Establishment and ecosystem effects of the alien invasive red king crab (*Paralithodes camtschaticus*) in the Barents Sea a review. *ICES Journal of Marine Science*, 68, 479–488.
FAO (2012). *The State of World Fisheries and Aquaculture 2012*. Rome: FAO.
FAO (2014). *FAO yearbook: Fishery and Aquaculture Statistics 2012*. Rome: FAO.
Finkl, C. W. (1996). What might happen to America's shorelines if artificial beach replenishment is curtailed: a prognosis for southeastern Florida and other sandy regions along regressive coasts. *Journal of Coastal Research*, 12(1), R3-R9.
Fleming, L. E., Broad, K., Clement, A. et al. (2006). Oceans and human health: emerging public health risks in the marine environment. *Marine Pollution Bulletin*, 53, 545–560.
Folke, C., Carpenter, S., Walker, B. et al. (2004). Regime shifts, resilience, and biodiversity in ecosystem management. *Annual Review of Ecology Evolution and Systematics*, 35, 557–581.
Folpp, H., Lowry, M., Gregson, M. and Suthers, I. M. (2013). Fish assemblages on estuarine artificial reefs: natural rocky reef mimics or discrete assemblages? *PLoS ONE*, 8(6), e63505.
Ford, J. K. B. and Ellis, G. M. (2006). Selective foraging by fish-eating killer whales *Orcinus orca* in British Columbia. *Marine Ecology Progress Series*, 316, 185–199.
Fowler, A. M. and Booth, D. J. (2013). Seasonal dynamics of fish assemblages on breakwaters and natural rocky reefs in a temperate estuary: consistent assemblage differences driven by sub-adults. *PLoS ONE*, 8(9), e75790.
Fox, H. E. (2004). Coral recruitment in blasted and unblasted sites in Indonesia: assessing rehabilitation potential. *Marine Ecology Progress Series*, 269, 131–139.
Frank, K. T., Petrie, B., Fisher, J. A. D. and Leggett, W. C. (2011). Transient dynamics of an altered large marine ecosystem. *Nature*, 477(7362), 86-U98.
Frederiksen, M., Edwards, M., Mavor, R. A. and Wanless, S. (2007). Regional and annual variation in black-legged kittiwake breeding productivity is related to sea surface temperature. *Marine Ecology Progress Series*, 350, 137–143.
Frederiksen, M., Wanless, S., Harris, M. P., Rothery, P. and Wilson, L. J. (2004). The role of industrial fisheries and oceanographic change in the decline of North Sea black-legged kittiwakes. *Journal of Applied Ecology*, 41(6), 1129–1139.
Frid, C. L. J. and Paramor, O. A. L. (2012). Feeding the world: what role for fisheries? *ICES Journal of Marine Science*, 69: 145–150.
Froese, R., Zeller, D., Kleisner, K. and Pauly, D. (2012). What catch data can tell us about the status of global fisheries. *Marine Biology*, 159(6), 1283–1292.
Garthe, S. and Huppop, O. (2004). Scaling possible adverse effects of marine wind farms on seabirds: developing and applying a vulnerability index. *Journal of Applied Ecology*, 41(4), 724–734.

Godfray, H. C. J., Beddington, J. R., Crute, I. R. et al. (2010). Food security: the challenge of feeding 9 billion people. *Science*, 327(5967), 812–818.

Gowen, R. J., Tett, P., Bresnan, E. et al. (2012). Anthropogenic nutrient enrichment and blooms of harmful phytoplankton. In *Oceanography and Marine Biology: An Annual Review*, Vol. 50, ed. R. N. Gibson, R. J. A. Atkinson, J. D. M. Gordon and R. N. Hughes. Boca Raton, FL: CRC Press, pp. 65–126.

Graham, M. H. (2004). Effects of local deforestation on the diversity and structure of southern California giant kelp forest food webs. *Ecosystems*, 7(4), 341–357.

Grosholz, E. D. and Ruiz, G. M. (1996). Predicting the impact of introduced marine species: Lessons from the multiple invasions of the European green crab *Carcinus maenas*. *Biological Conservation*, 78(1–2), 59–66.

Hall, C. M. (2001). Trends in ocean and coastal tourism: the end of the last frontier? *Ocean and Coastal Management*, 44(9–10), 601–618.

Hallegraeff, G. M. (1993). A review of harmful algal blooms and their apparent global increase. *Phycologia*, 32(2), 79–99.

Hall-Spencer, J. M. and Moore, P. G. (2000). Scallop dredging has profound, long-term impacts on maerl habitats. *ICES Journal of Marine Science*, 57(5), 1407–1415.

Hall-Spencer, J., Allain, V. and Fossa, J. H. (2002). Trawling damage to Northeast Atlantic ancient coral reefs. *Proceedings of the Royal Society B-Biological Sciences*, 269(1490), 507–511.

Hall-Spencer, J., White, N., Gillespie, E., Gillham, K. and Foggo, A. (2006). Impact of fish farms on maerl beds in strongly tidal areas. *Marine Ecology Progress Series*, 326, 1–9.

Harrold, C., Light, K. and Lisin, S. (1998). Organic enrichment of submarine-canyon and continental-shelf benthic communities by macroalgal drift imported from nearshore kelp forests. *Limnology and Oceanography*, 43(4), 669–678.

Hauser, L., Adcock, G. J., Smith, P. J., Bernal Ramírez, J. H. and Carvalho, G. R. (2002). Loss of microsatellite diversity and low effective population size in an overexploited population of New Zealand snapper (*Pagrus auratus*). *Proceedings of the National Academy of Sciences*, 99(18), 11742–11747.

HELCOM (2009). Eutrophication of the Baltic Sea: an integrated thematic assessment of the effects of nutrient enrichment in the Baltic Sea region. *Baltic Sea Environment Proceedings* 115.

Hoarau, G., Boon, E., Jongma, D. N. et al. (2005). Low effective population size and evidence for inbreeding in an overexploited flatfish, plaice (*Pleuronectes platessa* L.). *Proceedings of the Royal Society B: Biological Sciences*, 272(1562), 497–503.

Holling, C. S. and Meffe, G. K. (1996). Command and control and the pathology of natural resource management. *Conservation Biology*, 10, 328–337.

Holmer, M. and Kristensen, E. (1992). Impact of marine fish cage farming on metabolism and sulphate reduction of underlying sediments. *Marine Ecology Progress Series*, 80(2–3), 191–201.

Hooper, D. U., Chapin, F. S., Ewel, J. J. et al. (2005). Effects of biodiversity on ecosystem functioning: A consensus of current knowledge. *Ecological Monographs*, 75(1), 3–35.

Hooper, T. and Austen, M. (2014). The co-location of offshore windfarms and decapod fisheries in the UK: Constraints and opportunities. *Marine Policy*, 43, 295–300.

Hoyt, E. and Hvenegaard, G. T. (2002). A Review of whale-watching and whaling with applications for the Caribbean. *Coastal Management*, 30(4), 381–399.

Hussy, K., St John, M. A. and Bottcher, U. (1997). Food resource utilization by juvenile Baltic cod *Gadus morhua*: a mechanism potentially influencing recruitment success at the demersal juvenile stage? *Marine Ecology Progress Series*, 155, 199 208.

Isacch, J. P., Daleo, P., Alberti, J. and Iribarne, O. (2004). The distribution and ecological effects of the introduced Pacific oyster *Crassostrea gigas* (Thunberg, 1793) in northern Patagonia. *Journal of Shellfish Research*, 23:765–772.

Jahncke, J., Checkley, D. M. and Hunt, G. L. (2004). Trends in carbon flux to seabirds in the Peruvian upwelling system: effects of wind and fisheries on population regulation. *Fisheries Oceanography*, 13(3), 208–223.

Jennings, S. and Kaiser, M. J. (1998). The effects of fishing on marine ecosystems. *Advances in Marine Biology*, 34, 201–212, 212a, 213–266, 266a, 268–352.

Johnston, E. L. and Roberts, D. A. (2009). Contaminants reduce the richness and evenness of marine communities: a review and meta-analysis. *Environmental Pollution*, 157(6), 1745–1752.

Jonsson, B. and Jonsson, N. (2006). Cultured Atlantic salmon in nature: a review of their ecology and interaction with wild fish. *ICES Journal of Marine Science*, 63(7), 1162–1181.

Kaiser, M. J., Clarke, K. R., Hinz, H. *et al.* (2006). Global analysis of response and recovery of benthic biota to fishing. *Marine Ecology Progress Series*, 311, 1–14.

Kaiser, M. J., Collie, J. S., Hall, S. J., Jennings, S. and Poiner, I. R. (2002). Modification of marine habitats by trawling activities: prognosis and solutions. *Fish and Fisheries*, 3(2), 114–136.

Karakassis, I., Hatziyanni, E., Tsapakis, M. and Plaiti, W. (1999). Benthic recovery following cessation of fish farming: a series of successes and catastrophes. *Marine Ecology Progress Series*, 184, 205–218.

Kideys, A. E. (2002). The comb jelly *Mnemiopsis leidyi* in the Black Sea. In *Invasive Aquatic Species of Europe. Distribution, Impact and Management*, ed. E. Leppakoski, S. Gollasch and S. Olenin. Dordrecht, The Netherlands: Kluwer Academic Publisher, pp. 56–61.

Kite-Powell, H. L., Fleming, L. E., Backer, L. C. *et al.* (2008). Linking the oceans to public health: current efforts and future directions. *Environmental Health*, 7, (Suppl. 2), S6.

Korpinen, S., Honkanen, T., Vesakoski, O. *et al.* (2007a). Macroalgal communities face the challenge of changing biotic interactions: review with focus on the Baltic Sea. *Ambio*, 36(2–3), 203–211.

Korpinen, S., Jormalainen, V. and Honkanen, T. (2007b). Effects of nutrients, herbivory, and depth on the macroalgal community in the rocky sublittoral. *Ecology*, 88(4), 839–852.

Krone, R., Gutow, L., Brey, T., Dannheim, J. and Schröder, A. (2013). Mobile demersal megafauna at artificial structures in the German Bight: likely effects

of offshore wind farm development. *Estuarine, Coastal and Shelf Science*, 125, 1–9.
Laffoley, D. and Grimsditch, G. (2009). *The Management of Natural Coastal Carbon Sinks*. Gland, Switzerland: IUCN.
Laist, D. W. (1987). Overview of the biological effects of lost and discarded plastic debris in the marine environment. *Marine Pollution Bulletin*, 18(6, Supplement B), 319–326.
Lang, A. C. and Buschbaum, C. (2010). Facilitative effects of introduced Pacific oysters on native macroalgae are limited by a secondary invader, the seaweed *Sargassum muticum*. *Journal of Sea Research*, 63(2), 119–128.
Lappalainen, A., Westerbom, M. and Heikinheimo, O. (2005). Roach (*Rutilus rutilus*) as an important predator on blue mussel (*Mytilus edulis*) populations in a brackish water environment, the northern Baltic Sea. *Marine Biology*, 147(2), 323–330.
Lewison, R. L., Crowder, L. B., Read, A. J. and Freeman, S. A. (2004). Understanding impacts of fisheries bycatch on marine megafauna. *Trends in Ecology and Evolution*, 19(11), 598–604.
Lindemann-Matthies, P., Junge, X. and Matthies, D. (2010). The influence of plant diversity on people's perception and aesthetic appreciation of grassland vegetation. *Biological Conservation*, 143(1), 195–202.
Ling, S. D., Johnson, C. R., Frusher, S. D. and Ridgway, K. R. (2009). Overfishing reduces resilience of kelp beds to climate-driven catastrophic phase shift. *Proceedings of the National Academy of Sciences*, 106(52), 22341–22345.
Link, J. S. and Garrison, L. P. (2002). Trophic ecology of Atlantic cod *Gadus morhua* on the northeast US continental shelf. *Marine Ecology Progress Series*, 227, 109–123.
Lovelock, C. E., Ball, M. C., Martin, K. C. and Feller, I. C. (2009). Nutrient enrichment increases mortality of mangroves. *PLoS ONE*, 4(5), e5600.
Lye, C. M., Frid, C. L. J., Gill, M. E. and McCormick, D. (1997). Abnormalities in the reproductive health of flounder, *Platichthys flesus*, exposed to effluent from a sewage treatment works. *Marine Pollution Bulletin*, 34, 34–41.
MacLeod, C., Santos, M., Reid, R. J., Scott, B. and Pierce, G. J. (2007). Linking sandeel consumption and the likelihood of starvation in harbour porpoises in the Scottish North Sea: could climate change mean more starving porpoises? *Biology Letters*, 3, 185–188.
McClanahan, T. R., Donner, S. D., Maynard, J. A. et al. (2012). Prioritizing key resilience indicators to support coral reef management in a changing climate. *PLoS ONE*, 7(8), e42884.
McKinley, A. and Johnston, E. L. (2010). Impacts of contaminant sources on marine fish abundance and species richness: a review and meta-analysis of evidence from the field. *Marine Ecology Progress Series*, 420, 175–191.
McLeod, E., Chmura, G. L., Bouillon, S. et al. (2011). A blueprint for blue carbon: toward an improved understanding of the role of vegetated coastal habitats in sequestering CO_2. *Frontiers in Ecology and the Environment*, 9(10), 552–560.
McMahon, T. A., Halstead, N. T., Johnson, S. et al. (2012). Fungicide-induced declines of freshwater biodiversity modify ecosystem functions and services. *Ecology Letters*, 15(7), 714–722.

McManus, J. W. and Polsenber, J. F. (2004). Coral–algal phase shifts on coral reefs: ecological and environmental aspects. *Progress in Oceanography*, 60, 263–279.

Mermillod-Blondin, F. (2011). The functional significance of bioturbation and biodeposition on biogeochemical processes at the water–sediment interface in freshwater and marine ecosystems. *Journal of the North American Benthological Society*, 30(3), 770–778.

Mermillod-Blondin, F. and Rosenberg, R. (2006). Ecosystem engineering: the impact of bioturbation on biogeochemical processes in marine and freshwater benthic habitats. *Aquatic Sciences*, 68(4), 434–442.

Micheli, F., Mumby, P. J., Brumbaugh, D. R. et al. (2014). High vulnerability of ecosystem function and services to diversity loss in Caribbean coral reefs. *Biological Conservation*, 171, 186–194.

MA (2003). *Ecosystems and Human Well-being: A Framework for Assessment*. Washington DC: Island Press.

Mitchell, I. (2010). Marine birds. In *Charting Progress 2: Healthy and Biologically Diverse Seas Evidence Group Feeder Report*, ed. M. Frost. London: UK Marine Monitoring and Assessment Strategy, Defra. Available at: http://chartingprogress.defra.gov.uk/chapter-3-healthy-and-biologicaly-diverse-seas, accessed 1 August 2014.

Moberg, F. and Folke, C. (1999). Ecological goods and services of coral reef ecosystems. *Ecological Economics*, 29(2), 215–233.

Myers, R. A. and Worm, B. (2003). Rapid worldwide depletion of predatory fish communities. *Nature*, 423(6937), 280–283.

Naylor, R. L., Goldburg, R. J., Primavera, J. H. et al. (2000). Effect of aquaculture on world fish supplies. *Nature*, 405(6790), 1017–1024.

Nesheim, M. C. and Yaktine, A. L. (2006). *Seafood Choices: Balancing Benefits and Risks*. Washington DC: National Academy Press.

O'Connor, S., Campbell, R., Cortez, H. and Knowles, T. (2009). Whale watching worldwide: tourism numbers, expenditures and expanding economic benefits. A special report from the International Fund for Animal Welfare, Yarmouth, MA, prepared by Economists at Large.

O'Neill, J. and Spash, C. L. (2000). Conceptions of value in environmental decision-making. *Environmental Values*, 9(4), 521–536.

OSPAR (2010). *Quality Status Report 2010*. London, OSPAR Commission.

Osterblom, H., Hansson, S., Larsson, U. et al. (2007). Human-induced trophic cascades and ecological regime shifts in the Baltic Sea. *Ecosystems*, 10(6), 877–889.

Páez-Osuna, F. (2001). The environmental impact of shrimp aquaculture: a global perspective. *Environmental Pollution*, 112(2), 229–231.

Park, J. S., Jung, S. Y., Son, Y. J. et al. (2011). Total mercury, methylmercury and ethylmercury in marine fish and marine fishery products sold in Seoul, Korea. *Food Additives and Contaminants Part B-Surveillance*, 4(4), 268–274.

Parks, J. R. (2006). Shorebird use of smooth cordgrass (*Spartina alterniflora*) meadows in Willapa Bay, Washington. Unpublished thesis, The Evergreen State College.

Pauly, D. (1987). Managing the Peruvian upwelling ecosystem: a synthesis. In *The Peruvian Anchoveta and its Upwelling Ecosystem: Three Decades of Change*, ed. D. Pauly and I. Tsukayama. Manila, The Philippines: ICLARM.

Pauly, D., Christensen, V., Dalsgaard, J., Froese, R. and Torres, F. (1998). Fishing down marine food webs. *Science*, 279(5352), 860–863.

Piatt, J. F., Lensink, C. J., Butler, W., Kendziorek, M. and Nysewander, D. R. (1990). Immediate impact of the *Exxon Valdez* oil spill on marine birds. *The Auk*, 107, 387–397.

Principe, P., Bradley, P., Yee, S. *et al.* (2011). Quantifying Coral Reef Ecosystem Services. EPA/600/R-11/206 U.S. Environmental Protection Agency, Office of Research and Development, Research Triangle Park, NC.

Read, A. J., Drinker, P. and Northridge, S. (2006). Bycatch of marine mammals in US and global fisheries. *Conservation Biology*, 20, 163–169.

Reeve, M. R., Gamble, J. C. and Walter, M. A. (1977). Experimental observations on the effects of copper on copepods and other zooplankton: controlled ecosystem pollution experiment. *Bulletin of Marine Science*, 27(1), 92–104.

Reubens, J. T., Braeckman, U., Vanaverbeke, J. *et al.* (2013). Aggregation at windmill artificial reefs: CPUE of Atlantic cod (*Gadus morhua*) and pouting (*Trisopterus luscus*) at different habitats in the Belgian part of the North Sea. *Fisheries Research*, 139, 28–34.

Rice, J. C. and Garcia, S. M. (2011). Fisheries, food security, climate change, and biodiversity: characteristics of the sector and perspectives on emerging issues. *ICES Journal of Marine Science*, 68, 1343–1353.

Richardson, W. J., Miller, G. W. and Greene, C. R. (1999). Displacement of migrating bowhead whales by sounds from seismic surveys in shallow waters of the Beaufort Sea. *The Journal of the Acoustical Society of America*, 106(4), 2281–2281.

Riegl, B. (2001). Degradation of reef structure, coral and fish communities in the Red Sea by ship groundings and dynamite fisheries. *Bulletin of Marine Science*, 69(2), 595–611.

Roberts, C. M. (2002). Deep impact: the rising toll of fishing in the deep sea. *Trends in Ecology and Evolution*, 17(5), 242–245.

Rochman, C. M. and Browne, M. A. (2013). Classify plastic waste as hazardous. *Nature*, 494, 169–171.

Rodríguez, J. P., Beard Jr., T. D., Bennett, E. M. *et al.* (2006). Trade-offs across space, time, and ecosystem services. *Ecology and Society*, 11(1), 28.

Rossini, G. P. and Hess, P. (2010). Phycotoxins: chemistry, mechanisms of action and shellfish poisoning. *EXS*, 100, 65–122.

Rudnick, D. A., Chan, V. and Resh, V. H. (2005). Morphology and impacts of the burrows of the Chinese mitten crab, *Eeriocheir sinensis* H. Milne Edwards (Decapoda, Grapsoidea), in south San Francisco Bay, California, USA. *Crustaceana*, 78, 787–807.

Russell, D. J. F., Brasseur, S. M. J. M., Thompson, D. *et al.* (2014). Marine mammals trace anthropogenic structures at sea. *Current Biology*, 24(14), R638–R639.

Sandström, M. (1996). Recreational benefits from improved water quality: a random utility model of Swedish seaside recreation. Working Paper No. 121, Stockholm School of Economics.

Schmidt, A. L. and Scheibling, R. E. (2006). A comparison of epifauna and epiphytes on native kelps (*Laminaria species*) and an invasive alga (*Codium fragile* ssp. *tomentosoides*) in Nova Scotia, Canada. *Botanica Marina*, 49, 315–330.

Skinner, M. A., Courtenay, S. C. and McKindsey, C. W. (2013). Reductions in distribution, photosynthesis, and productivity of eelgrass *Zostera marina* associated with oyster *Crassostrea virginica* aquaculture. *Marine Ecology Progress Series*, 486, 105–119.

Smale, D. A., Burrows, M. T., Moore, P., O'Connor, N. and Hawkins, S. J. (2013). Threats and knowledge gaps for ecosystem services provided by kelp forests: a northeast Atlantic perspective. *Ecology and Evolution*, 3(11), 4016–4038.

Smith, P. J., Francis, R. and McVeagh, M. (1991). Loss of genetic diversity due to fishing pressure. *Fisheries Research*, 10(3–4), 309–316.

Speybroeck, J., Bonte, D., Courtens, W. et al. (2006). Beach nourishment: an ecologically sound coastal defence alternative? A review. *Aquatic Conservation-Marine and Freshwater Ecosystems*, 16(4), 419–435.

Springer, A. M., Estes, J. A., van Vliet, G. B. et al. (2003). Sequential megafaunal collapse in the North Pacific Ocean: an ongoing legacy of industrial whaling? *Proceedings of the National Academy of Sciences of the United States of America*, 100(21), 12223–12228.

Stevick, P. T., Incze, L. S., Kraus, S. D. et al. (2008). Trophic relationships and oceanography on and around a small offshore bank. *Marine Ecology Progress Series*, 363, 15–28.

Stewart, N. L. and Konar, B. (2012). Kelp forests versus urchin barrens: alternate stable states and their effect on sea otter prey quality in the Aleutian Islands. *Journal of Marine Biology*, 2012, doi:10.1155/2012/492308.

Taylor, T. and Longo, A. (2010). Valuing algal bloom in the Black Sea Coast of Bulgaria: A choice experiments approach. *Journal of Environmental Management*, 91(10), 1963–1971.

TEEB (2010). *The Economics of Ecosystems and Biodiversity: Ecological and Economic Foundations*, ed. P. Kumar. London and Washington DC: Earthscan.

Thrush, S. F. and Dayton, P. K. (2002). Disturbance to marine benthic habitats by trawling and dredging: implications for marine biodiversity. *Annual Review of Ecology and Systematics*, 33(1), 449–473.

Thrush, S. F., Hewitt, J. E., Dayton, P. K. et al. (2009). Forecasting the limits of resilience: integrating empirical research with theory. *Proceedings of the Royal Society B*, 276, 3209–3217.

Topelko, K. N. and Dearden, P. (2005). The shark watching industry and its potential contribution to shark conservation. *Journal of Ecotourism*, 4(2), 108–128.

Tougaard, J., Carstensen, J., Wisz, M. S. et al. (2006a). *Harbour Porpoises on Horns Reef in Relation to Construction and Operation of Horns Rev Offshore Wind Farm*. Roskilde, Denmark: National Environmental Research Institute.

Tougaard, J., Tougaard, S., Jensen, R. C. et al. (2006b). *Harbour seals on Horns Reef Before, During and After Construction of Horns Reef Offshore Windfarm*. Esbjerg, Denmark: Vattenfall A/S.

Twiner, M. J., Rehmann, N., Hess, P. and Doucette, G. J. (2008). Azaspiracid shellfish poisoning: a review on the chemistry, ecology, and toxicology with an emphasis on human health impacts. *Marine Drugs*, 6(2), 39–72.

UN Convention on Biological Diversity. (2000). Conference of the Parties (COP) 5 Decision V/6 Ecosystem Approach.

UNEP CEP (2013). Waste water, sewage and sanitation. Available at: http://www.cep.unep.org/publications-and-resources/marine-and-coastal-issues-links/wastewater-sewage-and-sanitation, accessed 8 January 2013.

Uyarra, M. C., Cote, I. M., Gill, J. A. *et al.* (2005). Island-specific preferences of tourists for environmental features: implications of climate change for tourism-dependent states. *Environmental Conservation*, 32(1), 11–19.

Wallace, N. (1985). *Debris Entanglement in The Marine Environment: A Review*, translated by R. S. Shomura and H. O. Yoshida, Honolulu, Hawaii, 27–29 November 1984, US Department of Commerce, pp. 259–277.

Wanless, S., Harris, M. P., Redman, P. and Speakman, J. (2005). Low energy values of fish as a probable cause of a major seabird breeding failure in the North Sea. *Marine Ecology Progress Series*, 294, 1–8.

Waycott, M., Duarte, C. M., Carruthers, T. J. B. *et al.* (2009). Accelerating loss of seagrasses across the globe threatens coastal ecosystems. *Proceedings of the National Academy of Sciences of the United States of America*, 106(30), 12377–12381.

Wen, C. K. C., Pratchett, M. S., Shao, K. T., Kan, K. P. and Chan, B. K. K. (2010). Effects of habitat modification on coastal fish assemblages. *Journal of Fish Biology*, 77(7), 1674–1687.

Westerbom, M. (2006). Population dynamics of blue mussels in a variable environment at the edge of their range. PhD thesis, Faculty of Biosciences, Department of Biological and Environmental Sciences, University of Helsinki.

Wilhelmsson, D., Malm, T. and Ohman, M. C. (2006). The influence of offshore windpower on demersal fish. *ICES Journal of Marine Science*, 63(5), 775–784.

Willette, D. A. and Ambrose, R. F. (2012). Effects of the invasive seagrass *Halophila stipulacea* on the native seagrass, *Syringodium filiforme*, and associated fish and epibiota communities in the Eastern Caribbean. *Aquatic Botany*, 103, 74–82.

Wilmers, C. C., Estes, J. A., Edwards, M., Laidre, K. L. and Konar, B. (2012). Do trophic cascades affect the storage and flux of atmospheric carbon? An analysis of sea otters and kelp forests. *Frontiers in Ecology and the Environment*, 10(8), 409–415.

Worm, B. (2000). Consumer versus resource control in rocky shore food webs: Baltic Sea and NW Atlantic Ocean. *Berichte aus dem Institut fuer Meereskunde Kiel*, 316, 1–147.

Worm, B., Barbier, E. B., Beaumont, N., *et al.* (2006). Impacts of biodiversity loss on ocean ecosystem services. *Science*, 314(5800), 787–790.

York, P. H., Booth, D. J., Glasby, T. M. and Pease, B. C. (2006). Fish assemblages in habitats dominated by *Caulerpa taxifolia* and native seagrasses in south-eastern Australia. *Marine Ecology Progress Series*, 312, 223–234.

12 · Conclusions

TASMAN CROWE, DAVE RAFFAELLI AND
CHRISTOPHER FRID

12.1 Introduction

Marine ecosystems provide a range of essential benefits to society, including food and other products, waste assimilation, coastal protection and climate regulation as well as less tangible, but no less important cultural and aesthetic benefits (Chapter 2). To a considerable degree, those benefits are underpinned by ecosystem services dependent on the efficient functioning of the ecosystems, although detailed understanding of relationships between particular services and particular functional processes is currently being developed (Chapter 2). In turn, the efficient functioning of ecosystems has been linked to the number and identity of species present, as well as the prevailing environmental conditions (Chapter 5).

Society derives many of its benefits from ecosystems via sectoral activities and industries, such as fishing, construction, energy, shipping, leisure and tourism. These activities can impose pressures on ecosystems, such as removal of biomass, inputs of nutrients and other contaminants and the introduction of artificial structures and non-indigenous species (Chapter 3). Such pressures act through a range of mechanisms affecting different levels of biological organisation to modify biodiversity and ecosystem functioning (Chapters 3 and 4). The chapters in this book have reviewed current knowledge of how the spectrum of human activities and pressures (collectively referred to as stressors) affect biodiversity, ecosystem functioning and the provision of services and benefits to society. A key objective was to provide a synthesis of evidence for policy makers and managers to facilitate trade-offs between sectors of activity based on the benefits they provide weighed against the degree to which they may compromise the delivery of ecosystem services now and into the future (see Chapter 11).

Marine Ecosystems: Human Impacts on Biodiversity, Functioning and Services, eds T. P. Crowe and C. L. J. Frid. Published by Cambridge University Press. © Cambridge University Press 2015.

In this chapter, we first synthesise and summarise the inferences presented in the book about the range of impacts of different activities and pressures on biodiversity and ecosystem processes. We then ask how this kind of knowledge can help policy makers and managers, particularly in the achievement of the international targets laid down in the Millennium Development Goals (Chapter 1) and whether the ecosystem approach and the concept of ecosystem services can provide an effective framework for facilitating the achievement of those goals.

12.2 Overview of impacts of human activities on marine ecosystems

The processes by which stressors affect marine ecosystems are inherently complex and impacts are highly dependent on the nature of the stressor, the regime with which it is imposed on the system and characteristics of the receiving environment (Chapter 4). Nevertheless, the extensive review of current knowledge presented in Chapters 6–10 revealed an overall tendency for some stressors to reduce species richness in comparatively predictable ways in given contexts. For example, contaminants have been shown by meta-analysis to reduce species richness by approximately 50% on average, initially reducing abundance of less tolerant species (although there are notable exceptions, Chapter 9). Similarly, excessive inputs of nutrients lead to a predictable change from coastal ecosystems based on perennial macroalgae or seagrass to eutrophic systems based on fast-growing ephemeral algae and ultimately to phytoplankton dominance. High levels of eutrophication lead to severe oxygen deficiency and widespread mortality of fish and invertebrates (Chapter 8). One of the main effects of fishing is to remove commercial fish species from the system in large numbers, particularly the larger individuals. These are often replaced by smaller, faster growing, shorter lived and often gelatinous species (Chapter 6), the dominance of which is also a predictable consequence of eutrophication (Chapter 8). Physical disturbances like trawling particularly affect erect and fragile growth forms, including biogenic reefs, and reduce species richness in the surface layers of sedimentary habitats (particularly those which were comparatively undisturbed prior to fishing).

Other stressors generally have more complex, unpredictable impacts. Invasive species, for example, interact in complex ways with ecosystems of which they become a part, but may often cause reduced abundance or local extinction of inferior competitors (Chapter 10). The effect of some stressors is further complicated because they cause an increased likelihood of others arising. For example, the likelihood of establishment and spread of invasive species in an area can be increased by the presence of artificial structures, contaminants and eutrophic conditions (Chapter 10). As such, the influence of these stressors can include indirect effects – those caused by invasive species – as well as direct effects that they may themselves cause. Similarly, the indirect effects of removing large predatory fish through fishing can also be complex and unpredictable, potentially resulting in trophic cascades and regime shifts (Chapter 6).

Not all effects of human activities on biodiversity and functioning are negative (in the numerical, ecological sense of *reducing* species richness and rates of functioning; see Chapter 1 and Section 12.4 for consideration of societal definitions of positive and negative effects). Indeed, the chapters highlight a number of positive effects of putative stressors: artificial structures can provide valuable habitat for marine organisms, particularly when ecological engineering is used to enhance suitability for colonisation (Chapter 7); offshore wind farms also protect nearby habitats from fishing, effectively contributing to networks of marine protected areas (Chapter 7); changes to habitat structure caused by invasive species can often lead to overall enhancement of diversity, particularly at low and intermediate density of the invader (Chapter 10).

The effects of some stressors remain poorly understood, even in broad terms, because they have not yet been studied in any detail, e.g. the influence of microplastics derived from marine litter (Chapter 9), the impacts of invasive microbes on biodiversity and ecosystem functioning (Chapter 10). Uncertainty remains about long-term impacts of even the best studied stressors, partly because of the inherent complexity of biological, physical and chemical processes that govern ecological systems. Throughout the book, chapter authors have highlighted key research needs.

Some environments tend to be inherently more sensitive to particular stressors than others. For example, although eutrophication has been documented in all kinds of marine ecosystems, sheltered coastal systems respond faster to nutrients and may be more vulnerable to eutrophication than systems with greater water exchange (Chapter 8). Similarly, biogenic habitats such as seagrass and maerl beds can be particularly vulnerable to a range of stressors. It is notable that comparatively little work has been done on impacts of stressors in highly modified environments (Chapter 7). Thus, research in so-called emergent or novel ecosystems is urgently needed as these habitats are becoming ever more prevalent

(Hobbs et al., 2013). Assessment of the sensitivity of a given environment should encompass not just the likelihood and severity of damage, but also the capacity of the system to recover its resilience (Chapter 3). Resilience is a complex property of ecosystems, however, and if a system passes a threshold and tips into an alternative stable state, it can be very difficult for it return to its previous state, even if the stressors that caused the change are removed ('hysteresis' – see Chapter 3). Reversing the state of eutrophic ecosystems can be slow, for example, because of large accumulations of phosphorus in sediments (Chapter 8). Recovery may also be hampered by a lack of availability of propagules from nearby habitats, which may also be degraded, or because new species have become fully established, modifying the environment and preventing the return of key species that had previously occupied it. Indeed some ecosystems may never return to the previous or desired state through natural processes. In some cases, however, where natural recovery is slow, it is possible to restore habitats artificially (Warren et al., 2002; Rodney and Paynter, 2006; Chapman, 2013). The appropriate target state (or baseline) against which to assess restoration or recovery to good environmental status needs to be identified with care given that many systems have been progressively modified by decades of human activity (Pauly, 1995; Chapter 3).

It is clear that most marine ecosystems are affected simultaneously by multiple anthropogenic stressors (Halpern et al., 2008). Unfortunately, however, most of our understanding of impacts is based on individual stressors, and we currently lack the detailed knowledge or a well-developed overarching framework to predict combined effects of multiple stressors. In principle, additive effects, in which the effects of two or more stressors is equal to the sum of their individual effects, could be predicted on the basis of their individual effects, which are often quite well understood. Unfortunately, meta-analyses of the limited evidence currently available have shown that more complex interactive (synergistic/antagonistic) effects are more common than additive effects (Crain et al., 2008; Chapter 4). For example, eutrophication processes are enhanced by exploitation of large fish and other top predators (Chapter 8). As such, mitigation of eutrophication impacts may not be possible without managing fisheries and other human uses of marine resources. Conversely, however, some combinations of activities are already known to have antagonistic effects, counteracting each others' influences to some degree, presenting opportunities for complementary economic activity. This forms the basis for approaches such as Integrated Multitrophic Aquaculture, for example, in which cultivating seaweed and/or deposit feeders close to fish cages can reduce the net efflux of nutrients and organic matter into the environment (Troell et al., 2009). However, unless more detailed knowledge is available for impacts of combinations of stressors on particular receiving environments, the precautionary approach should be applied, under which it is assumed that the combined effect of two or more stressors in an area will be greater than the sum of their individual effects.

Global changes to climate and ocean physico-chemistry will inevitably occur over the coming decades, with complex regional and local effects (Chapter 3). As well as directly affecting marine ecosystems, these changes will also modify the influence of the more localised and regional stressors described in this book – pollution, nutrient enrichment, and those caused by fisheries and aquaculture, physical structures and bioinvasions. Although global changes cannot be directly managed at a regional level, adaptive management of local stressors can mitigate both their direct impacts and potentially synergistic interactions with future global changes. At present, however, we lack understanding of how combinations of multiple climatic and non-climatic changes will affect ecosystems, or the societies dependent on them. We therefore have only a weak basis for forward planning to minimise the combined effects of global and local stressors and to ensure sustainable benefits to society from marine ecosystems.

12.3 How can ecological knowledge help address the UN Millennium Development Goals?

The United Nations' Millennium Development Goals (MDGs; see Chapter 1) provide a framework, agreed by all the nations and major NGOs, that should see poverty decreased, health increased and social, economic and political inequality decreased by 2015. While the MDGs are widely seen has having had major impacts on policy development and implementation, there remain major challenges. It is estimated that the combined challenges of lifting everybody out of food poverty (around 1 in 7 people in the world is currently estimated to be suffering from hunger/malnutrition) and dealing with the growing population will require around 70% more food (measured by calorific content) to be available, sustainably, by 2050 (United Nations, 2013). In Chapter 1, two of the MDGs were singled out as having close links with ecological processes. Here, we consider the ways in which the ecological understanding reviewed in this book bears on the achievement of those goals.

12.3.1 Millennium Development Goal 1: to eradicate extreme poverty and hunger

As marine foodstuffs provide around about 7% of the global protein requirement there will be pressure for marine sources to help fill this gap. As outlined in Chapter 6, the scope for growth in capture fisheries is limited but shifting to a framework that seeks to deliver the maximum sustainable yield from the global fisheries is a major shift in policy that will deliver (slightly) more food but ensure sustainability in that supply and hence stability in terms of impacts/demands on the marine ecosystem. This gap will however also support the growing demand for marine aquaculture. Aquaculture is the fastest growing food production sector and marine aquaculture has scope for further growth (Frid and Paramor, 2012). However, increasing marine food production, both by harvesting and aquaculture, will lead to conflicts with other sectors and also with the MDG for biodiversity protection (Rice and Garcia, 2011).

The aim of harvesting populations of fish and shellfish at their maximum sustainable yield as a contribution to MDG1 is not a simple one to achieve. Leaving aside the scientific, social and political challenges that mean that most fisheries have a history of overexploitation, there is no single answer to what this outcome would look like. If, for example, capelin in the North Atlantic are harvested at their maximum sustainable level then there will be insufficient capelin available to act as food for juvenile cod (Jakobsson and Stefánsson, 1998). So cod populations will be held below their maximum potential size and so harvesting both capelin at MSY and cod at MSY is not possible. Therefore the management objectives must be refined by some societal input – do we wish to prioritise cod for human consumption as opposed to capelin that are converted to fishmeal? However, we may get more total protein or calories from aquaculture using capelin as part of a mixed (fish meal and vegetable protein) artificial diet. However we resolve such decisions about which species are fished at MSY and which are fished to some lower level, to provide forage or other ecological services, exploiting wild stocks at such levels will have ecological impacts on the dynamics of the system. At the most basic, the fishery is acting as an additional, very effective predator in the food web. In other cases the fishing process may alter habitat features, alter biogeochemical pathways or rates and cause additional mortality or injury to non-target species (see Chapter 6).

The impacts on the marine ecosystem of increasing aquaculture are likely to be complex and dependent on the species cultured, the methods used and the extent to which these are driven by unfettered 'market forces'. The UN FAO Code of Conduct for responsible fisheries (FAO, 1995) seeks to reverse the trends in wasteful overfishing and environmentally damaging fishing practices. The FAO (1999) identifies seven major challenges facing the aquaculture sector, these are:

- meeting growing demands for seed, feed and fertilisers, in terms of quantities and quality;
- reducing production losses through improvement in fish health management;
- increasingly severe competition with other resource (land/water/feed) users;
- deteriorating quality of water supplies resulting from aquatic pollution;
- successful integration of aquaculture with other farming activities, and promotion of small-scale, low-cost aquaculture in support of rural development;
- improvements in environmental management including reduction of environmental impacts and avoidance of risks to biodiversity through better site selection, appropriate use of technologies, including biotechnologies, and more efficient resource use and farm management; and
- assurance of food safety and quality of products.

At least four of these involve interactions with the wider ecosystem, either in using ecosystem services such as feed or seed from the wild or potentially compromising other ecosystem service delivery, i.e. conflicts over habitat use, impacts on water quality, production of pollution, biosecurity. Therefore delivery of MDG1 interacts with MDG7.

12.3.2 Millennium Development Goal 7: to ensure environmental sustainability

The concept of sustainable development was one of the key aspects of the Convention on Biological Diversity in 1992 and the adoption of Agenda 21 (the international commitment to sustainable development as a key principle) saw ecological, economic and social sustainability enshrined as the 'three pillars' of sustainability and this is the framework for sustainability that underpins the MDG7 and in particular Target 9 – 'Integrate the principles of sustainable development into country policies and programmes and reverse the loss of environmental resources'. Agenda 21 recognises that humanity and human uses of ecosystem services are an integral part of a functioning ecosystem but established the critical importance of understanding the limits of sustainable use. The UN Convention on Biological Diversity (United Nations, 1992) also initially signed at Rio in 1992 made a commitment not just to protecting biodiversity per se but biodiversity in the context of functioning systems with humans as part of the ecosystem. This led to the development and application of the 'ecosystem approach' to environmental management (see Chapter 1) and in conjunction with Agenda 21 will work to promote the establishment of management regimes aimed at producing functioning, biodiverse ecosystems to underpin human drawdown of ecosystem services.

There has been considerable debate about the relative importance of the three pillars of sustainability (Pope *et al.*, 2003), but it is increasingly recognised (at least by ecologists) that ecological (or environmental) sustainability has primacy, because if we use our environmental capital in an unsustainable manner then there is no scope to have economic activity or social structures supported by it (but see Littig and Griessler, 2005).

From the perspective of this volume, the key test of sustainability is whether the system can continue to support a particular level of use of an ecosystem service indefinitely. This may seem a simple idea but in practice it is extremely difficult to assess analytically. Most human activities exert a range of pressures on a variety of ecosystem components. In most areas, multiple activities occur leading to interactions between activities mediated through the response of the ecosystem components. For example, using a sandy embayment for recreation − sun bathing and swimming − may seem low impact and hence sustainable. But the level of activity may interfere with bird feeding or roosting on the shore, this plus the physical trampling might impact the infauna, which may reduce the bay's ability to assimilate nutrients carried in by runoff. This, in turn, may be limited by the ability to develop aquaculture nearby.

While some may advocate the development of ever more complex ecosystem–social–economic models to assess the ramifications of such management schemes, the more pragmatic approach is to take an incremental and adaptive approach to management. Activities are allowed to continue until evidence of impact arises. This requires strong and robust environmental monitoring and a management regime with the strength to actually restrict activities. In many jurisdictions the halting of an activity requires a very high burden of proof − a condition that is hard to meet given the difficulty of getting environmental data and establishing causality with a high level of certainty. This approach would also tend to be in contrast to a 'precautionary approach' where proof of 'no impact' is required before the licensing of an activity.

So it is clear that ecosystem science has a critical role to play in delivering MDG7 (and MDG1). It is far from clear, however, even 25 years after the Brundtland report (World Commission on Environment and Development, 1987), how environmental sustainability can be achieved.

12.4 How do the ecosystem approach and the ecosystem services framework contribute to conservation management and policy?

Two frameworks have come to dominate ecosystem services approaches to environmental management: the *Ecosystem Approach* and the *Ecosystem Services Framework*. The ecosystem approach provides a framework for accounting for ecological, economic and social dimensions where people are part of the system, not apart from the system (Convention on Biological Diversity, 1995), and has been formally expressed as: 'a strategy for the integrated management of land, water and living resources that promotes nature conservation and sustainable use in an equitable way recognising that humans with their cultural diversity are an integral part of ecosystems' (Convention on Biological Diversity, COP 7 Decision VII/11). The underlying principles of the CBD's ecosystem approach, the so-called Malawi Principles, are intended to reflect that thoroughly holistic perspective, and they represent a shift from a species or biological focus to a landscape or seascape focus for conservation and sustainable use, with emphasis on functionality rather than structure (Table 12.1). The ecosystem approach, together with related initiatives such as the Millennium Ecosystem Assessment (2005), have made governments, academic researchers and wider society more aware of the dependence of human well-being on natural systems and in particular Nature's ability to deliver benefits to people through ecosystem services.

The concepts enshrined in the Malawi Principles have indeed been applied in practice in many diverse situations, most often by NGOs such as The Nature Conservancy, Conservation International, the World Wildlife Fund and the Royal Society for the Protection of Birds, as well as some government agencies such as Natural England. But it is rare that all 12 of the Malawi Principles are considered and it is undeniable that the concepts of ecosystem services and managing systems within an economic context (Principles 4 and 5) have become prime foci for

Table 12.1 *The Malawi Principles that underpin the Ecosystem Approach (from Convention on Biological Diversity, 1995).*

1	The objectives of management of land, water and living resources are a matter of societal choices.
2	Management should be decentralised to the lowest appropriate level.
3	Ecosystem managers should consider the effects (actual or potential) of their activities on adjacent and other ecosystems.
4	Recognizing potential gains from management, there is usually a need to understand and manage the ecosystem in an economic context.
5	Conservation of ecosystem structure and functioning, in order to maintain ecosystem services, should be a priority target of the ecosystem approach.
6	Ecosystem must be managed within the limits of their functioning.
7	The ecosystem approach should be undertaken at the appropriate spatial and temporal scales.
8	Recognizing the varying temporal scales and lag-effects that characterise ecosystem processes, objectives for ecosystem management should be set for the long term.
9	Management must recognise the change is inevitable.
10	The ecosystem approach should seek the appropriate balance between, and integration of, conservation and use of biological diversity.
11	The ecosystem approach should consider all forms of relevant information, including scientific and indigenous and local knowledge, innovations and practices.
12	The ecosystem approach should involve all relevant sectors of society and scientific disciplines.

managers. This is probably because of their greater resonance and ease of operation, compared to the other principles. As a result, in some policy arenas, ecosystem services have become the sole thrust of those claiming to take an ecosystem approach. Turner and Daily (2008) have produced an Ecosystem Services Framework (Figure 12.1) which attempts to re-emphasise all of the other principles, at least implicitly, through linkages to decision making, but this implicitness is not always recognised by those making decisions on the ground. The limited take-up of all 12 Principles is not helped by the plethora of 'Ecosystem'-type frameworks to which seascape managers are exposed, including: the ecosystem approach, an Ecosystems Approach, the Ecosystem Services Framework, an Ecosystem Services Approach, etc., often allowing the Malawi Principles to be cherry-picked and considered in isolation. Nevertheless, it is clear that the concept of ecosystem services, and all that surrounds it, currently dominates developments in this area.

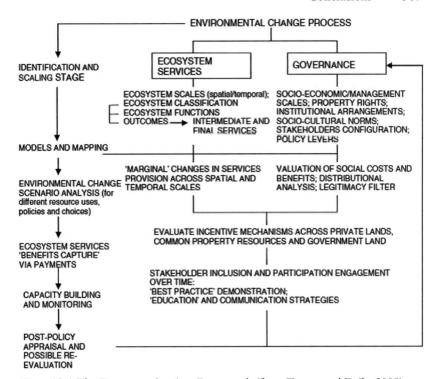

Figure 12.1 The Ecosystem Services Framework (from Turner and Daily, 2008).

12.4.1 Resonance of the ecosystem services concept with policy makers, conservationists and the wider public

The Ecosystem Approach and the Ecosystem Services Framework offer attractive solutions to all sectors involved in or affected by environmental management decisions. The interdependence of healthy ecosystems and human well-being is unarguable and forces all parties to consider and account for the full cost of any management intervention. In economic-speak, the externalities are internalised. For instance, the costs and benefits associated with the deployment of fields of offshore structures for energy generation should not be restricted to the engineering cost–benefit analyses, but extended to include such things as changes in shipping and navigation, changes in near-bed and nearshore hydrography, greenhouse gas emissions, commercial and recreational fishing, coastal erosion and deposition, opportunity costs for other ventures and amenity value to the general public (Chapters 2, 11). Similarly the traditional externalities of coastal catchment activities, such as nutrient loadings and sediment runoff into estuarine and coastal areas would need to be internalised within terrestrial catchment management decisions. Turner *et al.* (2014) provide an excellent exploration of this in the UK context. By providing balance sheets of all the pros and cons and costs and benefits of an intervention or policy, it should be possible to provide a more informed evidence base for decision making (Figure 12.1). In this respect, ecosystem services seem to offer a useful framework for exploring costs and benefits holistically through an understanding of how preferences for particular services might need to be traded off against other services (Chapter 11). Comprehensive typologies of services such as those in Chapter 2 provide a checklist of environmental assets that could be considered when thinking about possible implications of an intervention or new policy.

Nevertheless, however sensible and attractive such schemes might seem, not everyone is comfortable with the Ecosystem Approach and the Ecosystem Services Framework for informing management decisions. In particular, the exclusive application of monetary valuations of nature that so often characterise the decision-making process are resented in principle by some stakeholders, especially when they imply that economic and ecological values are equivalent and thus substitutable. Luck *et al.* (2012) recognise several dimensions to this ecosystem services backlash: (1) the anthropocentric view that nature is solely for the purposes of human well-being; (2) the use of the terms capital assets, services and related economic language implies an extension of 'the rationality of the profit calculus to the environmental domain'; (3) related to this, monetary valuation raises ethical issues, especially for species; (4) commodification whereby previously non-marketed dimensions of the environment are brought into market trading; (5) the socio-cultural impact of the ecosystem services concept, especially in arenas where external markets for environmental goods and services have not been part of those cultures, such as indigenous communities; (6) shifts in motivation for protecting the environment from moral to economic arguments; (7) failure to consider distributional issues when trading off services (who loses, who gains). Many of these issues reflect real differences in value systems held by different stakeholders which may not be reconcilable. Luck *et al.* (2012) argue that the ecosystem services paradigm raises many ethical issues and should be seen as only one of various approaches for capturing and packaging preferences about the way society would like to see the environment managed.

12.4.2 The Ecosystem Approach and the conservation of wild species

The attractiveness to many people of the ecosystem approach lies in part in its acknowledgement of the importance of societal choice and its focus on the maintenance of ecosystem services (Reid et al., 2006), and in part because traditional policies and instruments have failed to protect wild species so that a radically new approach is urgently needed (Black et al., 2010; Balmford et al., 2011; Gomez-Baggethun and Ruiz-Perez, 2011). Nevertheless, there are concerns in the conservation sector that the inevitable trade-offs required to manage services for a particular seascape, and the fact that similar services could be delivered by quite different sets of wild nature, could leave individual species of conservation importance out of the equation. Most of those species are probably not vital to most services, although some of them may be in the future (the insurance argument). Also, there are reservations about assigning some kind of value, especially a monetary value, to wild species, their habitats or other experiences of nature. Consequently, a mistrust is developing in the conservation world over the usefulness of the ecosystem approach in general and the motivations of those who promote it (McCauley, 2006). In addition, there is a fear that valuation is a prelude to the privatisation and sale of natural capital assets, including wild species.

These reservations and positions in the conservation community seem to be becoming more, not less, polarising since they were first aired several years ago. However, this growing mistrust is often based on fundamental misunderstandings of what a value is, as well as of the valuation agenda. There is also a surprising lack of acknowledgement of the ecosystem services and benefits provided by wild species of conservation interest, as well as understanding of how decisions about conservation and sustainable use are really made and the part played in those decisions by ecosystem services information.

12.4.3 Why should we value wild species at all?

All decisions about the conservation of wild species and habitats are an expression of our preferences for how we think the world should be and of the values we hold. The statutory policies that legally protect many wild species and habitats are essentially societal preferences and values codified, and such policies will not be negated or replaced if an

Ecosystem Services Framework is used to inform management. Not all species or aspects of nature are so protected, but stakeholders still express very strong preferences for their continued existence. Usually, these preferences resist quantification in any monetary sense, even though those species clearly have a net positive value to campaigning stakeholders, such as when predatory marine mammals come into conflict with fisherman and culls are suggested (of the mammals, that is). Whilst the financial (monetary) losses to society of maintaining those species might be estimated, for instance as lost catch, unless these are set against the value of the culled species as expressed by conservationists, then there can be no sensible analysis of the benefits and costs to inform decision-making processes. Those preferences need to be captured and expressed formally otherwise the value of those species will be assumed zero, whereas they are clearly not. Methods do exist for reflecting those preferences and values in monetary and non-monetary terms, but they are contentious and the estimates are open to debate (Gatto and Leo, 2000; Salles, 2011; Kontogianni et al., 2012). Nevertheless, refusing to engage at all in the process of estimating value will inevitably be to the detriment of those species. Importantly, even if all the preferences and values can be captured, there is no guarantee that these, along with ecosystem service evidence, will be given primacy in the decision-making process anyway. Those who have to make decisions may well be influenced by the scientific, economic and social evidence base, but they will tension that evidence against other quite different political considerations (Turner, 2011), including traditional species-based arguments put forward by conservationists. But refusing to engage with the broader valuation agenda and thus standing aloof of the process is hardly likely to improve matters.

12.4.4 Incorporating wild species in the Ecosystem Services Framework

A common misconception exists that wild species conservation cannot be sensibly incorporated into the Ecosystem Services Framework. Nothing could be further from the truth. Wild species underpin many ecosystem services, they are a service in their own right, and they are a good that is highly valued by society (Mace et al., 2011). Wild species pervade the Ecosystem Services Framework. Indeed, there is very real value of wild species and habitats in underpinning and delivering benefits to society through cultural pathways (Clarke et al., 2014; Chapters 2, 11). Such benefits are clear, although hard to quantify, but include economically quantifiable effects on physical and mental health and general well-being, presenting an economic rationale for species conservation (Shogren et al., 1999). Rather than ignoring these benefits, the conservation community needs to become better engaged with emerging research agendas so that their value is not assumed zero by decision makers.

12.4.5 Limitations of the Ecosystem Services Framework for conservation

The Ecosystem Approach and the Ecosystem Services Framework within it, along with the whole valuation agenda, will never be a panacea for all environmental ills. Decisions on future environmental options will embrace a whole raft of other evidences, analyses and epistemologies and decisions will always need to respect statutory legislative provisions (Turner, 2011). There are some environmental benefits and goods which will probably resist utterly attempts at quantification because of technical challenges or because of fundamental and irreducible belief barriers. For instance, the aesthetic and spiritual experiences associated with nature and with iconic species are so particular that they cannot be categorised or monitored in the same way as other ecosystem services and they simply resist commodification (Perlman and Adelson, 1997; Rodwell, 2013). But the reality is that the Ecosystem Services Framework has a momentum at all levels in the conservation world and is increasingly being used as a full-cost accounting approach in decision making. Rather than fighting against the approach, those detractors in the conservation community would do better to engage with researchers endeavouring to better represent those preferences and value systems in decision making.

12.5 Concluding remarks

If ecology can be thought of as a discipline that started with a consideration of how individuals interacted with their environment (home – the derivation of the term), then moved on to recognise the population, the community and then the ecosystem, then this volume has illustrated that the most recent step in that journey has been to recognise the fundamental role of humankind as a force acting on the system and in drawing services from the system. It is interesting to note that conservation has undergone a similar journey from a genesis in the 1960s which was holistic and owed much to the free thinking 'hippy' culture of the time, through a focus on species preservation (1970–80s), to habitat concerns

(1980s) to the need to protect functioning ecosystems (post-Rio 1992) and to the consideration of human society as having an integral relationship with the environment, in what is being described as social–ecological systems (Berkes *et al.*, 2003; Mace, 2014). This has lead to a convergence between ecological science and conservation and societal goals seen for example in the Aichi Biodiversity targets, which seek to ensure that biodiversity protection and maintenance of functioning ecological systems is recognised as essential for delivering the MDGs (Chapter 1).

This volume has explored the science that underpins this approach and how it links to the social and economic sciences. These linkages are still developing and new understanding in ecology and social–economic structures continues to drive an evolving agenda. In the marine environment, new initiatives such as marine spatial planning (Kidd *et al.*, 2011; Stelzenmüller *et al.*, 2013) provide a means of dealing with conflicting sectors and multiple pressures while providing scope for consideration of the ecology from a broader perspective than the traditional single species focus.

It must be remembered, as the cod–capelin example above illustrates, that there is rarely a single right, scientifically justified, solution to environmental management. Conflicts between sectors, between societal objectives and the uncertainty of the wider changing environment all conspire to make such ideas fanciful. Ecological science must engage with wider society to achieve the societal objectives set out in the MDGs. Ecology, and science in general, has two roles to play. First, informing the debate: society needs to make decisions about which options to take, or how priorities and trade-offs may play out. Science must provide the information, in an impartial way, to inform the debate. Once society has set out the targets, science is needed to monitor how management actions are performing. In a changing world filled with complex, nonlinear interactions, management must be responsive and adaptive to changes. Science must monitor the right things and, using and further developing the kind of knowledge presented here, interpret and translate the data into information of use to managers, regulators, policy makers and wider society.

References

Balmford, A., Fisher, B., Green, R. E. *et al.* (2011). Bringing ecosystem services into the real world: an operational framework for assessing the economic consequences of losing wild nature. *Environmental and Resource Economics*, 48, 161–175.

Berkes, F., Colding, J. and Folke, C. (2003). *Navigating Social–Ecological Systems: Building Resilience for Complexity and Change.* Cambridge: Cambridge University Press.

Black, J., Milner-Gulland, E. J., Sotherton, N. and Mourato, S. (2010). Valuing complex environmental goods: landscape and biodiversity in the North Pennines. *Environmental Conservation*, 37, 136–146.

Chapman, M. G. (2013). Constructing replacement habitat for specialist and generalist molluscs: the effect of patch size. *Marine Ecology Progress Series*, 473, 201–214.

Clark, N. E., Lovell, R., Wheeler, B. W. et al. (2014). Biodiversity, cultural pathways and human health. *Trends in Ecology and Evolution*, 29, 198–204.

Convention on Biological Diversity (1995). COP 7 Decision VII/11. Available at: www.cbd.int/decision/cop/?id=7748.

Crain, C. M., Kroeker, K. and Halpern, B. S. (2008). Interactive and cumulative effects of multiple human stressors in marine systems. *Ecology Letters*, 11, 1304–1315.

FAO (1995). *Code of Conduct for Responsible Fisheries.* Rome: FAO.

FAO (1999). *Into the next Millennium: Fishery perspective.* Bangkok, Thailand: UN FAO.

Frid, C. L. J. and Paramor, O. A. L. (2012). Feeding the world: what role for fisheries? *ICES Journal of Marine Science*, 69, 145–150.

Gatto, M., De Leo, G. A., 2000. Pricing biodiversity and ecosystem services: the never-ending story. *BioScience*, 50, 347–355.

Gomez-Baggethun, E., Ruiz-Perez, M. (2011). Economic valuation and the commodification of ecosystem services. *Progress in Physical Geography*, 35, 613–628.

Halpern, B. S., Walbridge, S., Selkoe, K. A., et al. (2008). A global map of human impact on marine ecosystems. *Science*, 319, 948–952.

Hobbs, R. J., Higgs, E. S., and Hall, C. M. (2013). *Novel Ecosystems: Intervening in the New Ecological World Order.* Oxford: Wiley-Blackwell.

Jakobsson, J. and Stefánsson, G. (1998). Rational harvesting of the cod–capelin–shrimp complex in the Icelandic marine ecosystem. *Fisheries Research*, 37, 7–21.

Kidd, S., Plater, A., and Frid, C. L. J. (2011). *The Ecosystem Approach to Marine Planning and Management.* London: Earthscan.

Kontogianni, A., Tourkolias, C., Machleras, A. and Skourtos, M. (2012). Service providing units, existence values and the valuation of endangered species: a methodological test. *Ecological Economics*, 79, 97–104.

Littig, B. and Griessler, E. (2005). Social sustainability: a catchword between political pragmatism and social theory. *International Journal for Sustainable Development*, 8, 65–79.

Luck, G. W., Kai, M. A., Chan, U. E. et al. (2012). Ethical considerations in on-ground applications of the ecosystem services concept. *Bioscience*, 62, 1020–1029.

Mace, G. M. (2014). Whose conservation? *Science*, 345, 1558–1560.

Mace, G. M., Bateman, I., Albon, S. et al. (2011). *The UK National Ecosystem Assessment Technical Report, UK National Ecosystem Assessment.* Cambridge: UNEP-WCMC.

McCauley, D. J. (2006). Selling out on nature. *Nature*, 443, 27.

Millennium Ecosystem Assessment (2005). *Ecosystems and Human Well-being: Biodiversity Synthesis*. Washington DC: World Resources Institute.
Pauly, D. (1995). Anecdotes and the shifting baseline syndrome of fisheries. *Trends in Ecology and Evolution*, 10, 430.
Perlman, D. L. and Adelson, G. (1997). *Biodiversity. Exploring Values and Priorities in Conservation*. Oxford: Blackwell Science.
Pope, J., Annandale, D. and Morrison-Saunders, A. (2003). Conceptualising sustainability assessment. *Environmental Impact Assessment Review*, 24, 595–616.
Reid, W. V., Mooney, H. A., Capistrano, D. *et al*. (2006). Nature: the many benefits of ecosystem services. *Nature*, 443, 749–749.
Rice, J. C. and Garcia, S. M. (2011). Fisheries, food security, climate change, and biodiversity: characteristics of the sector and perspectives on emerging issues. *ICES Journal of Marine Science*, doi: 10.1093/icesjms/fsr041.
Rodney, W. S. and Paynter, K. T. (2006). Comparisons of macrofaunal assemblages on restored and non-restored oyster reefs in mesohaline regions of Chesapeake Bay in Maryland. *Journal of Experimental Marine Biology and Ecology*, 335, 39–51.
Rodwell, J. (2013). Aesthetic and spiritual responses to the environment. A two-day BESS workshop at York, 22/23 January 2013. Available at: http://www.nerc-bess.net.
Salles, J.-M., 2011. Valuing biodiversity and ecosystem services: why put economic values on nature? *Comptes Rendus Biologies*, 334, 469–482.
Shogren, J. F., Tschirhart, J., Anderson, T. *et al*. (1999). Why economics matters for endangered species protection. *Conservation Biology*, 13, 1257–1261.
Stelzenmüller, V., Breen P, Stamford T. *et al*. (2013). Monitoring and evaluation of spatially managed areas: A generic framework for implementation of ecosystem-based marine management and its application. *Marine Policy*, 37, 149–164.
Troell, M., Joyce, A., Chopin, T. *et al*. (2009). Ecological engineering in aquaculture: potential for integrated multi-trophic aquaculture (IMTA) in marine offshore systems. *Aquaculture*, 297, 1–9.
Turner, K., Schaafsma, M., Elliott, M. *et al*. (2014). UK National Ecosystem Assessment Follow-on. Work Package Report 4: Coastal and marine ecosystem services: principles and practice. Cambridge: UNEP-WCMC.
Turner, R. K. (2011). A pluralistic approach to ecosystem assessment and evaluation. A Report to Defra, London. Available at: http://www.defra.gov.uk/naturalcapitalcommittee/files/ncc-assetcheck-03.pdf.
Turner, R. K. and G. C. Daily (2008). The Ecosystem Services Framework and natural capital conservation. *Environmental Resource Economics*, 39, 25–35.
United Nations (1992). *Convention on Biological Diversity*. New York: UN.
United Nations (2013). *World Resources Report: Creating a Sustainable Food Future*. New York: UN.
Warren, R. S., Fell, P. E., Rozsa, R. *et al*. (2002). Salt marsh restoration in Connecticut: 20 years of science and management. *Restoration Ecology*, 10, 497–513.
World Commission on Environment and Development (1987). *Our Common Future*. Oxford: Oxford University Press.

Index

additive effects, 90
agriculture, 208, 245
aquaculture, 49, 137, 145, 158, 170, 209, 213, 215, 275, 339, 342, 351, 361, 382
aquaculture feed, 154, 158
artificial reefs, 187
assessment and valuation of services, 36, 38, 157, 362, 385, 390, 391

Baltic Sea, 223, 230, 344, 352
benefits, 23, 28, 42, 52, 336
Beyond BACI, 61
biocides, 152, 153, 246, 250, 345
biodiversity, 6, 57, 173, 212, 252, 275, 293, 338
biodiversity–ecosystem functioning relationships, 88, 99, 111, 298, 309, 336
biological insurance, 112, 114
biological traits, 88, 100, 118, 126, 144, 212
bioturbation, 223, 291, 347, 348, 350, 373
body size, 141, 145

carbon sequestration, 51, 348, 358
Chesapeake Bay, 214, 227, 228, 230, 275
climate change, 16, 51, 56, 79, 169, 245, 263, 302, 304, 360, 381
climate regulation, 13, 25, 33, 34, 36, 37, 47, 50, 337, 347, 348, 349, 350, 377
coastal erosion, 170, 185, 347, 348
complementarity, 115
connectivity, 87, 125, 174, 177, 180, 182
conservation, 7, 21, 188, 311, 385, 389
contaminants, 50, 54, 82, 94, 147, 153, 244, 252, 302, 345, 349, 352, 355
Convention on Biological Diversity, 8, 383
cumulative effects, 78

density-dependent impacts, 308
detecting impacts, 59
discarding, 142

dispersal, 182, 275
disturbance regime, 74, 82, 93, 96, 178, 263
duration of disturbance, 77
dynamic energy budgets, 97

ecologically mediated interactions, 97
ecosystem approach, 140, 335, 384, 385, 386
ecosystem functioning, 5, 13, 58, 141, 145, 183, 252, 275, 289, 292, 295, 310
ecosystem service typology, 29
ecosystem services, 5, 13, 21, 48, 141, 144, 157, 233, 335, 385
endocrine disruptors, 345
endogenous stressor interactions, 94
energy, 54, 55, 169
environmental management, 36
eutrophication, 202, 206, 228, 303, 343, 348, 352, 359
evolutionary responses, 86, 180
exogenous stressor interactions, 94
experimental evidence, 60, 92, 124, 262, 282
extent of disturbance, 76
extraction of minerals, 53

fisheries, 48, 77, 87, 119, 137, 178, 228, 233, 303, 339, 351, 354, 360, 382
food security, 4, 7, 140, 145
food webs, 5, 51, 118, 140, 141, 142, 143, 150, 206, 227, 228, 337
functional redundancy, 88, 112, 114, 261
functional traits, 100, 118, 126

genetic diversity, 125, 180

habitat diversity, 125
habitat loss, 52, 53, 186, 281, 308, 350, 351

harmful algal blooms, 203, 227, 233, 307, 344
health, 346
hydrodynamics, 170, 174, 289, 347
hypoxia, 49, 55, 173, 202, 203, 219, 290, 301, 344, 352, 355

identity effects, 112, 113, 116
indicators, 15
indirect effects, 77, 140, 204, 251, 275, 280, 283, 284, 286, 342, 345, 354
intensity of disturbance, 76
interactions between stressors: synergisms and antagonisms, 89
intermediate disturbance hypothesis, 74
invasive ecosystem engineers, 284, 286, 292, 299, 353
invasive species, 82, 155, 178, 181, 223, 314, 346, 350, 353, 357

key species, 117, 120, 251, 257, 282, 283, 299, 337, 352

levels of biological organisation, 57, 78, 97
light pollution, 172, 175
local adaptation, 85, 180

macroalgal blooms, 148, 174, 203, 217, 232, 233, 251
Malawi Principles, 385
management, 12, 37, 73, 157, 169, 184, 234, 264, 309, 358, 381, 382, 384, 385
Marine litter/debris, 142, 244, 246, 251, 259, 342, 356
Marine Protected Areas, 188
marine spatial planning, 11, 157, 159, 362
Marine Strategy Framework Directive, 10
maximum sustainable yield, 141, 382
Mediterranean Sea, 219, 231, 274, 306, 312, 317, 321, 323, 324, 329
metals, 249, 304
microbial pathogens, 151, 307, 343
Millennium Development Goals, 7, 48, 381
multiple ecosystem functions, 123
multiple ecosystem services, 357
multiple stable states, 65, 76, 79, 89, 98, 210, 337
multiple stressors, 47, 88, 228, 263, 302, 380
multiplicative effects, 90

nonlinear responses, 75, 89
North Sea, 232, 267, 274, 323, 354, 357, 360, 368, 369, 372, 374, 376
novel ecosystems, 190
novel habitats, 173, 176, 379
nursery habitats, 174
nutrient cycling, 204, 206, 214, 289, 347, 348, 350
nutrients, 50, 147, 202, 208, 246, 348, 355

ocean acidification, 16, 57, 89, 94, 338, 381
oil, 54, 76, 83, 169, 246, 249, 250, 352, 356
organic material, 49, 50, 147, 202, 208, 290, 348

parasites, 151
phylogenetic diversity, 119
physical impacts, 140, 281
physical structures, 52, 55, 56, 87, 167, 304, 343, 349, 351, 355, 361
plastics, 251, 259
policy, 7, 21, 37, 55, 234, 362, 385
precautionary approach, 62, 311, 381, 385
pressure, 44, 46
productivity, 184, 202, 206, 225, 252, 337, 352

radionuclides, 251
rare species, 120, 126, 149
realistic extinction scenarios, 120
recreation, 52, 170, 338, 353, 355, 360, 384
regime shifts, 97, 210, 337, 355
renewable energy, 169, 189, 352
research, 59, 190, 191, 234, 260, 262, 339, 358, 363
resilience, 63, 65, 89, 149, 184, 217, 338, 363, 380
resistance stability, 62, 184
restoration, 185, 188, 361

sectoral activities, 44
sediment, 49, 55, 94, 143, 149, 170, 173, 174, 292, 304
selection effects, 115
sensitivity, 62, 64, 81, 85, 86, 379
shifting baseline, 66
shipping, 56, 249, 275, 304, 307
society, 7
socioeconomic impacts, 233

spatial and temporal scales of impact, 61, 76, 124, 264, 358
species distribution, 180, 305
stability, 6, 62, 114, 116, 118, 123, 183
stressors, 47
sustainability, 384

taxonomic distinctness, 119
temporal pattern and timing of disturbance, 79, 85, 96
tipping points, 66, 89, 217, 337, 361

tourism, 355, 359, 361
trade-offs, 14, 38, 336, 359, 389

urban environments, 188, 190, 209, 245

variation in impact, 73, 205, 210, 279, 342
variation in impact due to environmental context, 81, 122, 146, 280, 285, 308, 379
variation in impact due to societal context, 358
variation in service provision, 34, 37